Basic
Electricity and Electronics

Third Edition

Orla E. Loper
Arthur F. Ahr
Lee R. Clendenning

VNR VAN NOSTRAND REINHOLD COMPANY
NEW YORK CINCINNATI TORONTO LONDON MELBOURNE

Copyright © 1979 by Litton Educational Publishing, Inc.
Library of Congress Catalog Card Number 78-10781
ISBN 0-442-24883-0

Published in 1979 by Van Nostrand Reinhold Company
A division of Litton Educational Publishing, Inc.
135 West 50th Street, New York, NY 10020, U.S.A.

Van Nostrand Reinhold Limited
1410 Birchmount Road
Scarborough, Ontario M1P 2E7, Canada

Van Nostrand Reinhold Australia Pty. Ltd.
17 Queen Street
Mitcham, Victoria 3132, Australia

Van Nostrand Reinhold Company Limited
Molly Millars Lane
Wokingham, Berkshire, England

16 15 14 13 12 11 10 9 8 7 6 5 4 3 2

Library of Congress Cataloging in Publication Data

Loper, Orla E
    Basic electricity and electronics.

    Includes index.
    1. Electric engineering. 2. Electronics.
I. Ahr, Arthur F., joint author. II. Clendenning,
Lee R., joint author. III. Title.
TK146.L695 1979   621.3   78-10781
ISBN 0-442-24883-0

Preface

BASIC ELECTRICITY AND ELECTRONICS is designed to acquaint you with the behavior of electronics and to provide a thorough understanding of electron theory. This knowledge is then applied to learning the operations of electrical equipment, producing fundamental components, and using these components in working circuits. In all areas, the behavior of the electron remains the center of the presentation of principles.

Each chapter is concluded with basic "points to remember" consisting of formulas, rules, or concepts excerpted from the text for easy recognition and remembrance. These are followed by review questions that test your retention and understanding of the material presented in that chapter. Research and development material in the form of detailed experiments and projects is also provided at the end of each chapter. This inclusion enables you to utilize your cumulative knowledge of electricity and electronics in a hands-on approach for an even more complete understanding of this area of study.

The book is introduced by information pertaining to safe practices when working with electricity. Subsequent chapters are also dotted with highlighted safety precautions appropriately coordinated with the text material. Included as the last chapter of the book is a career guide for the benefit of those interested in a future in electricity and electronics. As a specific update, this edition presents metric equivalants of the English system where applicable.

BASIC ELECTRICITY AND ELECTRONICS is an "all-in-one" package. As a guide it provides easy access to a thorough understanding of the electrical world surrounding us.

Orla Loper has been a professor of physics and science for over 35 years. Twenty-five of these years were spent teaching at The University of New York at Oswego. Arthur Ahr also draws upon nearly 20 years of experience as a classroom industrial arts teacher on the junior and senior high school level. Mr. Ahr was a member of The Bureau of Industrial Arts, New York State Education Department for 15 years. He was the chief of that bureau for 6 years. Lee Clendenning is the Chairman of the Industrial Education Department at Berry College, Mt. Berry, Georgia. As an associate professor, he also teaches undergraduate professional and laboratory courses in industrial arts.

Introduction–Safe conduct in electricity-electronics shops

Electricity as a household utility and as a source of power is at our disposal in a convenient form. We need only to throw a switch, press a button, or connect an attachment plug, and electrons flow, providing power to produce light, heat elements, and run motors and other appliances. We tend to take this source of power for granted without considering that it is potentially dangerous if carelessly used.

When preparing to use electrical equipment, follow instructions carefully. Whenever there is doubt, consult an electrician. Never take a chance.

The best way to avoid costly accidents is to practice safe working habits. The following are some suggestions for safe procedures:

- Be serious. Avoid all fooling or horseplay.

- Wear suitable clothing according to shop standards and use safety devices when required.

- Use only tools and equipment that are in good condition.

- Keep work stations neat, clean, and free of unnecessary tools and equipment. Remember to keep extension and line cords out of aisles of travel.

- Never assume that an electrical circuit is dead. Always check to be certain.

- Always use proper instruments for testing circuits.

- Electricity is powerful. Although you cannot see it, it may be present. Handle it with care.

- Lamps, heating elements, soldering irons, and other electrical devices should be allowed to cool before handling.

- When wiring a circuit or connecting an appliance, use appropriate wire that is large enough to carry the current.

- When mixing acid and water, always pour the acid into the water. Wear safety glasses and avoid getting acid on your hands and clothing.

- When charging a storage battery, place it in a well-ventilated room. Follow the manufacturer's instructions to connect the proper leads from the charger to the battery terminals before power is applied. Always disconnect power before removing charger leads from battery terminals since a spark could ignite the highly explosive hydrogen gas which is usually present during the charging process.

- Be extremely careful when testing a storage battery with a hydrometer. Always hold the rubber tube on the hydrometer over the battery filler hole while taking the reading. Avoid getting drops of sulfuric acid on your hands or clothing. If acid comes in contact with your skin or clothes, neutralize it with baking soda or cold water.
- When working with heaters and resistors, remember that they remain hot (the same as a piece of hot metal) for a considerable time after the power is disconnected.
- Fuses and circuit breakers are safety devices. When a circuit breaker trips or a fuse blows, locate the cause and provide a remedy before resetting the circuit breaker or replacing the fuse.
- Never bridge a fuse with wire or other metal.
- When working with transformers, remember that some produce high voltages. Check with the instructor before connecting transformers into a circuit or attempting voltage measurements. Always use extreme caution when working with or around high voltages. They are dangerous.
- When testing high-voltage circuits, wear rubber soled shoes or stand on a rubber or plastic mat to avoid a direct path to ground. Never use two hands since current through one hand across your body to the other hand has a direct path through your chest and heart. Even a small current can be harmful and sometimes fatal.
- Equip all motor-driven hand tools made of metal with a three-wire cord. The third wire is connected to the ground screw in the plug and the metal frame of the tool.
- When it is necessary to use an electric power tool with a convenience outlet that does not have provision for three-prong plugs, an adapter is used. Attach the grounding lead on the adapter to the center screw of the receptacle.
- Do not use electric tools when standing on a wet surface.
- Electric power tools operate more efficiently when supplied with adequate power. When using an extension cord, select one with wire of sufficient size to avoid voltage drop. Low voltage causes tools to overheat and possibly burn out.
- Radios, televisions, and other electronic equipment have large capacitors. These capacitors retain their charge after the power is disconnected. A charged capacitor can give a severe shock. To protect the worker and test equipment, they are discharged before work is begun. This is done by shorting the capacitor terminals to the chassis with a screwdriver that has an insulated handle.
- Be aware of harmful radiations from electron or other electric discharge tubes.
- Always select a motor of sufficient horsepower to carry the anticipated load. An overloaded motor will become hot and possibly burn out or cause a fire.
- Replace all worn or deteriorated cords on lamps, appliances, and power tools. Use good quality cord that is large enough to carry the load. Frayed cords with deteriorated rubber are a fire hazard.
- Always disconnect the power, when finished using electric power tools.
- Remove a line cord plug by grasping the plug, not the cord.
- Never use electric power tools or apparatus near flammable vapors and gases, since sparks may cause an explosion or fire.
- Avoid touching wires that dangle from power lines.
- When installing or adjusting a TV antenna, take care not to come in contact with the electric service wires.
- When using a metal ladder near a building or in a tree, be careful to keep the ladder free of electric service wires and other wires that carry the same voltage.
- When servicing the battery in an automobile, follow the previous instructions for charging and testing a storage battery.

Contents

Chapter 1

Electricity–
The behavior of electrons

Electricity can be as gentle as the music it makes in a stereo or as violent as lightning. In an automobile, a battery supplies electrical energy to the starting motor which cranks the engine and electric sparks to ignite the gasoline which starts and keeps the engine running. This is not all the car owes to electricity: electric motors rolled its steel; electric welding joined it; electric current released aluminum from its ore, purified copper for wires and tubing, and formed the chromium plating. Electrically controlled machines made the glass, plastic, and rubber; machined the engine block; and cut the gears. Although these tasks appear diverse and unrelated, they are all understood or explained as a part of the unified body of knowledge about electricity.

Electrical energy has relieved many burdens of daily physical work. Electricity even operates calculating machines and controls devices which relieve people of mental drudgery. The computer or the pocket calculator performs mathematical operations with ease and accuracy, freeing people to think about goals, purposes, values, and other decisions that cannot be reduced to mechanical manipulation.

The ever-increasing use of electrical equipment in homes, factories, commerce, entertainment, transportation, and communication has opened a field of expanding employment opportunities.

The eye-opening discoveries that made it easy to understand the behavior of electricity were made in college laboratories in the years 1890-1900. The major discovery was that all kinds of materials contain, and in part consist of, tiny particles called *electrons*. The rest of this book is a description of electrons and their behavior and how they can be used and controlled to do useful work.

Electric current in a wire is the movement of electrons through the wire. Electrons do not move through the wire unless some special effort is made to make them move. This gives them *energy,* which is the ability to move and do work. One way to make electrons move in a wire is to place a coil of wire where it will be affected by moving magnets. The power company that supplies electrical energy over wires to homes has such machines, called *generators.*

Automobile engines are equipped with small generators, which supply the electrical requirements of the car while the engine is running. Combinations of materials that can supply energetic electrons can also be assembled. These combinations of materials that can supply energetic electrons are called *batteries*. Generators and batteries are only two ways of making electrons move. This text will aid in understanding many different electrical power sources and some useful applications of them. First, some basic concepts must be understood.

TWO KINDS OF CHARGES

You have probably found at some time, just after walking across a thick rug, sliding out of a car, or getting up from a plastic-covered chair, that you have the ability to cause an electric spark when you touch some other object. When this happens we say you are electrically charged, or that you carry an electrical charge. If you rub a plastic or hard-rubber object on your clothing (a wool or nylon sweater, for example) the plastic or hard rubber may get an electrical charge and produce a tiny spark that you can hear if you bring the plastic close to your ear. When the plastic is charged, it will attract light-weight objects, such as thread, hair, or scraps of paper. (It attracts heavy objects too, but not strongly enough to move them.)

Unlike charges attract, like charges repel.

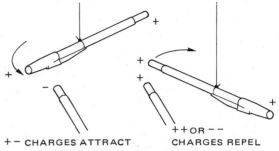

+ − CHARGES ATTRACT   + + OR − − CHARGES REPEL

Fig. 1-1 Electric Charges

Besides being able to attract ordinary uncharged objects, electrically charged objects may attract or repel each other. Whether they attract or repel depends on the materials. To see whether heavy charged objects, such as plastic pens, attract or repel, suspend a pen by a thread so it is free to move. Rub a second pen on wool or nylon and hold it close (not touching) to one end of the suspended pen as shown in figure 1-1.

A few hundred years ago men experimenting with charged objects classified the materials they knew about into two groups, figure 1-2. The materials listed in Group 1 repel each other after having been electrically charged by rubbing. Materials listed in Group 2 also repel each other. However, Group 2 materials attract the materials listed in Group 1 when both kinds of materials are charged.

Group 1	Group 2
Glass (rubbed on silk)	Hard rubber (rubbed on wool)
Glass (rubbed on wool or cotton)	Block of sulfur (rubbed on wool or fur)
Mica (rubbed on cloth)	Most kinds of rubber (rubbed on cloth)
Asbestos (rubbed on cloth or paper)	Sealing wax (rubbed on silk, wool or fur)
Cat's fur (rubbed with a block of sulfur)	Dry wood (rubbed with glass or mica)
Wool (rubbed with a stick of sealing wax)	Amber (rubbed on cloth)

Anything from Group 1 attracts anything from Group 2 (charged glass attracts charged rubber).
Any pair of materials in Group 1 repels each other (charged glass repels charged mica).
Any pair of materials in Group 2 repels each other (charged rubber repels charged rubber).

Fig. 1-2

If two pieces of the same kind of hard rubber are rubbed on the same cloth, they are charged alike and repel each other. If two glass tubes of the same kind are charged on the same piece of cloth, they are charged alike and repel each other. However, the charged glass and charged rubber attract each other. Although they are different materials, all of the items in Group 1 are said to have like charges because they repel each other. The charges on items in Group 2 are different from those on Group 1 items even though they are similar to each other. Therefore, two different kinds of like charges were identified. The pair of names that is used to describe these groups was suggested by Benjamin Franklin: *positive* for Group 1 and *negative* for Group 2. The words positive and negative are simply names, they do not explain anything. The positive terminal of a battery has the same kind of charge as glass rubbed on cloth; the negative terminal has the same kind of charge as hard rubber. The rubbing action causes electrons to move from one material to another. The material that loses electrons is left with a positive charge. The material that gains electrons has a negative charge.

The electric charges formed by rubbing materials such as listed in figure 1-2 are called static charges. *Static* means stationary, as contrasted to *current* electricity, which refers to charges in motion.

USEFUL STATIC CHARGES

Since we have observed that like charges repel and unlike charges attract, what are some useful applications of this principle? To cause attraction between two items, give one a negative charge (a surplus of electrons) and the other a positive charge (a lack of electrons). This is what has been done in an expanding field of applications known as electrostatics.

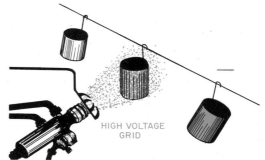

Fig. 1-3 Electrostatic Painting

ELECTROSTATIC APPLICATIONS
Electrostatic Painting

In some paint-sprayer machines used for mass production, the paint particles are given a positive charge as they leave the gun. The objects to receive the paint are given a negative charge. The objects then attract the paint spray, producing an even coat without wasting paint, figure 1-3.

Electrostatic Manufacture of Sandpaper

Sandpaper grit is forced to stand up and produce a sharper sandpaper by giving the sand particles a positive charge while the glue-covered paper is negatively charged. The like-charged grains of sand repel each other and stand apart, figure 1-4, page 4. Similarly, chopped fibers can be attracted to glue-covered backings, forming new types of nonwoven fabrics.

Electrostatic Precipitation

Smoky factory chimneys are sometimes cured by an electrostatic precipitator. This device puts a positive charge on the particles of soot and dust in the smoke. The particles are then attracted to negative-charged screens where they collect instead of going out the chimney to pollute the air.

Xerography (Dry Writing)

The application of electrostatic charging in an office copying machine, figure 1-5, page 4, came on the market in 1950.

The core of the machine is a rotating aluminum drum coated with a thin film of the element selenium.   Selenium is a semiconducting material whose electrical conductivity increases when light hits it.

(1) As the first step in the process, the selenium surface rotates past a highly positive-charged wire and loses electrons to the wire. The selenium is charged in darkness.

(2) Light reflected from the sheet of material to be copied is projected by lenses and mirrors to the surface of the drum and forms an image there, just as a camera or projector forms an image. The black form of the letter R, for example, reflects no light to the drum, while the white paper background reflects much light. Where light strikes the sele-

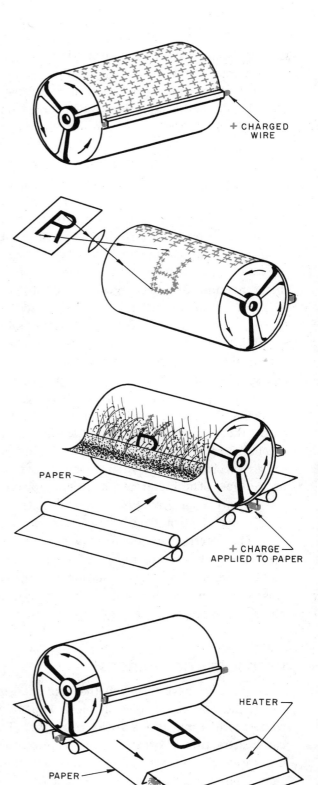

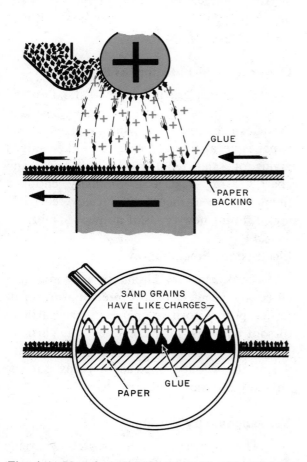

Fig. 1-4 Manufacturing Sandpaper by Electrostatic Process

Fig. 1-5 The Xerographic Process

nium, the selenium becomes a conductor. Electrons from the aluminum base flow into the selenium and neutralize its positive charge. The dark areas of the drum remain positively charged.

(3) A small conveyor belt (not shown in the sketch) pours a mixture of glass beads and toner (a black powder) over the previously illuminated surface of the drum. During the mixing of the glass beads and powder, the glass beads become positively charged, and the black dust particles become negatively charged. The black particles are attracted to the positive-charged areas left on the drum in step 2. The black particles do not attract to the neutralized areas of the polished selenium surface; they fall away along with the glass beads.

(4) The piece of paper, which is to receive the finished print, now passes under the drum. The paper is charged by a wire so that it attracts the toner from the surface of the drum.

(5) Finally, the paper carrying black copy of the image passes either under a radiant heater or over a heated roller where the black particles are melted into the paper, forming a permanent copy. A brush (not shown) removes any particles still on the drum, and the selenium surface is ready for recharging. One model of this device produces 40 copies per minute.

ELECTRICITY AND MATERIALS

These practical uses of electrical charges have been developed only in recent years. This is because only recently has there been enough information to produce and control these charges effectively. Production and control of electricity requires the use of special materials with special characteristics. Insulators, conductors, semiconductors, and magnetic materials are such materials. If these materials behave differently with respect to electricity, then they must differ from each other in terms of their basic structure or makeup. If it were known what things are really made of, the reasons why some materials conduct, some insulate, some can be made into amplifiers, and others might be found in electric motors could easily be understood.

Sometimes the question "what are things really made of?" is answered by saying "molecules." If a grain of sugar is divided into smaller and smaller specks, the smallest possible speck of sugar is a *molecule*. The molecule of sugar can be split apart further, but the fragments of this breaking of the molecule are not little pieces of sugar, they are specks of carbon, hydrogen, and oxygen. Carbon, hydrogen, and oxygen are examples of simple fundamental materials, called *elements*. There are hundreds of thousands of different kinds of complex materials, but all of these complex materials are combinations containing a few elements. There are 107 known elements.

The smallest possible specks of elements are called *atoms*. There are 107 different types of atoms because there are 107 different elements. The word atom is correctly used only as a name for the smallest particle of an element. Atoms of carbon, oxygen, copper, and iron can be discussed because these materials are elements. The term "atom of sugar" cannot be used because there is no such thing; sugar is not an element. The term "molecule of sugar" is the name of the smallest particle of sugar.

ELECTRONS, PROTONS, AND NEUTRONS

Each of these 107 kinds of atoms consists of three kinds of still smaller particles. These particles are so completely different from any known material that no picture of them can be imagined.

Atoms of ordinary hydrogen gas are the simplest in structure of all atoms. They consist of a single positive-charged particle in the center, with one negative-charged particle revolving around it at high speed. The positive-charged particle has been given the name *proton;* the negative-charged particle is called an *electron.*

Figure 1-7 is not drawn to scale since the diameter of the atom is several thousand times greater than the diameter of the particles in it. A more exact sketch would illustrate the electron as pinhead size, revolving in an orbit 150 feet (48 meters) across, to better show relative dimensions. There is relatively as much open space as in our solar system. However, this electron is not properly represented by a pinhead, for it is highly indefinite in shape, a fuzzy wisp that ripples and spins and pulses as it encircles the proton in the center. The electron is too small to have the properties that ordinary materials have. In fact, an electron is not an ordinary material, but a particle of negative electricity.

The proton that forms the center of the hydrogen atom is smaller than the electron, but 1 840 times as heavy. A proton's most important property is its positive charge.

There are two more terms used to describe an atom. The *nucleus* of the atom is the tightly packed, heavy central core where the protons of the atom are assembled. Together with the protons are other particles called neutrons.

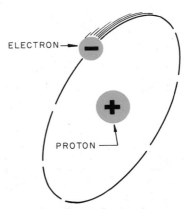

Fig. 1-7 The Hydrogen Atom

The name *neutron* indicates that these heavy particles are electrically neutral; weight is their most important property. A neutron is probably a tightly collapsed combination of an electron and a proton.

It may be hard to realize that these three kinds of particles, electrons, protons, and neutrons make up all materials. All electrons, protons, and neutrons are alike, regardless of the material they come from or exist in, figure 1-9. An atom has no outer skin other than the surface formed by its whirling electrons, a repelling surface comparable to the whirling surface that surrounds a child skipping rope.

The kind of element is determined only by the number of protons in the atom nucleus. If there was a way to assemble 29 protons as tightly as they are packed in the nucleus of

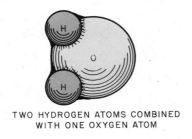

TWO HYDROGEN ATOMS COMBINED
WITH ONE OXYGEN ATOM

Fig. 1-6 A Water Molecule

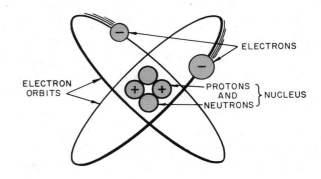

Fig. 1-8 The Structure of an Atom

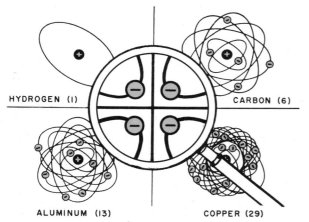

Fig. 1-9 All Electrons are Identical

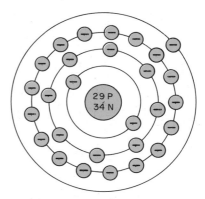

Fig. 1-10 A Copper Atom

an atom, 34 neutrons would also be needed to hold those protons together. This nucleus would attract 29 electrons making an electrically neutral atom (the positive charge of a proton is equally as strong as the negative charge of an electron). This assembly of 29 protons, 29 electrons, and 34 neutrons is an atom of copper. All copper atoms contain 29 protons, and any atom with 29 protons is a copper atom, figure 1-10.

The number of protons in the nucleus determines what the element is. The arrangement of electrons around the nucleus determines most of the physical and chemical properties and behavior of the element. The electrons of the atom are arranged in distinct layers, or shells, around the nucleus. The innermost ring or shell contains no more than two electrons, the next is limited to 8, the third can have 18; and the fourth, 32.

CONDUCTORS

Referring to the copper atom diagrammed in figure 1-10, 29 electrons are arranged in four layers. Two electrons in the shell nearest the nucleus, 8 in the next, and 18 in the third account for 28 electrons. The single 29th electron circulates alone in the fourth layer. In this position, relatively far from the positive nucleus and screened from the

attracting positive charge by the other electrons, this single electron is not tightly held to the atom, and is fairly free to travel. Inside a piece of copper, these outside single electrons of atoms often change places with one another, sliding easily from one atom to another. This easy mobility of the outside electron of copper atoms accounts for the good electrical conductivity of copper. Electrical conduction in a wire is simply a drift of electrons, sliding from one atom to another along the wire.

By examining diagrams of the electron arrangement in all kinds of atoms, it is found that most of them have one, two, or three electrons in an outer ring, shielded from the positive nucleus by one or more inner shells of electrons. These elements are all called metals. They are fairly good conductors because one of those outermost electrons is free to wander if it can be replaced by an identical electron from a nearby atom, figure 1-11.

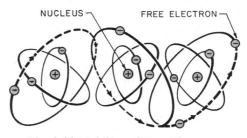

Fig. 1-11 Mobility of Free Electrons

INSULATORS

To contrast with metals, atoms of two nonmetallic elements, sulfur and iodine, are shown in figure 1-12. (Both are solids; the iodine used to treat cuts is a small amount of pure iodine dissolved in alcohol.)

Elements with 5, 6, or 7 electrons in their outermost ring are classed as nonmetals. They are poor conductors for the following reasons:

1. Their outside electrons are not as well shielded from the attracting force of the nucleus because a larger percentage of the electrons of the atom is in the outside ring, not helping to screen any individual electron from the force of the nucleus.

2. The behavior of atoms indicates that it is desirable to have eight electrons in their outer rings. The presence of 8 electrons in the outer ring seems to stabilize the atom and

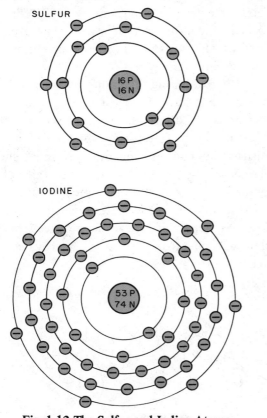

Fig. 1-12 The Sulfur and Iodine Atoms

isolate it from further chemical activity. This tendency to complete an outer shell of 8 electrons causes atoms which usually have 5, 6, or 7 electrons in the outer shell to pick up and hold the additional electrons needed to reach 8. If one tries to force electrons to move through a block of sulfur (which has 6 outer electrons), the electrons become trapped in empty places in the outer shell and are held fast. In sulfur there are no free electrons ready to slide over into the next atom and make room for a newcomer. Therefore, sulfur and *all elements with 5 or more electrons in the outer shell are generally good insulators. Insulators* are materials which do not allow electrons to move through them easily. Air, glass, rubber, and certain plastics, although not pure elements, are the most common insulating materials.

SEMICONDUCTORS

Elements with four electrons in the outer shell, such as silicon and germanium, can be made into *semiconductors*. Atoms with less than four electrons in their outer shell readily lose electrons, while atoms with five or more electrons easily pick up more. The former are conductors, the latter insulators. To reach a stable state, it would seem logical to assume those elements with four electrons in the outer shell could either gain four more electrons, or lose the four they already have. In one case they would be behaving like conductors, in the other they would be behaving like insulators. However, the atoms in a solid semiconductor crystal neither give up electrons nor take electrons from their neighboring atoms. Instead, the atoms in the crystal structure share their electrons with each other. By sharing, each atom always has eight electrons associated with it and is relatively stable. The conduction capability of the semiconductor can be controlled by adding impurity atoms to the crystal or affecting the energy level of the electrons by heat, light, or electric fields. Usually, a combination of heat and impurity atoms is used in semiconductor applications. Chapter 17 gives detailed information about semiconductor action.

INERT GASES

A short look at helium and neon atoms, figure 1-13, will complete this discussion of atom structure.

Helium has two electrons, but it is not considered a metal because those electrons are not screened from the nucleus by other electrons. Both of its two electrons are held tightly in place.

Neon has eight electrons in its outer ring that are tightly held, also. The chemist limits the term *nonmetallic* element to those with 5, 6, or 7 electrons in the outer ring.

There is another classification for elements like helium and neon, in which the outer ring is filled. These elements are *inert gases*. (Chemically, they are called inert because they do not ordinarily form combinations with other elements.)

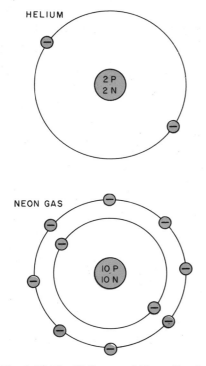

Fig. 1-13 The Helium and Neon Gas Atoms

These inert gases can be made conducting, however, by bombarding the gas at low pressure with high-speed electrons violently enough to jar electrons entirely loose from the atom. This is the condition in a neon sign.

MAGNETIC MATERIALS

An electron in motion has a magnetic field associated with it. In fact, one cannot find a magnetic field without an associated electron motion. Conversely, if a magnetic field is moved across a conductor which contains electrons, the electrons will try to move. It would seem logical that all materials would have magnetic properties, since all materials contain electrons in motion in their atoms. However, in most materials the magnetic field created by the motion of one electron is cancelled by the opposite motion of another electron. Only in iron, cobalt, and nickel are there enough uncancelled electron motions to make the materials have enough magnetic effects. Magnetic effects and forces are discussed further in Chapter 6.

To return to the discussion that began the atom-structure problem: our purpose was to secure "a further understanding of what is occurring in materials when they are electrically charged." There are now two charged particles with which to explain electrical behavior:

Electrons: Negative-charged, lightweight, and movable.

Protons: Positive-charged, heavy, and immovable in solids.

Because protons are immovable in solid materials, all electrical behavior in solid materials is explained by movements of electrons.

Refer again to Groups 1 and 2 listed on page 2. When a block of sulfur is rubbed with fur, the sulfur and the fur become oppositely charged; the sulfur negative, and the fur positive. Electrons are removed from the fur and transferred to the sulfur. When hard rubber or sealing wax receives a negative charge from wool, it is actually receiving electrons from the wool. At the same time, we say the wool is getting a positive charge, but it is actually losing some of its electrons.

> An object becomes positively charged by losing some of its electrons.
> An object becomes negatively charged by gaining electrons.

During these electric-charging operations, relatively few atoms are affected, considering the huge number of atoms in a piece of material. The addition of a billion electrons to a hard-rubber comb will give it noticeable attraction force, but a billion is a small number in this case. The sleeve on which the comb was rubbed had so many atoms that if only one atom out of every 1 000 000 000 000 000 lost one electron, the billion could be supplied.

Electricity is the study of how electrons behave. The loss or gain of a few electrons changes the properties of a material.

> The motion of electrons through a material is electric current.

TRUTH ABOUT ELECTRONS

Electrons cannot be seen directly. However, scientists can extend and surpass the range of our senses by the use of instruments, just as every day people use instruments and machines to accomplish tasks that cannot be done by hand and eye alone. Many of these instruments offer new senses and abilities to get information.

Is this belief fact or theory? Theories change.

The best evidence of truth is that devices built in accordance with these theories actually work. The existence of electrons is evident. We have learned how to make elec-

trons travel where we want them to go, doing useful work as they travel.

ELECTRIC LINES OF FORCE

How two objects can push or pull on each other when they are some distance apart with no material connection between them remains unexplained.

Two oppositely charged objects attract each other in a vacuum. Light and heat do not cause their attraction. There is no evidence of anything concrete going back and forth between them, no strings pulling them together if they attract, no sticks pushing them apart if they repel.

Therefore, the attraction or repulsion is due to *invisible lines of force*. The space near a charged object is an *electric force field*. The electric lines of force are pictured as stretched threads or bands going from a positive charge to a negative charge. These lines not only have a lengthwise pull but also push sideways against each other. The pattern of these imaginary lines may be seen with the help of strong static charges produced by a generator, as shown in figure 1-14. If a piece

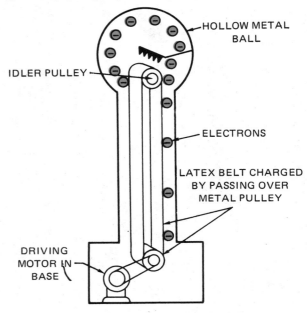

HOLLOW METAL BALL

IDLER PULLEY

ELECTRONS

LATEX BELT CHARGED BY PASSING OVER METAL PULLEY

DRIVING MOTOR IN BASE

Fig. 1-14 The Van de Graaff Generator

of glass is placed over two strong unlike charges, and short, chopped fibers, grass seed, or crumbled shredded wheat are sprinkled on the glass, the short fibers become charged and arrange themselves in the pattern shown, figure 1-15.

Although the lines are imaginary, they can represent graphically a real force.

ELECTRON ENERGY

The ability of electrons to do useful work, such as running a motor or producing heat, is called *energy*. The most common sources of energetic electrons are batteries and generators because they can produce useful amounts of energy that can easily be controlled. Electrostatic charges have energy, but no commercially practical way of producing energy by friction devices has yet been found that would compete with generators or batteries.

In all generators, magnets are rotated near coils of wire or coils of wire are rotated near magnets. In either case, the electrons in the wire are given energy so they can move and

Fig. 1-15 Electric Lines of Force

do work. The details of how generators work are discussed in a later chapter.

There are several other ways to produce small amounts of electrical energy. (1) Photocells and solar cells can convert the energy of light directly into electron motion. (2) The pickup of a record player usually contains a small, mineral-like crystal. When this crystal is mechanically vibrated, it produces electron vibrations which are made powerful enough to operate a speaker by an amplifier. (3) A metallic device called a thermocouple produces electric current when it is heated. Thermocouples can measure temperature and are used in the controls on furnaces and water heaters.

All these devices develop *electromotive force* (emf). More accurately, it is described as energy that makes electrons move. This emf energy is measured in volts.

DIRECT CURRENT AND ALTERNATING CURRENT

The electric current produced by cells and batteries is a steady one-way flow, called *direct current* (dc). The power company wires that supply homes with energy provide an alternating current (ac) which is a vibrating back-and-forth motion of electrons. Both types of motion, direct and alternating, effectively deliver energy. To produce and transmit large amounts of energy, it is easier and more efficient to use alternating than direct current. Lamps operate equally well when supplied with electrons of the proper amount of energy, whether the motion is dc or ac; some motors require ac, some require dc and some (universal motors) operate on either type of current.

POINTS TO REMEMBER

- Electricity is explained by the behavior of electrons.
- Like charges repel, unlike charges attract.

- A negative-charged object is one that has gained some extra electrons. It will repel charged hard rubber.

- A positive-charged object is one that has lost some of its electrons. It will repel charged glass.

- All materials can be electrically charged.

- An atom is the smallest particle of an element; a molecule is the smallest possible portion of a compound. A molecule consists of two or more atoms fastened together.

- All atoms consist of various numbers of electrons, protons, and neutrons. The number of protons determines the kind of element.

- Electrons are negative-charged, light-weight, and movable.

- Protons are positive-charged, heavy, and immovable in solids.

- Neutrons are not charged, and are as heavy and immovable in the atom as protons.

- Electrons are arranged in rings or shells in the atom: the number of electrons in the outer ring determines most of the electrical properties of the atom and the element.

- Charged objects are surrounded by an electric field.

- Conductors have 1 to 3 loosely held electrons in the outer shell which are free to move from atom to atom.

- Nonconductors (insulators) have no free electrons in their outer shells.

- Semiconductors are crystals made with atoms having 4 electrons in the outer shell. These 4 electrons are *shared* with other atoms in the crystal. Conduction in a semiconductor is controlled by outside energy forces and the addition of impurity atoms to the crystal.

- Magnetic effects are always associated with an electron in motion.

REVIEW QUESTIONS

1. How are the different kinds of electricity classified? Define each kind.

2. a. How were the words positive charge and negative charge originally defined in terms of materials?
   b. Using a knowledge of electrons, how are the words positive charge and negative charge defined?

3. Using electrons, explain what happens to an object when it becomes positively charged. What happens to an object when it becomes negatively charged?

4. State the law of attraction and repulsion of electrical charges.

5. What do each of these words mean: atom, molecule, elements, compound?

6. What is the difference between an electron and a proton? A proton and a neutron?

7. What factor determines most of the physical and chemical properties and behavior of an element?

8. Tell how atoms of metals differ from atoms of nonmetallic elements in their electron arrangement. Why are metals good conductors?

9. How is electrical behavior in solid materials explained?

10. What is electricity? What is electric current?

11. How do we know that electrons exist?

12. Define electromotive force (emf).

13. How does an object become positively charged?

14. What is energy?

15. Explain why copper and other metals are good conductors of electricity.

RESEARCH AND DEVELOPMENT

Experiment and Projects on the Behavior of Electrons

INTRODUCTION

The experiment and projects for research and development in this unit require application of the content presented in Chapter 1. Perform the experiment suggested, and develop, assemble, or construct (as required) the project or projects which you select with the approval of your instructor. Use graph paper to develop some idea sketches. Proceed to develop working drawings, and describe the manner in which the project is completed by writing a report of Materials Needed, Procedure, and Evaluation.

EXPERIMENT

1. Investigate the electrostatic behavior of various kinds of materials.

PROJECTS

1. Study the electronic principle involved and the construction and operation of an electrostatic flue precipitator, electrostatic device for making sandpaper, electrostatic device for separating materials, and electrostatic copier.

2. Design and build an electroscope.

3. Devise a way to illustrate the structure of an atom.

4. Construct a model of a molecule for a specific compound.

EXPERIMENT

Experiment 1   ELECTROSTATICS

OBJECT
To study the electrostatic behavior of various kinds of materials.

APPARATUS

Wire stirrup
  (to hold charged rods)
Silk or nylon thread
  (to support stirrup)
Rods or bars of:
  plastics
  glass
  hard rubber
  sulfur
  sealing wax
  varnished wood

Swatches of clean cloth:
  wool
  silk
  cotton
  nylon
  orlon
Sheet plastic:
  cellophane
  cellulose acetate
  polyethylene

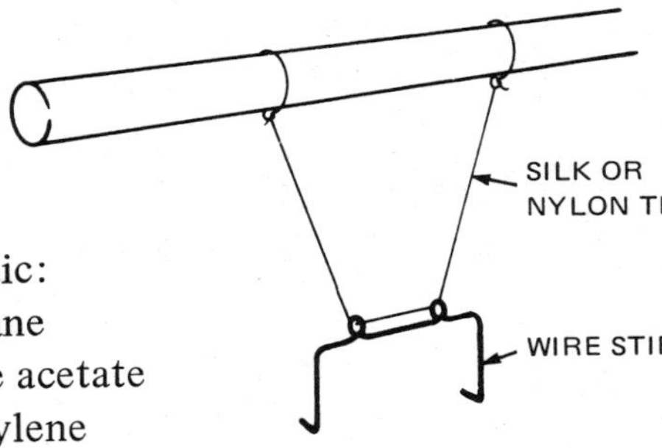

SILK OR NYLON THREAD

WIRE STIRRUP

Fig. 1-16

PROCEDURE

1. Charge a hard-rubber rod by rubbing it on wool, fur, or silk.

2. Test its ability to attract lightweight scraps of foil, paper, cloth, etc. Find out, by such experimenting, what material to rub the rod on to get the strongest possible charge.

3. Charge the hard-rubber rod strongly and place it in the wire stirrup to serve as a standard of comparison. Presumably the hard rubber will be negatively charged.

4. Charge each of the materials you have available by rubbing it on cloth or plastic.

5. Determine the charge on the object by bringing it near the suspended hard rubber. If it repels hard rubber, it is negative. Record your results. (Probably you will need to recharge the hard rubber while these trials are being made.)

6. Bring your finger near the suspended rubber rod. If it attracts, that does not indicate that your finger was charged before you brought it to the rod; the charged rod attracts neutral objects. To indicate that an object is positively charged, it should attract the negative rod more strongly than your finger does.

7. Replace the rubber rod in the stirrup with a charged glass or plastic rod which has shown the ability to strongly attract hard rubber.

8. Bring charged objects near the suspended glass (or plastic) rod. Record the effect. (Some materials may show no special behavior worth listing.)

9. Wrap one end of a smooth piece of metal (knife blade or tubing) with plastic to insulate it. Rub the uncovered portion with fur or plastic. Are you able to charge this conductor?

OBSERVATIONS

Use a table similar to the one shown here to record your observations.

MATERIAL TESTED	ACTION OBSERVED (Repel or Attract)		CHARGE ON MATERIAL (Positive or Negative)
	WITH RUBBER	WITH GLASS	

QUESTIONS

1. In general, do your results agree with the Law of Attraction and Repulsion, "unlike charges attract; like charges repel"?

2. Would the charge on a piece of varnished wood be characteristic of the varnish or the wood? Explain.

3. Name three materials that have a tendency to pick up electrons from other materials.

4. Do these experiments give you any visual proof of the fact that electrons are movable and protons immovable in solids?

PROJECTS

DESIGN AND BUILD AN ELECTROSCOPE

The electroscope is an interesting device. It can be used to show the presence of an electrical charge in a material and to demonstrate the law of charges: Unlike charges attract, like charges repel.

If you are interested in the phenomenon of static electricity and the law of charges, you will want to make an electroscope with which to experiment.

MATERIALS

1 — Glass jar

1 — Wooden stopper

1 — Brass rod, 1/2″ to 5/8″ (12 mm to 16 mm) diameter

1 — Copper rod or substitute, 1/8″ to 1/4″ (3 mm to 6 mm) in diameter

1 — Pc. sheet copper or brass 3/4″ (20 mm) square, #22 – #24 gauge

2 — Pcs. foil, 3/4″ x 1 1/2″ (20 mm x 40 mm) (dimensions approximate). Foil to be as thin as possible; gold or silver leaf would be ideal. Epoxy cement.

PROCEDURE

Obtain a suitable bottle.

Make sketches, with dimensions of each part.

Make a wooden stopper.

Turn the ball or elongated knob on the engine lathe. Drill a hole in the piece while it is in the lathe. *Note:* The rod should fit the hole tightly.

Make the foil support of sheet metal.

Cut the rod and face ends to finished length. Drill a hole of the proper size in one end to receive a small, self-tapping screw.

Drill a hole in the center of the stopper. This should be a tight sliding fit.

Mount the foil support on the end of the rod.

Cut the foil pieces and cement them in place on the support.

Assemble the electroscope and test.

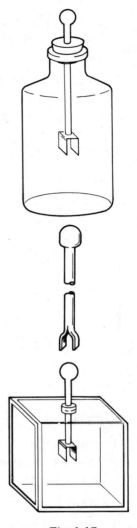

Fig. 1-17

Chapter 2
Controlling electron motion

In Chapter 1, it was suggested that electrons can usefully carry energy when they move. The motion of electrons, called *electric current*, must be controlled so that energy can be usefully carried from one place to another. One way to control electron flow is by using a property of conductors called resistance. This chapter will be mainly concerned with the use of resistance in electric circuits.

OPEN AND CLOSED CIRCUITS

One of the first things that must be learned about electric current is that a complete (or closed) conducting path must exist if electrons are to flow.

> Electricity is powerful. Although it cannot be seen it may be present. Handle it carefully.

A dry cell maintains a supply of electrons on its negative terminal. These electrons are attracted to the positive terminal of the cell, but they do not have enough energy to jump through the air to get there. They flow through a wire from the (–) to the (+) terminal. Figure 2-1 shows a lamp connected to the (–) terminal of the battery. The lamp does not light because elec-

trons cannot flow through it. The open gap in the wires must be closed if electrons are to flow through the wires and the lamp. A *switch* is a device for closing and opening a circuit, thus starting or stopping the electron flow.

ELECTRICAL MEASUREMENTS

It is hard to imagine any industry which has no use for measurement or arithmetic. Long ago, units of measure of weight, length, volume, and money were set up for use in trade. Before the electrical industry could begin to grow, units of measure for electrical quantities had to be established so what occurs in an electrical circuit can be described accurately.

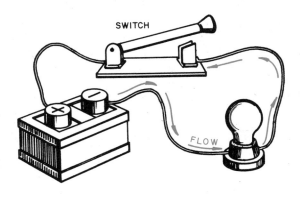

Fig. 2-1 A Switch Controls Electron Flow

Quantity of Electric Charge

The simplest electrical unit is the *coulomb.* One coulomb is 6 240 000 000 000 000 000 electrons. If that number of surplus electrons accumulated on a metal ball, we would say that the ball has a negative charge of one coulomb. If there are twice as many electrons, the charge is two coulombs. If 6 240 000 000 000 000 000 electrons are missing from an object, the object has a positive charge of one coulomb. The coulomb was named after French scientist, Charles A. Coulomb, who experimented with electrical charges and discovered many of the principles and laws related to them.

Rate of Electron Flow

Flow rates are expressed as a quantity passing by according to units of time. For example, water may stream from a fire hose at 50 liters per second, the flow in a small river may be 1 000 cubic feet per second. a person may earn three dollars per hour, or grain may be loaded into a ship at a rate of 80 tons per minute. Similarly, a quantity of electrons measured in coulombs measures the rate of electron flow in coulombs per second. If 600 coulombs pass through an electric heater in 60 seconds, the rate of flow is 10 coulombs each second. If 50 coulombs pass through a lamp in 100 seconds,

the rate is 0.5 coul./sec. Read the slanting line / as "per," 0.5 coul./sec. is said as "0.5 coulomb per second." Measurement of flow rate is one of the most often used electrical measurements. Measurement of flow rate is so important that one word is now used to replace the longer term "coulombs per second." The word used is *ampere* (amp). For example, instead of saying, 5 coulombs per second, the term 5 amperes, which means the same thing, is used. The ampere is named for Andre Ampere (1776–1836), a French mathematician and scientist who investigated magnetic forces due to currents and set up the mathematical theory of electromagnetism.

A measurement of rate of flow: One ampere = one coulomb/second.

To review the use of the terms ampere and coulomb, try to answer the following questions:

1. A current of 2 amperes was maintained in a lamp for one hour. How many coulombs passed through the lamp? Think of 2 amperes as 2 coulombs per second, and one hour as 60 min. x 60 sec./min. = 3 600 seconds. 2 coulombs each second. . .for 3 600 seconds. . .2 x 3 600 = 7 200 coulombs.

2. When the current in a wire is 4 amperes, how much time is needed for 100 coulombs to pass by? 4 amp, is 4 coul./sec. For 100

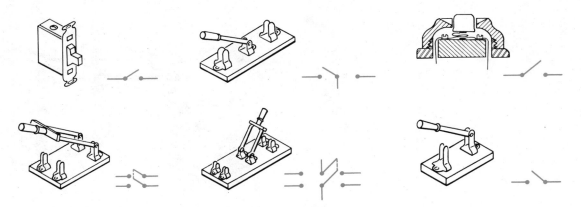

Fig. 2-2 Control Devices

CHARGE: COULOMB
FLOW RATE: AMP
FORCE: VOLT

coul. to pass by, it would require 100/4 = 25 seconds. (If it is not clear that 4 should be divided into 100, try another question that is like the electrical problem: If a man gets 4 dollars per hour, how many hours must he work to get 100 dollars?)

3. A battery pushed 240 coulombs through a circuit in 30 seconds. Find the current (in amperes). Given 240 coul. in 30 sec., how many coul. in each one second?

240/30 = 8 coul./sec. = 8 amperes.

For this type of problem, the following formula can be used:

$$\text{Amperes} = \frac{\text{coulombs of charge}}{\text{seconds of time}}$$

PRESSURE ENERGY OF ELECTRONS

It is often helpful to compare the flow of electrons through a wire-circuit to the flow of water through a pipe-circuit. The rate of flow of water (gallons or liters per second) through a pipe depends on the force or pressure on the water and on the size and frictional resistance of the pipe. When the pressure increases, the flow increases. Greater resistance causes less flow, figure 2-3. Similarly, the rate of flow of electrons through a wire depends on the resistance of the wire and on the electrical pressure or electromotive force (emf). The term, pressure, is inaccurate because it cannot be measured in pounds per square inch (or kilopascals) as we measure water pressure or air pressure. Yet, the pressure idea may be helpful in forming a mental picture of what determines the rate of electron flow in electrical equipment.

The unit of measure of emf is called a *volt*. This was named after Alessandro Volta, an Italian scientist who discovered how to make batteries.

Resistance

Electrical resistance is the name given to the internal friction involved in the passage of electrons through a wire, or through any material. The unit of measure of resistance is named the *ohm* after the German scientist, G.S. Ohm, who discovered that the current in a wire is proportional to the emf. One ohm of resistance is defined as exactly enough resistance to allow one volt of applied pressure (emf) to cause a current of one ampere.

1. When the emf applied to a device is increased, the current increases.

2. When the resistance is increased without changing the emf, there is less current.

The two facts above can be combined into one statement written as a formula that explains the relationship between the units of measure.

$$\text{Ampere of currents} = \frac{\text{emf in volts}}{\text{ohms resistance}}$$

This formula is called *Ohm's Law*. To shorten it further, it may be written either as

$$I = \frac{E}{R} \text{ or } E = IR.$$

E represents emf measured in volts. I stands for intensity of current, the electron flow rate, measured in amperes. R represents resistance, measured in ohms.

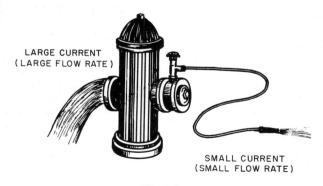

LARGE CURRENT
(LARGE FLOW RATE)

SMALL CURRENT
(SMALL FLOW RATE)

Fig 2-3

To illustrate how Ohm's Law, E = IR, is used, following are three examples:

1. If a lamp of 8-ohm resistance is connected to a 12-volt battery, how much is the current in figure 2-4?

E = IR

12 = I x 8

I = $\frac{12}{8}$ = 1.5 amps, ans.

2. How many volts are needed to cause a current of 1.5 amp in an 8-ohm lamp, figure 2-5?

E = IR

E = 1.5 x 8

E = 12 volts, ans.

3. Find the resistance of a lamp that permits a current of 1.5 amp when connected to a 12-volt battery as in figure 2-6.

E = IR

12 = 1.5 x R

R = $\frac{12}{1.5}$ = 8 ohms, ans.

Resistors

A *resistor* is any piece of material having a known electrical resistance which is used to control current or produce heat, figure 2-7. The heating element in toasters and irons is a wire or ribbon made of a nickel-chromium alloy. Such resistors have relatively few ohms

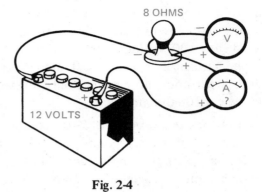

Fig. 2-4

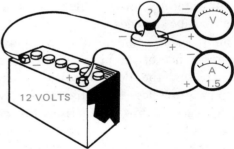

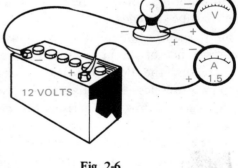

Fig. 2-5

Fig. 2-6

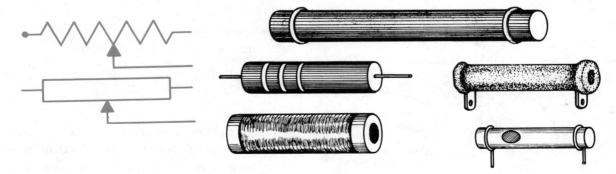

Fig. 2-7 Schematic and Pictorial Representation of Resistors

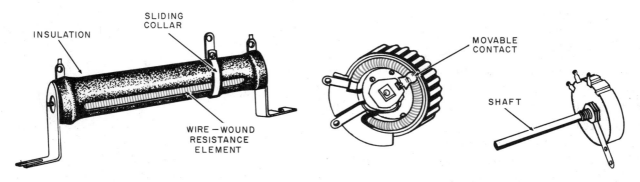

Fig. 2-8 Types of Variable or Adjustable Resistors

resistance (10-20) so they can allow a relatively large current which produces plenty of heat. The resistor in an incandescent lamp is a coil of fine wolfram (tungsten) wire of a size that will become white-hot. The color-banded resistors in radio and TV receivers are made in relatively small standard sizes. Their actual ohms values are independent of their physical size and may be small or great. For example, a 1-watt 47-ohm resistor will be the same physical size as a 1-watt 47 000-ohm resistor. The intended ohms value of this type of resistor is found by interpretation of the color code marking bands on the resistor. Refer to "Resistor Color Code Marking" in the appendix for information regarding the interpretation of resistor codes. These resistors are used to control amounts of current, and any heat they produce is undesirable.

The selection of wire used in making motors and generators, in home wiring or in the power-line wires along the street, is originally based on a calculation of resistance of wire. If the wire is too small, it wastes energy, if it is too large, it wastes copper.

Effects of Length, Material, and Cross-Section Areas

1. Resistance is directly proportional to the length of a wire. For example, if 50 feet (15 meters) of wire has one ohm resistance, 100 feet (30 meters) of the same wire has

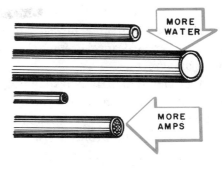

Fig. 2-9

2 ohms resistance. The 100 foot (30 meter) length has the resistance of 50 feet + 50 feet (15 meters + 15 meters).

2. Resistance depends on the kind of material, as shown in figure 2-13, page 23. Silver has the lowest resistance, copper is next.

3. Resistance is inversely proportional to cross-section area. This is a way of saying that the larger the wire, the less its resistance, provided we are comparing wires that differ only in thickness. A thick wire allows many electrons to move through easily, just as a wide road can carry many cars per hour, or a large pipe can easily let much water flow through, figure 2-9.

These three ideas are combined into one useful formula, using the following numerical representations for length, cross-section area, and kind of material.

Length: For ordinary use, feet or millimeters is the convenient measure.

Area: Because the use of square inches, square feet, or square millimeters would require inconveniently small numbers for ordinary sizes of wire, a more workable small unit of area is used.

Let this circle in figure 2-10 represent the end of a wire that is one-thousandth of an inch thick. The dimension, 0.001 inch, is often called one *mil.*

The area of this circle, one mil in diameter, we call one circular mil. A *circular mil* (C.M.) is a unit of area measurement. It is the same type of measurement as a square foot or an acre.

A circle of 0.001 inch diameter by definition has one circular mil area. What is the area of a circle of 0.002 inch diameter? (It would be found from $A = \pi r^2$. The answer would be in square inches . . . . much useless work.)

A circle 0.002 inch in diameter has 4 times as much area as a circle 0.001 inch in diameter. Comparing the two circles, recall that doubling the dimensions of any flat surface multiplies its area by four. Or, using area of a circle = πr^2, when r is 5, r^2 is 25; when r is 10, r^2 is 100. Comparing the 25 and the 100, we see that the radius-10 circle is four times as large in area as the radius-5 circle. The areas of circles may be compared by comparing the squares of their radii or the squares of their diameters. Notice that the same relationship exists in a square, figure 2-11.

What is the area of 0.002 inch diameter circle? It is four times as much as a one-mil circle, or 4 circular mils.

A circle 0.003 inch in diameter has how much area? 0.003 inch is 3 mils. 3^2 is 9. This circle is 9 times as large in area as a one-mil circle. Its area, therefore, is 9 circular mils. The convenience of the circular mil

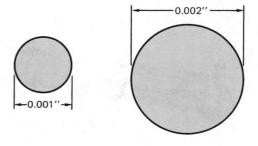

Fig. 2-10

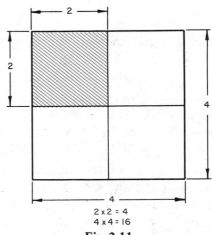

2 x 2 = 4
4 x 4 = 16

Fig. 2-11

area unit is that, by its use, we reduce the job of finding the area of a circle to this:

(1) Write the diameter of the circle in mils.

(2) Square this number (multiply by itself) and the result is the area of the circle, in circular mils.

For example, the diameter of 12-gauge wire is 0.081 inch or 81 mils. 81 x 81 equals 6 561. The cross-section area of the wire is 6 561 circular mils. (Even this calculation is not always necessary. Tables that tell the diameters of wires often also tell the circular mil cross-section area.)

Wire diameter is measured in terms of thousandths of an inch and the size is designated by gauge number. The larger the gauge number, the smaller the wire size. For instance, No. 12 wire is larger than No. 24 wire. The American wire gauge (AWG) system, which is the same as B & S, is used in the electrical industry in the United States.

As industry converts to the use of SI metric measurements, the size of wire will be specified according to its diameter. For example, a wire size of 1.25 mm is approximately the same as 16-gauge wire in the AWG system.

Returning to the resistance calculation, in which we were to decide how to represent length, area, and kind of material numerically, we find that length is in feet, and the cross-section area is to be expressed in circular mils, figure 2-12. Next, how should the type of wire be expressed numerically?

To simplify the proposed formula, the number used as a standard is the resistance of a piece of wire one foot long and one circular mil in cross-section area (mil-foot).

For example, find the resistance of a copper wire 20 feet long and 50 circular mils in end area.

The resistance of a mil-foot of copper is 10.4 ohms according to the table in figure 2-13.

We multiply the 10.4 by 20 (because resistance increases with length) and then divide by 50 (because, the larger the area, the less the resistance).

$$R = \frac{10.4 \times 20}{50} = \frac{208}{50} = 4.16 \text{ ohms, ans.}$$

The preceding calculation can be put in formula form:

$$R = \frac{K \times L}{C.M.}$$

In this formula, K is the ohms resistance of one mil-foot of the kind of wire used, L is the length of the wire in feet, and C.M. is the circular mils cross-section area of the wire.

The value of K for common wire materials is given in figure 2-13. Ohms per mil-foot for other metals can be calculated from the resistivity table, in the Appendix. The resistivity figure given in that table is the resistance of a cubic centimeter of material.

The C.M. area for various wire sizes may be found from the wire table in the Appendix.

USE OF THE WIRE-RESISTANCE FORMULA

$$R = \frac{K \times L}{C.M.}$$

1. Find the resistance of 150 feet of #20 aluminum wire. (From table below, K is 17. The wire table in the Appendix gives 1 022 C.M. cross-section area for #20 wire.)

$$R = \frac{17 \times 150}{1\,020} = 2.5 \text{ ohms, ans.}$$

2. How long a piece of #20 nichrome will have 30 ohms resistance? (Using the same formula, put 30 for R, 600 for K, C.M. is 1 020.)

$$30 = \frac{600\,L}{1\,020}$$

$$30\,600 = 600\,L$$

$$L = 51 \text{ ft., ans.}$$

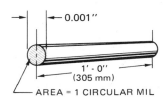

Fig. 2-12

OHMS RESISTANCE PER MIL-FOOT (at 70°F.) "K"	
Aluminum	17
Brass	42
Bronze (Cu, Sn)	108
Chromel (Ni, Cr)	420-660
Copper	10.4
German silver	200
Gold	14.6
Graphite	4 800
Iron, pure	59
Lead	132
Mercury	575
Nickel	42
Nichrome (Ni, Cr)	550-660
Platinum	60
Silver	9.6
Steels	72-500
Tungsten (wolfram)	33

Fig. 2-13

3. A two-wire power line 1 200 feet long (2 400 feet of wire) is to be erected. The resistance of the line should not be more than 1.5 ohms. What size copper wire should be used?

$$1.5 \ = \ \frac{10.4 \ \times \ 2 \ 400}{(C.M.)}$$

$$1.5 \ (C.M.) \ = \ 24 \ 960$$

$$C.M. \ = \ 16 \ 640$$

This is close to #8 wire (refer to wire table, page 500).

In using this formula, as in most electrical calculations, it is sensible to round off figures. A good rule is to round off numbers after the first three figures. For example, if the arithmetic problem is 10.4 x 2 317, multiply 10.4 x 2 320 and round off the answer to 24 100.

Measuring instruments, such as ammeters and voltmeters, can seldom be read to more than 3-figure accuracy. When an inexpensive voltmeter reads "118 volts," it means only that the voltage is somewhere between 116 and 120. According to the tables, #24 wire has 404-C.M. area, but wiremakers are also allowed a tolerance. A given sample of #20 wire might be 401 C.M. or 408 C.M.

Arithmetic processes deserve to be done with no more accuracy than the original measurements possess, or the final answer requires. In calculating a wire size, for example, whether the correct answer is 1 483 C.M. or 1 552 C.M. is of no concern because #18 wire (1 624 C.M.) will have to be used.

When wiring a circuit or connecting an appliance, use appropriate wire large enough to carry the current.

USE OF WIRE TABLES

Notice that the tables in the Appendix give ohms per 1 000 feet for copper and aluminum. This information makes it easy to find resistance of given lengths by comparison.

Example 1

Find the resistance of 3 000 feet of #6 copper at 68°F. (20°C).
The table gives 0.395 ohm per 1 000 feet. Three times as much wire will have 3 x 0.395 = 1.19 ohm, ans.

Example 2

Find the resistance of 200 feet of #18 copper at 68°F. (20°C). The table says that 1 000 feet has 6.39 ohms. 200 feet is one-fifth of 1 000, so we will take one-fifth of 6.39 ohms =1.28 ohms, ans.

There are conductors larger than 0000 (also written 4/0). Stranded cables larger than 4/0 are rated directly in circular mils. For flexibility, stranded wire is available. No. 16 stranded has the same amount of copper and the same current-carrying area as No. 16 solid. Solid rectangular conductors (bus bars) are used for large currents. They are often easier to assemble and more economical of space than large round conductors, and the flat shape provides more surface area from which heat can radiate.

Resistances of wires of various materials may be found by calculating information from the copper-wire table in the Appendix, then multiplying by factors given in figure 2-14.

Effect of Temperature on Resistance

Resistance depends not only on length, area, and kind of material, but also on the temperature of the material. The wire table in the Appendix gives two sets of values for resistance of copper at different temperatures. (The higher-temperature values are used in calculation of motor and transformer windings, which are intended to operate warm.) These figures illustrate the fact that

RESISTANCES OF METALS COMPARED TO COPPER (Copper = 1)	
Aluminum	1.59
Brass.	4.40
Gold	1.38
Iron	6.67
Lead	12.76
Nichrome.	60.00
Nickel	7.73
Platinum	5.80
Silver	0.92
Steel	8.62
Tin	8.2
Tungsten	3.2
Zinc	3.62

(See also Nichrome Wire Table, page 81 and Resistance of Metals and Alloys, Appendix, page 496.)

Fig. 2-14

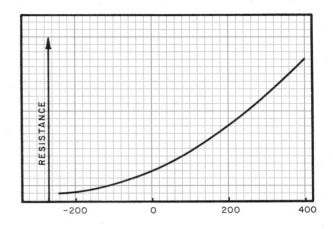

Fig. 2-15

as the temperature rises, the resistance of metals increases, figure 2-15.

The graph in figure 2-15 shows how the resistance of most metals increases as temperature goes up. At higher temperatures, not only is resistance higher, but it is also increasing at a faster rate.

The resistance of carbon decreases slightly with rise in temperature, the resistance of conducting liquid solutions decreases rapidly with temperature rise, and the resistance of semiconductors, such as germanium and metal oxides, decreases very rapidly as the temperature goes up.

Not all metals increase in resistance at the same rate, but for most metals the amount of increase is about 0.003 times the original resistance for each degree Celsius of temperature rise. For example, if a wire had 100 ohms resistance at 40°C, when heated to 50° it adds 100 x 0.003 x 10° = 3 ohms, so its resistance becomes 103 ohms at 50°C.

The resistance of a device when operating may be very different from its resistance when cold. For example, the resistance of an ordinary incandescent lamp in use is many times its cold resistance, and a measurement of cold resistance would not be usable to predict its behavior when heated.

The temperature of an inaccessible coil may be estimated from its resistance, found from voltmeter and ammeter readings. Occasionally overheating of a coil in a motor or transformer may be detected in this way.

POINTS TO REMEMBER

- Large quantites of electrons are measured in coulombs. A coulomb is a quantity or amount similar to liters of gasoline, bushels of grain, or pounds of sugar.

- Amperes measure rate of flow of electrons. One ampere is a rate of one coulomb per second.

- Electrons start moving at all points in an electric circuit when the switch is closed so the effect of closing the switch is practically instantaneous, although the average drift speed of individual electrons is slow.

- Volts measure electron-moving force, emf, electrical pressure.

- Ohms measure resistance, the opposition of atoms in the material to the passage of electrons.

- E = IR. This is Ohm's Law. E is emf, in volts. I is current, in amperes. R is resistance, in ohms.

- $R = \dfrac{K \times L}{C.M.}$

- Find K from the table in figure 2-13, C.M. from wire table in the Appendix, page 500. Use formula to find resistance length, or wire size.

- C.M. area = (diameter in mils)^2
  Square-mil area = C.M. area x 0.7854.

- Resistance is more
  for a long wire than a short one.
  for a thin wire than a thick one.
  for a hot wire than a cold one.
  for iron than it is for copper.

- Change in resistance, due to temperature change, is original resistance x coefficient x degrees change.

- Resistance of copper compared with temperature:

$$\frac{R_1}{234.5 + T_1} = \frac{R_2}{234.5 + T_2}$$

REVIEW QUESTIONS

1. How is the motion of electrons controlled?

2. In what direction do electrons flow in a resistor? Positive to negative, or negative to positive?

3. What is the purpose of a switch?

4. Define a coulomb and an ampere, and explain their relationship in the measurement of electron movement?

5. What is a volt? An ohm?

6. What happens when the resistance in an electrical circuit is increased without increasing the emf?

7. How is wire measured and the size designated?

8. In wire measure, what is a mil? What is a circular mil? What is a mil-foot?

9. What factors determine the total resistance of a wire?

10. How much is the cross-section area of a wire 0.012″ in diameter? Of a wire 0.0155″ in diameter?

11. Find the diameter of a wire that has 81 C.M. area.

12. Find the resistance of these wires using $R = \dfrac{KL}{C.M.}$

    a. 100 feet of #14 aluminum
    b. 25 feet of #20 nichrome
    c. One mile of #8 iron
    d. 6 inches of #18 copper

13. The emf applied to a 50-foot line connecting the power panel and a garage is 120 volts. Calculate the size copper wire needed to deliver a current of 30 amperes. Make the same calculations for an aluminum wire.

14. If an emf of 120 volts causes a current of 3 amperes in a lamp, calculate the resistance.

RESEARCH AND DEVELOPMENT

Experiments and Projects on Controlling Motion

INTRODUCTION

Perform as many of the six experiments listed as agreed with your instructor. Study the rules below before attempting to plan and construct electrical circuits. Of the projects suggested, two have been partially developed for your guidance.

EXPERIMENTS

1. Use Ohm's Law to measure amount of resistance.

2. Investigate how resistance varies with the kind of material used.

3. Show how resistance of a conductor varies with its length.

4. Determine the relationship between resistance of a conductor and its cross-sectional area.

5. Observe the effect of temperature of a conductor upon its resistance.

6. Measure resistance by using an ohmmeter.

PROJECTS

1. Measure solid copper wire in several sizes from No. 8 to No. 30 with a wire gauge and with a micrometer. Also measure several pieces of stranded wire.

2. Examine receptacles, switches, and other devices used in circuits to control electric current.

3. Plan several simple circuits to operate and control devices such as bells, buzzers, and lamps.

4. Plan and assemble a test board to use in learning how to measure volts, amperes, and ohms.

5. Disassemble an old iron, toaster, or some other electric heating appliance and examine the heating element. Notice the design and construction details of the unit.

6. Design and build a project in which resistance wire is used as a heating element.

RULES for planning a conducting path or circuit in which electrons are to flow and be controlled:

1. The high potential side of the current supply should be connected to the switch or other control device.

2. The remaining side of the switch or control device should be connected to the apparatus to be energized and controlled.

3. The remaining side of the apparatus should be connected to the neutral side of the current supply.

    Note: In certain applications, it is necessary to determine whether one side of a current source is connected to ground. Examples: A storage battery on a bench is not grounded, one terminal of a battery in an automobile is grounded to the chassis, and one side of the 120-volt ac line is connected to the earth. This ground, which is usually a white wire, should not be confused with bare wire used in safety grounding.

4. Colored wires should be used for the potential current supply and control parts of a circuit. The potential and control wires are usually black and other colors, such as red. The neutral return wires are white.

Always check the type and rating of a meter before installing it in a circuit.

EXPERIMENTS

Experiment 1 OHM'S LAW

OBJECT

To use Ohm's Law to find amount of resistance.

APPARATUS

1 - AC voltmeter (0-150 V)
1 - AC ammeter (0-1 A)
1 - AC ammeter (0-10 A)
1 - Lamp, 50-watt, 120-volt
1 - Lamp, 100-watt, 120-volt
1 - Lamp, 200-watt, 120-volt
1 - Electric toaster or iron

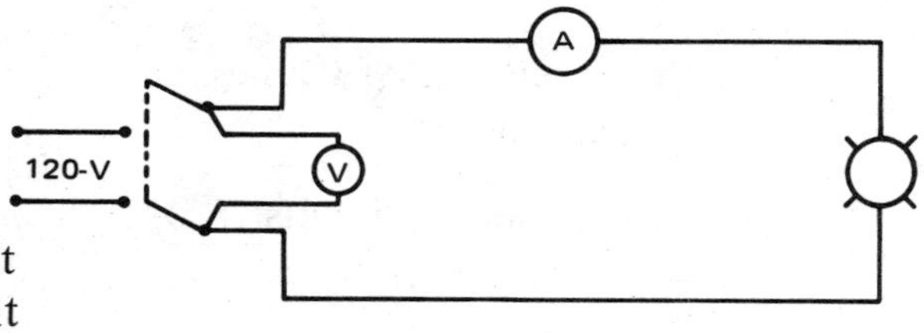

Fig. 2-16

PROCEDURE

1. Connect apparatus as shown, figure 2-16. Have the instructor check the circuit before closing the main switch.

2. For Observation No. 1, use the 50-watt lamp as a load and the 0-1 ammeter.

3. Energize the circuit, using 117-volt alternating current and record readings of the ammeter and voltmeter.

4. In Observation No. 2, use the 100-watt lamp as a load and record readings of the ammeter and voltmeter. (Start with high scale on ammeter.)

5. In Observation No. 3, change to the 0-10 ammeter and use the 200-watt lamp as a load. Record simultaneous readings of the ammeter and voltmeter.

6. In Observation No. 4, use the toaster or iron as the load and record simultaneous readings of the ammeter and voltmeter.

OBSERVATIONS

Use a table similar to the one shown here to record your observations.

OBS. NO.	LINE VOLTS	LOAD AMPERES	CALCULATED RESISTANCE IN OHMS
1			
2			
3			
4			

Sample Calculations

Resistance in ohms equals Electromotive force in volts divided by Intensity of current in amps.

$$R = \frac{E}{I}$$

QUESTIONS

1. Did the 200-watt lamp used in Observation No. 3 take twice the current that the 100-watt lamp required in Observation No. 2? Explain, using Ohm's Law to clarify your answer.

2. What would happen to the current required by the toaster in Observation No. 4 if the line voltage was decreased to 60 volts? Explain, using Ohm's Law as a reference.

3. What kind of devices have the highest resistance? Which device had the lowest resistance? Could you estimate which would have the highest resistance?

Experiment 2 FACTORS DETERMINING RESISTANCE

OBJECT

To learn how electrical resistance varies with the kind of material used.

APPARATUS

1 - Dry Cell

1 - DC ammeter (0-1 A)

1 - Board, 2' (600 mm) long with two terminals

1 - Miniature lamp (1.5 V)

1 - Miniature lamp receptacle

2' (600 mm) #22 wire of each of the following kinds: copper, aluminum, iron, German silver, manganin, nichrome, or other available material.

> When working with heaters and resistors, remember that they remain hot (the same as a piece of hot metal) for a considerable time after the power is disconnected.

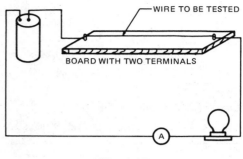

Fig. 2-17

PROCEDURE

1. Connect the terminals of the board, a dry cell, and a miniature lamp receptacle, as shown in figure 2-17. Fasten a piece of copper wire between the two terminals. Observe the brightness of the lamp and record your observation.

2. Connect the ammeter in the circuit and record its reading.

3. Remove the copper wire and try each of the remaining wires in turn. Record the brightness of the lamp and the ammeter reading in each case.

OBSERVATIONS

Use a table similar to the one shown here to record your observations.

KIND OF WIRE	BRIGHTNESS OF LAMP	AMMETER READING
COPPER		
ALUMINUM		
IRON		
GERMAN SILVER		
MANGANIN		
NICHROME		

QUESTIONS

1. Should a good conductor of electrons have a high or low resistance? Why?

2. Which will have greater resistance: a piece of iron wire or a piece of copper wire?

3. Which has the higher resistance: aluminum or copper? Nichrome or silver?

4. Name two good conductors of electrons.

5. Why is silver not generally used for an electric conductor?

6. Will a low-resistance conductor cause the lamp to burn brightly?

7. Name two materials which have a high resistance.

8. Name two materials which are used for resistors.

9. Explain why the resistance depends on the material of a conductor.

10. Why is aluminum sometimes used on high tension lines?

11. Why is copper used for most electrical conductors?

12. What materials might be used in making a heating element?

Experiment 3 LENGTH OF A CONDUCTOR

OBJECT

To learn how the resistance of a conductor varies with its length.

APPARATUS

1 - Dry cell

1 - DC ammeter (0-30 A)

1 - Board, 2′ (600 mm) with two terminals

1 - Miniature lamp (1.5 V)

1 - Miniature lamp receptacle

1 - Test clip

2′ (600 mm) #22 Nichrome wire

PROCEDURE

1. Connect the board with the resistance wire, dry cell, and miniature lamp, figure 2-18. Fasten the test clip to the left binding post. Record the brightness of the lamp.

2. Connect the test clip on the resistance wire 1/8 of the total length from the left binding post. Record the brightness of the lamp.

3. Repeat the process with the test clip at the 1/4, 1/2 and full-length positions. Record the results.

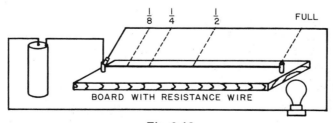

Fig. 2-18

4. Insert an ammeter in the circuit in place of the lamp. Repeat all the positions of the test clip taken before and record the readings of the ammeter.

OBSERVATIONS

Use a table similar to the one shown here to record your observations.

LENGTH OF RESISTANCE WIRE	BRIGHTNESS OF LAMP	AMMETER READING
0		
$\frac{1}{8}$		
$\frac{1}{4}$		
$\frac{1}{2}$		
FULL		

QUESTIONS

1. Does a long conductor pass more or less current than a short one?

2. Does the current vary directly or inversely with resistance?

3. If 1 000 feet (305 m) of #28 wire have a resistance of 65 ohms, what is the resistance of 2 000 feet (610 m) of #28 wire?

4. If 100 feet (30 m) of wire have a resistance of 4 ohms, what will be the resistance of 50 feet (15 m) of the same wire?

5. Name two factors which affect the resistance of a conductor.

6. How does the resistance of a wire vary with its length?

Experiment 4 CROSS-SECTIONAL AREA OF A CONDUCTOR

OBJECT

To learn how the resistance of a conductor depends upon its cross-sectional area.

APPARATUS

1 - Dry cell
1 - DC ammeter (0-30 A)
1 - Miniature lamp (1.5 V)
1 - Miniature lamp receptacle
1 - Board, 2' (600 mm) long with two terminals
    10' (3 m) copper wire of each of the following sizes: #24, #28, #32, #36

PROCEDURE

1. Connect the board with two terminals: the dry cell and the miniature lamp receptacle as shown in figure 2-19. Place the 10 foot (3 m) length of #24 copper wire between the two terminals, allowing it to loop around. Record the brightness of the lamp.

2. Repeat Step 1, using #28, #32, and #36 wire. Note and record lamp brightness.

3. Connect the ammeter in place of the lamp in the circuit. Place the 10-foot (3 m) length of #24 copper wire between the two terminals of the board, allowing it to loop around. Take the ammeter reading and record it.

4. Repeat Step 3, using #28, #32, and #36 wire. Note and record the ammeter reading.

5. Measure diameter of wire with wire gauge and with a micrometer. Look up dimension of gauge and measure with a micrometer. Look up dimension of gauge number in table.

OBSERVATIONS

Use a table similar to the one shown here to record your observations.

WIRE SIZE	BRIGHTNESS OF LAMP	AMMETER READING
#24		
#28		
#32		
#36		

QUESTIONS

1. Does an increase of resistance allow more or less current?

2. Does a large-diameter wire have more or less resistance than a smaller one?

3. Would three wires 1/3 as large in diameter as a single wire have more or less resistance? Explain your answer.

4. Does a bright lamp indicate more or less current than a dim one?

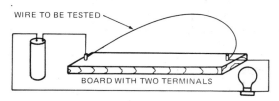

WIRE TO BE TESTED

BOARD WITH TWO TERMINALS

Fig. 2-19

5. How does the lamp used in the experiment indicate the relative amount of resistance in the wire to be tested?

6. How does the reading of the ammeter indicate the amount of resistance in the wire to be tested?

7. What effect does the cross-sectional area have on the resistance of a conductor?

8. Explain why it is more difficult for electrons to flow through a small wire than through a large wire.

9. Tell how to measure the size of a wire with a wire gauge.

10. How would you check the size of a wire with micrometers?

Experiment 5 TEMPERATURE

OBJECT

To observe that the resistance of a conductor depends upon its temperature.

APPARATUS

1 - Dry cell
1 - DC ammeter (0-30 A)
1 - Bunsen burner
1 - Support
2' (600 mm) #28 iron wire
3' (900 mm) #18 bell wire

PROCEDURE

1. Wind the 2-foot length of iron wire around a dowel to form a coil. Remove the dowel. Hang this coil on a support so that it is directly above an unlighted Bunsen burner or a propane torch. Connect the coil of wire to the dry cell and the ammeter as shown in figure 2-20. Take the reading of the ammeter and record.

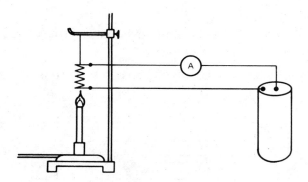

Fig. 2-20

2. Heat the wire with the Bunsen burner or propane torch. Notice the deflection of the pointer. Record the reading of the ammeter several times while the wire is heating.

OBSERVATIONS

Use a table similar to the one shown here to record your observations.

DESCRIPTION OF WIRE	RELATIVE TEMPERATURE	AMMETER READING	RELATIVE RESISTANCE
2 FEET OF #28 IRON WIRE	COLD		
	WARM		
	HOT		
	VERY HOT		
	RED HOT		

QUESTIONS

1. How does heating a metal change the movement of its atoms?
2. Why does heating a conductor limit the flow of electrons?
3. Will heating an iron wire increase or decrease its resistance? Why?
4. How does increasing the temperature of a conductor affect its resistance?
5. What does a decrease of current indicate in this experiment?
6. See if you can get information about one material, the resistance of which does not increase appreciably with an increase of temperature.
7. Explain how the increased movement of the atoms increases the resistance of a conductor.
8. How does an increase of temperature affect the resistance of the purest metals?

Experiment 6 RESISTANCE MEASUREMENT — USE OF OHMMETER

OBJECT

To learn how to measure resistance by using an ohmmeter and to learn color code.

APPARATUS

1 - Ohmmeter
1 each of the following resistors: ten color-coded resistors of widely different values

PROCEDURE

1. Short-circuit the terminals of the ohmmeter with a short piece of wire, and adjust the resistance until the pointer gives full-scale deflection.
2. Measure and record the resistance of each resistor.

OBSERVATIONS

Use a table similar to the one shown here to record your observations.

	RESISTORS									
	100	200	300	400	5 000	2 000	3 000	4 000	10 000	20 000
OHMMETER READINGS										

QUESTIONS

1. How is resistance measured with an ohmmeter?

2. Which of the following meters is used as a basic instrument in the ohmmeter: millivoltmeter, ammeter, wattmeter, milliammeter, or galvanometer?

3. Why should an ohmmeter read zero resistance when its terminals are connected with a short piece of wire?

4. Why should an ohmmeter read infinite resistance when its terminals are disconnected?

5. Why is a variable rheostat necessary on an ohmmeter?

6. What is the significance of color bands on resistors?

PROJECTS

DESIGN AND CONSTRUCT A TEST BOARD

This project may be constructed by an individual for personal use or assigned as a team project to be built as an equipment item for the shop.

The test board can be used when measuring current, emf and resistance in a circuit. It can also be used to test for continuity in a circuit or to locate a short circuit.

Note: When it is used for the latter purpose, a lamp should be substituted for the fuse.

MATERIALS

1 — Porcelain lamp receptacle
1 — Toggle switch and cover plate
1 — 4" octagon box
2 — 2" x 4" handy boxes for switch and convenience outlet
1 — Single convenience outlet mounted in 4" octagon cover plate
1 — Double convenience outlet and cover plate
4 — Binding posts
1 — AC voltmeter (0-150 V)
1 — AC ammeter, (0-50 A)
1 — Plug-type fuse, 15-ampere
1 — Prong-type lamp adapter

1 — Base, 18″ x 24″ (450 mm x 600 mm) (fiberboard or wood)
8 — Screw, kind and size determined by type of base selected

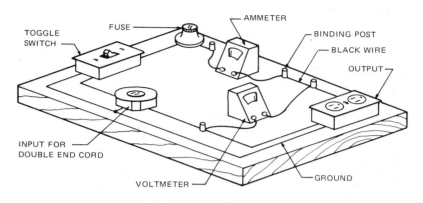

Fig. 2-21

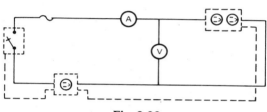

Fig. 2-22

PROCEDURE

Cut base to desired size and finish.

Make a 1/4″ = 1′ scale drawing of placement of parts and a circuit diagram. If metric dimensions are used, make the drawing to the scale of 1:50.

Mount parts on base.

Connect wires according to the circuit diagram, figure 2-22, and test circuit.

DESIGN AND BUILD AN ELECTRIC FEATHER TRIMMER

The electric feather trimmer is a device which utilizes the heating characteristic of electron flow through a suitable resistance wire.

MATERIALS

2 — Porcelain lamp receptacles
2 — 2″ x 4″ handy boxes for switch and convenience outlet
1 — Toggle switch and cover plate
1 — Convenience outlet
1 — Fuse

1 — Glocoil
1 — Pc. transite board
1 — Pc. sheet metal
2 — Pcs. wood
1 — Base, fiberboard or wood
    Assortment of screws

PROCEDURE

Study the pictorial diagram, figure 2-24.

Make some idea sketches of your own with approximate dimensions of parts.

Select your best sketch and consult your instructor for approval.

Make a scale drawing of the plan.

Make a circuit diagram.

Make a sketch with dimensions of each part.

Calculate the size and amount of resistance wire required for the cutting element. Form the cutting element.

Calculate the resistance necessary to control the heat in the element. Select a heater coil of the required size to control the amount of heat needed in the cutting wire.

Make all the pieces; assemble, test, and evaluate the unit.

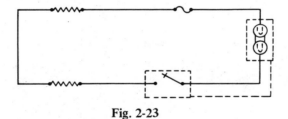

Fig. 2-23

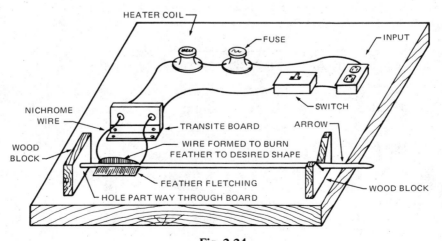

Fig. 2-24

Chapter 3
Electrical circuits

When someone connects an electric lamp to a wall outlet and turns it on, he or she is assembling and using a simple electric circuit. A circuit is any arrangement of materials that permits electrons to flow. It must contain a source of energy, connecting wires, and, to be useful, some device that makes use of electron energy, along with a switch for opening and closing the circuit.

A *circuit* implies a complete conducting path leading from the source, to the energy-using device, and back to the source again. Electrons are not used up as they flow through a lamp or a motor, and they can flow out of a source no faster than they are permitted to return to the source.

The necessity of a complete conducting path is illustrated by what may be done if the lamp fails to light: closing the switch, screwing the lamp bulb into the socket tightly to make metallic contact, replacing the bulb if an open (broken or burned-out) filament is suspected, and examining connections at the plug and lamp socket.

SERIES CIRCUITS

If electrons are provided only one possible route to follow from the source through several devices and back to the source,

this path is called a *series circuit*. Electrons must flow through each device in series at the same rate.

You may have seen a string of lights wired in series; when one lamp burns out, all of the lamps fail to light. The burnout of one lamp removes its filament from the circuit, which has the same effect as opening a switch in the circuit. When electrons cannot flow through one lamp, they cannot flow through any of the other lamps.

TOTAL RESISTANCE IN A SERIES CIRCUIT

A single 30-ohm resistor connected to a 120-volt source contains a current of 4 amps (Ohm's Law) figure 3-2, page 40.

$$I = \frac{E}{R} = \frac{120}{30} = 4 \text{ amps}$$

If two 30-ohm resistors are connected in series, electrons have twice as much opposition to fight through as they complete their

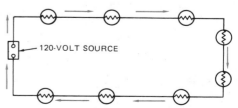

Fig. 3-1 A Series Circuit

circuit. That is, they have to overcome a total resistance of 60 ohms.

$$I = \frac{E}{R} = \frac{120}{60} = 2 \text{ amps}$$

Total resistance of a series circuit is the sum of the individual resistances.

If the symbols R_1, R_2, R_3, etc., are used for the ohms value of each resistor in a circuit, and the symbol R_T is used for the total resistance of the combination, a formula for finding the total resistance in a series circuit would be:

$$R_T = R_1 + R_2 + R_3 \ldots + R_n$$

The right-hand side of the equation has as many terms as there are resistors in the circuit.

CURRENT IN A SERIES CIRCUIT

In a series circuit, the one path through each resistor must be traveled by all of the electrons. Electrons must move through every part of the circuit at the same time. Therefore, to calculate the current in the circuit, all voltage sources and all resistances must be taken into account. When applying Ohm's Law to such a circuit, the voltage figure used in the calculation is the algebraic sum of the voltage sources in the circuit, and the resistance figure is the total resistance (R_T) described above.

Problem:

Eight lamps, 30 ohms each, are in series in a string on 120 volts, figure 3-3. Find the current. Ohm's Law is to be used for the whole circuit in this way:

Current in the circuit =

Voltage applied to the whole circuit
Total resistance of the whole circuit

$$I = \frac{120 \text{ volts}}{240 \text{ ohms}} = 0.5 \text{ amp, ans.}$$

This 0.5 amp is the current in any one lamp, and it is also the current in the whole circuit. There is only one current in a series circuit.

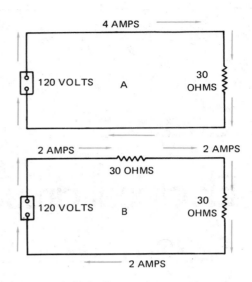

Fig. 3-2 Resistors in a Series Circuit (A) One 30-ohm Resistor (B) Two 30-ohm Resistors

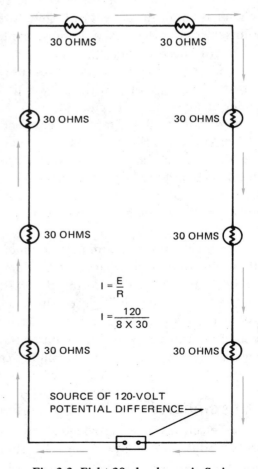

Fig. 3-3 Eight 30-ohm lamps in Series

> *To find current in a series circuit, use total resistance and line voltage applied to the circuit.*

$$I = \frac{\text{line voltage}}{\text{total resistance}}$$

Series circuits are common in electrical equipment. The tube-filaments in small radio receivers are usually in series. When one filament burns out, no tubes light. Some TV sets have a few tube filaments in series. For economy in wiring, incandescent street lamps are usually in series. When a filament is broken, a device in the lamp-holder closes the circuit to permit current to the other street lights in the series group.

Current-controlling devices are wired in series with the controlled equipment. A thermostat switch is in series with the heating element in an electric iron. Fuses are in series with the equipment they protect. Automatic house heating equipment has a thermostat, electromagnetic coils, and safety cutouts in series with a voltage source. Rheostats are placed in series with the coils in large motors to control motor current. All on and off switches, fuses, and circuit breakers must be wired in series with the devices controlled or protected. Some of these applications will be explored further after basic mathematical concepts of circuits are introduced.

AMPERES, VOLTS, AND OHMS IN SERIES CIRCUITS

Ohm's Law, $E = IR$, applies to each part of a circuit, as well as to the entire circuit. Using as an example five radio-tube filaments in series, with values as given in figure 3-4, find (1) current in each resistor, and (2) voltage across each resistor.

1. To find the current in each resistor, first find the total resistance of the circuit. $80 + 80 + 80 + 230 + 330 = 800$ ohms. Then $I = \frac{E}{R} = \frac{120}{800} = 0.15$ amp, ans. This is the

current everywhere in the circuit. It is also the current in each 80-ohm resistor; it is also the current in the 230- and the 330-ohm resistors.

2. We might connect a voltmeter to the 230-ohm resistor as shown in the diagram. The meter will tell us the voltage used by the 230-ohm resistor. This voltage can be calculated from the known current, 0.15 amp, and the 230 ohms, using Ohm's Law for just this one part of the circuit:

$$E = IR$$

Volts on one resistor = Current x ohms of one resistor

$E = 0.15 \times 230 = 34.5$ volts, figure 3-5.

Similarly, we can find the voltage for any one of the 80-ohm filaments:

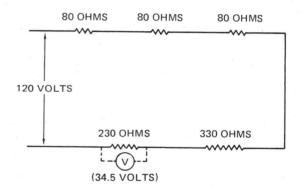

Fig. 3-4 Five Tube Filaments in Series

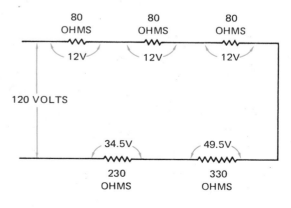

Fig. 3-5

E = 0.15 x 80 = 12 volts for each 80-ohm filament.

The voltage for the 330-ohm filament is 0.15 x 330 = 49.5 volts.

Do these five separate voltages have any relation to the 120 volts applied to the circuit? Yes, it is no accident that these five voltages, 34.5, 12, 12, 12, and 49.5 add up to 120 volts. These *part voltages* show how energy is divided up among the series resistors.

> *The total of individual series voltage drops equals the applied line voltage.*

In the case of a series circuit with more than one voltage source, the net voltage of the circuit must be determined. The sum of the voltage drops on series resistances or other devices equals this net voltage. For example, the two circuits shown in figure 3-6 are identical in terms of circuit current, power developed by R_1 and R_2, and voltage drops on each resistor regardless of the visual differences in their diagrams and the different physical arrangement of components. The reader should verify this equivalence mathematically and experimentally if there is any doubt about it.

Note that in both circuits shown in figure 3-6, the voltage sources are both attempting to move electrons around the circuit in the same direction. This connection is called *series aiding*. Net voltage in a series

aiding connection is the sum of the individual voltage sources so connected. If the polarity of one of the batteries were reversed, the net voltage would be 0 and there would be no current in the circuit. Voltage sources connected so as to force electrons in two different directions around the series circuit is called a *series bucking* condition. In such cases, the *net voltage* is the difference between the sum of sources of one polarity and the sum of sources of the opposite polarity.

The following problem is a more difficult one using these series-circuit principles. With the information shown in figure 3-7, find (A) current in R_2, (B) voltage across R_2.

(A) To find current in R_2, one might first try Ohm's Law, but E and I are not yet known. Observe from the diagram that the current in R_2 must be the same as the current in R_1, which can be found from I = E/R, using volts and ohms for the single resistor:

    I = 20 volts/10 ohms = 2 amps (for all parts of the circuit).

(B) To find volts across R_2, again try Ohm's Law, E = IR, but we still lack the value of R_2 in ohms. Recall that individual voltages must add to equal the applied line voltage, 120. The voltage across R_2 must therefore be 100 volts, to add with the 20 volts on R_1 to make the 120-volt total. Now that we have the current in R_2, 2 amps, and the voltage on R_2,

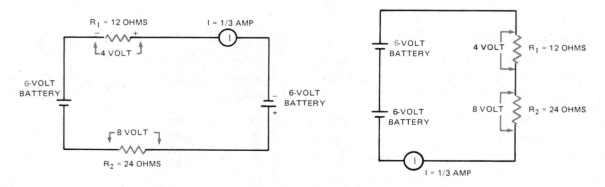

Fig. 3-6 Equivalent Series Circuits

100 volts, we can also find the ohms for R_2; R = 100/2 = 50 ohms.

Applications of Series Circuits

Since the current in a series circuit depends upon the continuity and the resistance of all devices in the system, the series connection is widely used for purposes of switching, continuous control, circuit protection, and voltage division.

Figure 3-8 illustrates a switch in series with the lamp it controls.

Figure 3-9 illustrates a fuse in series with the rest of the circuit. In either illustration if the switch is turned off, or the fuse wire melts, the rest of the circuit is off.

Figure 3-10 illustrates a variable resistor (sometimes called a rheostat) in series with a small dc motor. By varying the resistance of the rheostat, one can control the speed of the motor. This is a form of continuous control where the amount of power used by a device is adjustable over a wide range, not simply turned on or off. The brightness of lamps, or the amount of heat produced by a heater can be controlled in the same manner.

Continuous control by a series resistor, however, can be a waste of electrical energy. If the lamp in figure 3-8 usually draws 1 ampere on a 120-volt line and you wish to reduce this current to 1/4 of its original value, on what ohms value should the rheostat be set? To solve this problem, first find the total resistance needed by this circuit to limit the current to 1/4 amperes when 120 volts is applied.

$$R_T = \frac{E_T}{I} = \frac{120}{0.25} = 480 \text{ ohms.}$$

But, the total resistance is equal to the resistance of the rheostat plus the resistance of the lamp. If the resistance of the lamp were known, it could be subtracted from the 480-ohm total to find the rheostat setting. Resistance of the lamp, if it usually draws 1

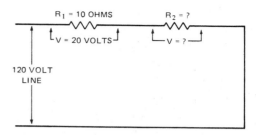

Fig. 3-7 A Series Circuit Problem

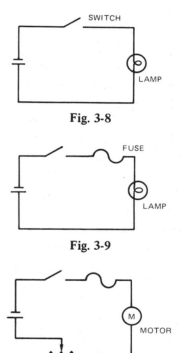

Fig. 3-8

Fig. 3-9

Fig. 3-10

ampere when 120 volts are applied, is

$$E = \frac{120 \text{ volts}}{1 \text{ ampere}} = 120 \text{ ohms.}$$

Therefore, the setting of the rheostat must be:

480 ohms - 120 ohms = 360 ohms.

If the circuit is studied further, one discovers that in this controlled position the rheostat has 90 volts across it while the lamp has only 30 volts across it. The current through both the lamp and the rheostat are equal even though the rheostat has three times as much resistance. Therefore, we must be using three times as much energy in the rheostat for control purposes as we are using in the lamp. This waste of energy in order to gain control is unavoidable with a simple series rheostat. In later chapters, more efficient power controls will be discussed.

Voltage division applications of series circuits are special cases of the continuous control discussed above. If it were desirable to operate a 9-volt transistor radio from a 12-volt source (for example the cigarette lighter in an automobile), the source must be divided into 3-volt and 9-volt components. Since the applied voltage to a series circuit is equal to the individual voltage drops on the components, it is only necessary to connect the right value of resistor in series with the radio and the 12-volt power source.

Figure 3-11 illustrates such an arrangement. The real problem is finding the right resistance. Before the problem can be solved, one must know how much current the radio draws on its proper 9-volt source. An ammeter determines this current requirement by connecting it in series with one of the wires between a 9-volt battery and the radio. A particular radio may draw 150 milliamperes. (A *milliampere* (ma) is one thousandth of an ampere.) When the radio is connected in series with the 12-volt source and a resistor, it should still experience a 9-volt drop across it with 150 milliamperes of current through it. The resistor selected must experience a 3-volt drop with the same flow of 150 milliamperes. Therefore, the ohms value of the resistor will be given by Ohm's Law.

$$R = \frac{3 \text{ volts}}{0.150 \text{ amperes}} = 0.150 \overline{)3.000}^{\,20 \text{ ohms}}$$

Note: *Your* transistor radio may draw more or less than 150 milliamperes. Therefore, measure the current load and recalculate the resistance if you intend to save money on batteries using this circuit.

Often the technician needs a particular value of resistor for a circuit repair or an experiment and discovers that it is not in

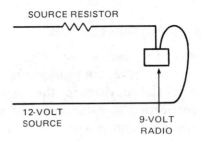

Fig. 3-11

stock. The technician simply remembers that total resistance is equal to the sum of the individual resistors and selects two or more smaller resistors to accumulate the value needed. For example, two 75 ohm resistors in series are just as good as a 150 ohm resistor for most applications.

As proven by the examples discussed, series circuits and series connections have many applications. Mastery of the principles involved and a lively imagination can solve many problems which the reader may encounter. Additional applications of series circuit principles should be recognized throughout the rest of this text. Summarizing the three important rules for series circuits:

1. Total resistance equals the sum of individual resistances.
2. Current is the same everywhere.
   $$I = \frac{\text{Line voltage}}{\text{Total resistance}}$$
3. Line voltage equals total of individual voltages.

PARALLEL CIRCUITS

When two or more devices are connected to a source of energy in a way that divides the total current causing the electrons to flow through each device in a separate path, the devices are said to be connected in *parallel*.

Parallel circuits are frequently used. Most lamps and home appliances are connected in parallel, so that each one can be operated independently. In series circuit either everything operates or nothing operates. If devices are to be turned on and off separately without affecting other devices, they are wired in parallel.

A parallel circuit can be compared to a circuit of water. The water will represent the electrons; the water pressure, the electrical pressure, or voltage; the branches from the water main represent the branches of the electrical circuit.

Trace through the circuits in figures 3-12 and 3-13, noticing how the current divides and recombines.

Electrons

- The wires are full of electrons, whether there is a current or not.

- Electrons in the supply side have more potential energy than those on the return side. This energy difference (volts) drives electrons through each device; each operates at the same voltage.

- Branches A, B, and C can have separate currents. A could have 2 amps; B, 3 amps; and C, 4 amps, with a total of 9 amps through the generator.

Water

- The water main and ditch are full of water, whether there is current or not.

- Water in the supply main has more pressure energy than water in the ditch. This energy drives water through each device; each operates at the same pressure.

- Branches A, B, and C can have separate currents. A can have 2 gal/sec.; B, 3 gal/sec.; and C, 4 gal/sec., with a total of 9 gal/sec. through the pump.

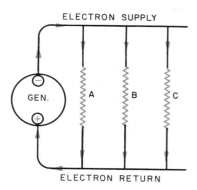

Fig. 3-12 A Generator and Three Resistors in Parallel

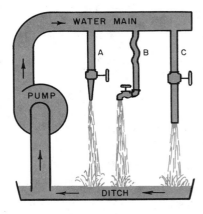

Fig. 3-13 A Pump and Three Water Valves in Parallel

Calculating Currents in Parallel Circuit Using Ohm's Law

The following discussion explains how Ohm's Law applies to parallel circuits in solving some typical problems of current and resistance.

In figure 3-14, page 46, each switch controls only one device. If all switches are open, there is no current, but the 120-volt potential energy difference exists between the two line wires.

In a simple parallel circuit, all devices operate on the same line voltage.

When the switches are closed, the voltage applied to each device is the same because each device has its own direct wire connection to the 120-volt source. Therefore, the current through each device is independent of the other currents and depends only on the line voltage and the individual resistance.

The current in each device is found by Ohm's Law, $I = \frac{E}{R}$.

For A: $\frac{120}{240} = 0.5$ amp

For B: $\frac{120}{72} = 1.67$ amps

For C: $\frac{120}{300} = 0.4$ amp

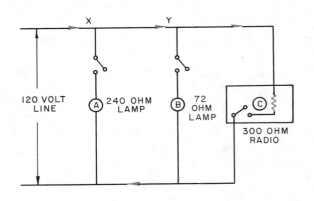

Fig. 3-14 Three Devices in Parallel

To find the current in the line at X: All electrons pass point X, that is, the current at X is 0.5 + 1.67 + 0.4 = 2.57 amps.

The total current in a parallel circuit is the sum of the individual currents.

How much current is there at Point Y? The diagram shows that electrons must pass point Y on their way to B and C, so the current at Y is 1.67 + 0.4 = 2.07 amps.

Figure 3-15 shows another example, using the same principles: R can be found by Ohm's Law, if we can first find the current in R.

How much current can the 30-ohm resistor on 120 volts have? I = 120/30 = 4 amps. If the total current is 10 amps, then the unknown resistance must contain a current of 6 amps.

$$R = \frac{120}{6} = 20 \text{ ohms}$$

The example above, with 6 amps in a 20-ohm resistor, 4 amps in the 30-ohm resistor, and 10 amps total current, illustrates a few facts about parallel circuits:

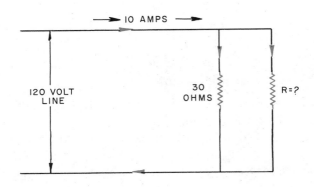

Fig. 3-15 Two Devices in Parallel

1. With both resistors in use, there is more total current than if only one were connected. Each resistor is another opportunity for electrons to move from one line wire to the other. This combination of two resistors allows a total current of 10 amps from a 120-volt source, so the 30-ohm resistor and the 20-ohm resistor together have the same effect as a single 12-ohm resistor.

As more resistors are connected in parallel to a line, the total effective resistance of the combination is reduced and line current increases.

2. There is more current in a low-resistance branch than in a high-resistance branch, but each has a current that is determined by $I = \dfrac{E}{R}$.

3. When too many additional appliances are added to a parallel circuit, the total current is sometimes increased to a dangerous level. This causes a fuse to blow or a circuit breaker to disconnect the circuit.

Finding Total Resistance of a Combination of Parallel Resistors

There are three ways of finding the combined resistance of a group of resistors in parallel. The method used depends partly on the type of problem and partly on personal preference. Method (2) is often recommended because it will solve most practical problems without using too much arithmetic. Method (3) is easiest when more than two resistors are involved and a pocket calculator can be used.

1. If all of the resistors in the group have the same ohms resistance, divide the resistance of one by the number of resistors in the group.

Three resistors, 60 ohms each, are in parallel. It is three times easier for electrons to go through the group than it is to go through only one, so the total effective resistance is 20 ohms.

Four resistors, each rated 12 000 ohms, are in parallel. Their total resistance is 12 000 ÷ 4 = 3 000 ohms.

If a technician needs a 15 000-ohm resistor but only has a drawerful of 10 000-ohm resistors, how can they be combined to make 15 000 ohms? One way: two 10 000-ohm resistors in parallel equal 5 000 ohms. Another 10 000 added in series to the parallel pair makes the total 15 000, figure 3-16.

2. If the resistors are not alike, the total resistance of two resistors in parallel can be

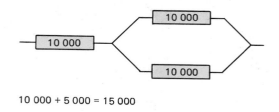

10 000 + 5 000 = 15 000

Fig. 3-16 A Series-Parallel Arrangement of Resistors

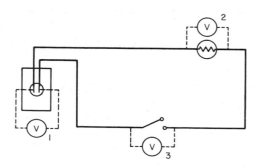

Fig. 3-17 Location of Voltage

found by dividing their product by their sum.

A 30-ohm and 20-ohm resistor are paralleled. Find their total resistance.

$$R = \frac{product}{sum} = \frac{30 \times 20}{30 + 20} = \frac{600}{50} = 12 \text{ ohms}$$

This method can be used for only two resistors at a time. Given the problem of finding the total resistance of 30 ohms, 20 ohms, and 10 ohms in parallel, the 30 and 20 can be combined as before, giving 12 ohms. Then this 12 ohms can be combined with the 10 ohms.

$$R = \frac{12 \times 10}{12 + 10} = \frac{120}{22} = 5.45 \text{ ohms}$$

3. Another method, useful for unlike resistors, is the formula $\dfrac{1}{R} = \dfrac{1}{r_1} + \dfrac{1}{r_2} + \dfrac{1}{r_3}$, where R is the total resistance and each smaller r is the resistance of one of the resistors

in the group. The formula uses as many 1/r terms as there are single resistors.

Find the total resistance of resistors of 30, 20, and 10 ohms.

$$\frac{1}{R} = \frac{1}{30} + \frac{1}{20} + \frac{1}{10} \; ; \quad \frac{1}{R} = \frac{2}{60} + \frac{3}{60} + \frac{6}{60}$$

$$\frac{I}{R} = \frac{11}{60} \; ; \frac{R}{I} = \frac{60}{11} \; ; \quad R = 5.45 \text{ ohms}$$

If one has access to a pocket calculator, this type of problem is solved by dividing 1 by each ohmic value, adding the results, and then dividing 1 by the sum of the results. For example:

$$1/30 = 1 \div 30 = 0.033\,33$$
$$1/20 = 1 \div 20 = 0.050\,00$$
$$1/10 = 1 \div 10 = \underline{0.100\,00}$$
$$\text{Sum} = 0.183\,33$$
$$R = 1 \div 0.183\,33 = 5.45 \text{ ohms}$$

The necessary divisions and additions are easily performed on the calculator. Hand calculation using this method is tedious.

In all cases, the total combined resistance of a group of resistors in parallel is less than the resistance of the smallest resistor in the group. Adding resistors to a parallel circuit is not a matter of increasing the opposition to current, but rather adding opportunities for electrons to flow from one line wire to the other.

SHORT CIRCUIT

A parallel path of very low resistance, often caused accidentally, is given the name *short circuit.*

For example, damaged insulation on an appliance cord may permit the two copper wires to touch each other, forming a path of nearly zero resistance. This permits an unreasonably large current in the wires leading to the place of contact. This large current could overheat the wires, starting a fire. To prevent this, fuses are used in series in each house circuit. Excessive current melts the fuse wire, thus opening the circuit and stopping the flow.

The term *open circuit* applies to a circuit containing an open switch, a burned-out fuse, or any separation of wires which prevents the flow of electrons.

OPEN CIRCUITS

When a lamp is turned off, where does the current go? That is an easy question, for it is similar to asking "when a water faucet is turned off, where does the water-flow go?" It does not go anywhere. The flow stops. There is still water in the pipe; its motion has ceased. When we turn off the lamp, the electrons are there, but not moving. Current is motion.

When a lamp is turned off, where does the voltage go? It is still there.

(1) If a voltmeter is connected to the source, it reads 120 volts. At the source, electrons still have energy, even though it may not be in use. (2) If the voltmeter is connected across the lamp, it reads zero, for V = IR, and I = zero. (3) Do you expect zero at the switch? When you try it, the meter reads 120 volts. Think of the switch and lamp as two elements in series. The total of their individual voltages should equal the applied total line voltage. The lamp uses up no volts now, so 120 at the switch plus 0 at the lamp equals the 120 volts supplied at the source. Does V = IR apply to the open switch? Yes, but we cannot learn much this way: V = 120, I = 0, then R = 120/0 which we cannot calculate because it is so large.

VOLTAGE DROP ON A LINE

There are two related reasons why lights in a house dim when a motor is started. First, the starting current requirement on a motor is generally much greater than that required to maintain its running speed. Therefore, the dimming effect may be of a short duration as the motor starts. Second, the wires taking electrical energy to and from the home do have a small value of ohms resistance. This resistance is necessarily in series with the rest of the house electrical load. Series and parallel circuit principles and Ohm's Law are all that is needed to explain the dimming situation.

Assume that each wire leading to the house has 1/2-ohm resistance, and lamps in the house cause a 2-amp current in the line. We then have a series circuit, and can calculate the voltage at the house.

Each line wire is, in effect, a 1/2-ohm resistor with 2 amps in it, figure 3-18.

E = IR = 2 x 1/2 = 1 volt which is used in forcing 2 amps through the 1/2 ohm of wire. One volt is used on each wire. From the 120, this leaves 118 volts between wires at the house.

If a motor is turned on, so that the current in the line becomes 20 amps instead of 2 amps, more volts will be used up on the line leading to the house.

E = IR = 20 x 1/2 = 10 volts for one wire, and another 10 volts for the other. Subtracting this 20 volts from 120 volts leaves 100 volts delivered at the house.

With 2 amps in the line, voltage at the house was 118 volts. With 20 amps in the line, voltage at the house is 100 volts. Since all the rest of the home fixtures, including the lamps, are wired in parallel and parallel-wired devices receive the same voltage, then the lamps only receive 100 volts. Lights are dimmer on 100 volts than they are on 118 volts because with less voltage, there is less current in the lamps. This 2-volt or 20-volt loss is called *voltage drop* and, by Ohm's Law, depends on the resistance of the line and current in the line.

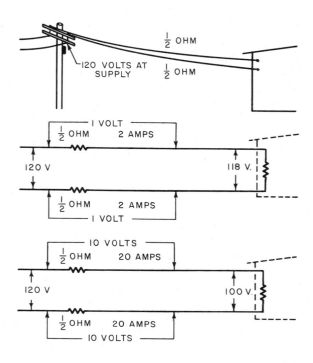

Fig. 3-18 Calculating Voltage Drop

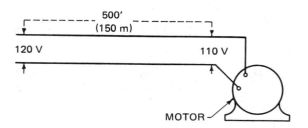

Fig. 3-19 Problems in Voltage Drop

Example of a Line-Voltage Drop Calculation

A certain electric motor requires at least 12 amps at 110 volts to operate properly. It is to be used 500 feet from a 120-volt power line. What size copper wire must be used for the 500-foot extension?

Solution: 120 volts is available at the start. 110 volts must be delivered to the motor, so allow 10 volts to be used on the wires when current is 12 amps.

Knowing the 10 volts and 12 amps, we can find resistance of the wires of the extension:

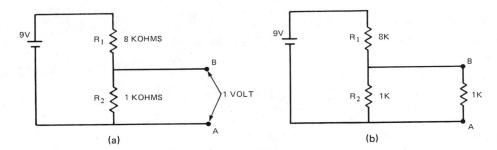

Fig. 3-20 Voltage Divider Circuits (1K = 1 000, 8K = 8 000)

$R = E/I = 0.83$ ohm. (These wires are a resistance in series with the motor. In Ohm's Law, we used the 10 volts to find the resistance of the wire that is responsible for the 10-volt potential difference.)

Wire size can now be found from the tables, page 500. The 500 foot extension uses 1 000 feet of wire, with not more than 0.83-ohm resistance. Using the "Ohms per 1000 feet" column, we find the closest wire size is No. 9. (If No. 9 copper is not available, the line should be made of No. 8.)

Since the resistance is directly proportional to the length of the wire, it is important to consider the distance from the source to the apparatus using current. You must be certain that the wire size is adequate to deliver the required current without excessive voltage drop. This is especially important with motors because at reduced voltage, the motor may fail to start or run properly. If the motor stalls due to low voltage, it overheats, causing serious damage to the motor or wiring, with the added possibility of causing a fire.

> Electric power tools operate more efficiently when supplied with adequate power. When using an extension cord, select one with wire of sufficient size to avoid voltage drop. Low voltage causes tools to overheat and possibly burn out.

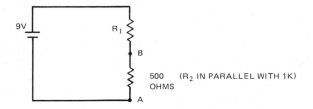

Fig. 3-21

SERIES-PARALLEL CIRCUITS

Often it is advisable to operate a device requiring a low voltage from a high-voltage source. Assume that one volt is needed from a 9-volt source. Two series resistors can be used to accomplish the division as in figure 3-20.

Simple circuit analysis shows that one volt is available at terminals A and B. Such a situation is called an *unloaded voltage divider* because the voltage has been divided, but the real power user (load) has not been connected. If one attempts to use the one volt obtained from the unloaded voltage divider, it is found that the voltage falls below one volt the instant the load is connected. Analysis of figure 3-20, where a 1 K ohm resistor is connected to terminals A and B, will show why the voltage falls. The parallel combination of R_2 and R_1 is 500 ohms. The equivalent circuit of the loaded divider showing R_2 and R_1 combined appears in figure 3-21.

The total circuit resistance is now 8.5 K ohms. The circuit current is:

$$I = \frac{9 \text{ volts}}{8\ 500 \text{ ohm}} = 0.001\ 06 \text{ amp} = 1.06 \text{ ma.}$$

The voltage A to B is

V = 500 ohms x 1.06 milliamperes (ma) = 0.53 volts.

The addition of a load lowers the resistance of that portion of the voltage divider, and therefore lowers the voltage that had been across that portion of the circuit.

Loaded voltage dividers are only one of a number of series-parallel circuit combinations that a technician might have to design or analyze. To use this analysis technique, combine parallel resistor portions into single resistor equivalents and then use basic series and parallel analysis procedures to determine voltage drops. Once voltage drops are determined for each resistor, Ohm's Law can be used to find individual currents. An example is given in figures 3-22 through 3-24 where each circuit is mathematically equivalent to the others.

R_3 and R_4 are in series, figure 3-22, and therefore can be combined into a single 150-ohm resistor. The new equivalent circuit is shown in figure 3-23. The combination of R_2 with R_3 - R_4 to equal 75 ohms is shown in figure 3-24. The total resistance of the circuit is 100 ohms. Therefore, total current from the source is:

$$I = \frac{10}{100} = .1 \text{ amp}$$

This .1 amp flows through R_1; therefore, voltage drop on R_1 is 25 x .1 = 2.5 volts. The equivalent circuit, figure 3-24, indicates that the same .1 amp flows through the 75 ohms of combination resistance R_2, R_3, and R_4. Therefore, the voltage drop across this portion of the circuit is 75 ohms x .1 amp = 7.5 volts. Since R_2 is directly in parallel with this part of the circuit, its voltage drop must be 7.5 volts. R_2 current will be

$$I = \frac{7.5 \text{ v}}{150 \text{ ohms}} = 0.05 \text{ amp.}$$ The current

through combination R_3, R_4 must also be

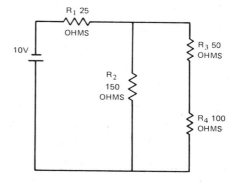

Fig. 3-22

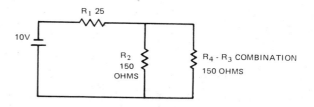

Fig. 3-23

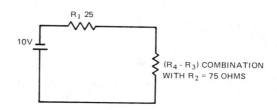

Fig. 3-24

0.05 amp (I = $\frac{7.5 \text{ v}}{150 \text{ ohms}}$).

The voltage drop on R_3 will be V = 50 ohms x 0.05 amp = 2.50 volts. The voltage drop on R_4 will be V = 100 ohms x 0.05 amp = 5 volts.

Unfortunately, series-parallel combination problems do not come in standard format to the technician. It is not possible to suggest a specific set of rules and procedures to apply in each case. Mastery of the general series and parallel relations and a little creativity will generally suffice.

POINTS TO REMEMBER

For Series Circuits

- There is only one current in a series circuit. Each device has the same current as the other devices.

$$I_L = I_1 = I_2 = I_3 = \ldots I_N$$

- Current depends on applied voltage and total resistance.

$$I_L = \frac{E}{R_T}$$

- Total resistance is the sum of individual resistances.

$$R_T = R_1 + R_2 + R_3 + \ldots R_N$$

- The sum of the individual voltages is equal to the total applied voltage.

$$E_L = V_1 + V_2 + V_3 + \ldots V_N$$

- Individual voltages are found from $E = IR$, used for individual resistances.

For Parallel Circuits

- Line current divides, so electrons pass through each separate device, independent of other devices.

- Total line current is the sum of the individual currents in the parallel paths.

$$I_T = I_1 + I_2 + I_3 + \ldots I_N$$

- If several equal resistors are in parallel, their combined resistance equals resistance of one ÷ number of resistors.

$$R_T = \frac{R \text{ (of one)}}{N}$$

- The combined resistance of any two resistors in parallel is equal to product ÷ sum.

$$R_T = \frac{R_1 \times R_2}{R_1 + R_2}$$

- Total resistance of any number of resistors in parallel is found by solving this equation for R_T.

$$\frac{I}{R_T} = \frac{1}{R_1} + \frac{1}{R_2} + \frac{1}{R_3} + \ldots \frac{1}{R_N}$$

- Combination circuits are best analyzed by drawing more simple equivalent circuits.

- A short circuit is an accidental path of too-low resistance.

REVIEW QUESTIONS

1. Name examples of useful series circuits. Would it be reasonable to connect the two headlights of an automobile in series?

2. In a circuit like figure 3-1, if the current is 5 amps entering the first resistor, how much current is there leaving the last resistor? And how much current at a point half-way around the circuit?

3. What is used up in an electric circuit — electrons, current?

4. Resistors of 500, 5 000 and 4 500 ohms are connected in series to a 100-volt source. Calculate the current in each, and find the potential difference (volts) across each.

5. Two wires lead from a pole to a house. The resistance of each wire is 0.4 ohm, the current is 10 amps, and the voltage between wires at the pole is 122 volts. Calculate:
   (a) Voltage drop in the line (both wires combined).
   (b) Voltage between wires at the house.

6. If the current is 25 amps in the same line described in Question 5, how much is the voltage drop in the line, and voltage at the house?

7. A certain motor needs at least 8 amps at 209 volts to operate properly. It is to be used 600 feet from a 225-volt source. How much resistance can the line have? (Suggestion: Find voltage drop first.)

8. A 10-ohm resistor is connected in series with another resistor (of unknown resistance) to a 120-volt source. The voltage measured across the 10-ohm resistor is 48 volts. Find (a) current in each resistor; (b) resistance of the second resistor.

9. In a series string of eight lamps used on a 120-volt line, how much is the voltage across each single lamp? If two more lamps are connected in the string, making a series of ten lamps, how much voltage is across each? What effect will this have on the current in the lamps? On the life of the bulbs?

10. What features distinguish a parallel circuit from a series circuit?

11. Three resistors, one 4-ohm, one 10-ohm, and one 6-ohm are connected in parallel to a 12-volt battery. Calculate the current in each resistor. Calculate the current in the battery.

12. Four resistors, 2 200 ohms each, are connected in parallel. This group is connected to a 110-volt line. Find the current in each single resistor. Find the total current in the line leading to the group of resistors. Find the combined resistance of the group of resistors, by two different methods.

13. Three lamps of resistances 48, 72, and 240 ohms are connected in parallel to a 120-volt line. Find the current in the line.

14. Three resistors are connected in parallel to a 120-volt line. Total current is 12 amps. One resistor is 20 ohm, one is 30 ohm. Calculate ohms for the third resistor.

15. A 50-ohm resistor, carrying 0.48 amp, is in parallel with a 60-ohm resistor. Find the current in the 60-ohm resistor.

16. A certain string of lamps consists of eight lamps in series. Lamp #3 fails to light, but the other seven light brightly. When lamp #3 is removed from the socket, the other seven remain bright. Is there any fault in the circuit? Explain.

17. Secure the data on the plate of several motors — 1/4, 1/2, 1, 2 and 3 to 5 horsepower, ac 115/230-V single phase. Compare the ampere rating for operation at 115 V and at 230 V. Assume that the motor will be placed 100 feet from the distribution panel. Calculate the wire size required to operate each motor on 115 volts and on 230 volts.

18. You wish to use a flexible rubber-covered extension cord 30 feet long to operate an electric power tool. The extension cord is made with No. 18 copper wire. Calculate the voltage drop if the tool takes 10 amps at 115 volts. Calculate the drop if the tool takes 5 amps at 230 volts.

| RESEARCH AND DEVELOPMENT |

Experiments and Projects on Electrical Circuits

INTRODUCTION

Electrical circuits form the path electrons travel to provide electrical energy where it is needed to do work. Included in the circuit is a source of energy. Connecting wires control a device and an appliance that makes use of the electron energy. In Chapter 3, you studied about the relationships of volts, amperes, and ohms in a circuit. The experiments and projects in this unit provide an opportunity for you to use, control, and measure electrical energy in series, parallel, and series-parallel circuits.

EXPERIMENTS

1. To investigate the relationships of current, voltage, and resistance in a series circuit.

2. To show voltage loss in a conductor and to prove that voltage drop is directly proportional to the wire length.

3. To use series circuits to determine resistance by the voltage drop method.

4. To investigate the relationships of current, voltage, and resistance in a parallel circuit.

5. To study the path an electron may take through a parallel-multiple circuit.

6. To check and verify calculated resistance values of a series-parallel circuit using an ohmmeter.

PROJECTS

1. Nine low-voltage circuit problems are suggested to provide experience in planning wiring diagrams and in methods of connecting component parts to form a circuit. Review "Rules for planning a conducting path or circuit in which electrons are to flow and be controlled.", Chapter 2.

Project 1. Plan, wire, and test a circuit containing a buzzer, a pushbutton, and current supply.

Project 2. Plan, wire, and test a series circuit containing a buzzer, two pushbuttons, and current supply.

Project 3. Plan, wire, and test a parallel circuit containing a buzzer, two pushbuttons, and current supply.

Project 4. Plan, wire, and test a series circuit containing two bells, a pushbutton, and current supply.

Project 5. Plan, wire, and test a parallel circuit containing two bells, a pushbutton, and current supply.

Project 6. Plan, wire, and test a circuit containing a bell and a pushbutton for the front door, a buzzer and pushbutton for the rear door.

Project 7. Plan, wire, and test a circuit containing a chime, a pushbutton for the front door, a pushbutton for the rear door, and current supply.

Project 8. Plan, wire, and test an apartment door lock circuit containing a buzzer, door lock device, two pushbuttons, and current supply.

Project 9. Plan, wire, and test a circuit containing a switch, a lamp, and a pilot light.

2. Using a double-pole, double-throw switch, devise a circuit so that when the switch is one way, two lamps will light in series. When the switch is engaged the other way, two lamps will light in parallel. Wire and test.

3. Examine a string of lights to determine if they are connected in series or in parallel. If they are connected in series, measure the voltage. Cut the line and add one or two lights and measure the voltage. What happened?

EXPERIMENTS

Experiment 1 SERIES CIRCUIT RELATIONSHIPS

OBJECT

To study the relation of current, voltage, and resistance of the entire combination to current, voltage, and resistance of the separate parts of a series circuit.

APPARATUS

1 - AC ammeter (0-10 A)
1 - AC voltmeter (0-150 V)
1 - Heater coil, 600-watt
1 - Lamp, 200-watt, 120-volt
1 - Lamp, 300-watt, 120-volt
1 - DPST knife switch
3 - SPST knife switches

PROCEDURE

1. Connect the apparatus as shown in the circuit diagram, figure 3-25, page 56.
2. Have the instructor check the circuit before closing the main switch.
3. Close the main switch and single-pole, single-throw switches 2 and 3 and take simultaneous readings of voltmeter and ammeter.

4. Open main switch, disconnect ammeter and insert it between R_1 and R_2. Connect the voltmeter across R_2.

5. Close main switch and single-pole, single-throw switches 1 and 3. Take simultaneous readings of voltmeter and ammeter.

6. Repeat Steps 4 and 5 for R_3.

OBSERVATIONS

Use a table similar to the one shown here to record your observations.

OBS. NO.	E-LINE	E_1	E_2	E_3	I-LINE	I_1	I_2	I_3
1								

CALCULATIONS

Use a table similar to the one shown here to record your calculations.

OBS. NO.	R_1	R_2	R_3	R TOTAL = $R_1+R_2+R_3$	R TOTAL = $\dfrac{\text{LINE E}}{\text{LINE I}}$

QUESTIONS

1. What is the relation between the line current and the current through each resistor?

2. What is the relation between the total voltage and the voltage across the individual resistors?

3. What is the relation between total resistance and individual resistances?

4. What would happen if one of the parts of this series burned out?

5. Which is brighter in the series circuit: the low-resistance lamp or the high-resistance lamp?

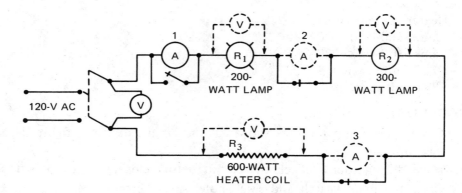

Fig. 3-25

Experiment 2 VOLTAGE LOSS IN CONDUCTORS

OBJECT

To show that voltage is lost when there is a current in a conductor and to prove that the voltage drop is directly proportional to the length of the wire.

APPARATUS

1 - AC ammeter (0-5 A)
1 - AC voltmeter (0-150 V)
4 - Coils, 50′ (15 m) #24 AWG magnet wire
1 - Lamp base
1 - Lamp, 100-watt
6 - Binding posts
1 - Board, 9 1/2″ x 36″ (250 mm x 900 mm)
1 - SPST switch

PROCEDURE

1. Connect apparatus on a board as shown in figure 3-26. This now represents a lamp operating one hundred feet (30 m) away from the source of current.

2. Close the switch and measure voltage between points marked 1-1, 2-2, and at lamp 3-3.

3. From ammeter reading and voltage drop between 2 and 1, calculate the resistance of the 100 feet (30 m) of wire between points 1 and 2.

4. From ammeter reading and voltage drop between 3 and 1, calculate the resistance of the 200 feet (60 m) of wire between source and the lamp.

OBSERVATIONS

Use a table similar to the one shown here to record your observations.

LINE VOLTS		DROP VOLTS
VOLTAGE AT	1,-1	
" "	2,-2	
" "	3,-3	

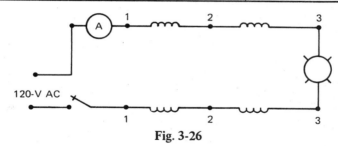

Fig. 3-26

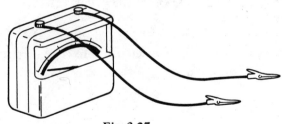

Fig. 3-27

To simplify observations, a voltmeter with flexible test probes may be used, figure 3-27.

QUESTIONS

1. Why is the voltage reading at 2 less than at 1?

2. Why is the voltage reading at 3 less than at 2?

3. What causes voltage drop?

4. Was the voltage drop uniform at each point?

5. What effect does the voltage drop have on the voltage at the source?

Experiment 3 VOLTAGE-DROP METHOD

OBJECT

To use series circuits in finding resistance.

APPARATUS

1 - AC voltmeter (0-150 V)
1 - SPST switch
Known resistance
Unknown resistance

PROCEDURE

1. Connect two resistors in series to the current source as shown in figure 3-28.

2. Close the switch and read the voltage, first across the known resistance and then across the unknown resistance.

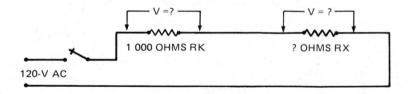

Fig. 3-28

3. Calculate current in RK from voltmeter reading on RX and known resistance. $I = \dfrac{E}{R}$. Do you expect current in RX to be the same as in RK?

4. If you know the current in RX and you know the voltage on RK, then you can find out how many ohms resistance in RX. $R = \dfrac{E}{I}$

OBSERVATIONS

Use a table similar to the one shown here to record your observations.

OBS. NO.	E KNOWN	EX	RX	CALCULATED RX	RES. SAMPLE
1					
2					
3					

QUESTIONS

1. What effect do the variations between the unknown and the known resistance have on the accuracy of the measurement?

2. What effect does the value of voltage and current have on the accuracy of the measurement?

3. Compare this method of resistance measurement with other methods.

Experiment 4 PARALLEL CIRCUITS

OBJECT

To study the relation of current, voltage, and resistance of the entire parallel combination to the current, voltage, and resistance of each separate part.

APPARATUS

1 - AC voltmeter (0-150 V)
1 - AC voltmeter (0-150 V)
1 - AC ammeter (0-20 A)
1 - AC ammeter (0-10 A)
1 - Lamp, 300-watt
1 - Lamp, 100-watt
1 - Heater coil, 600-watt
4 - SPST switches
1 - DPST switch

PROCEDURE

1. Connect apparatus as shown in the circuit diagram, figure 3-29, page 60, with ammeter in series with R_1 and voltmeter across R_1. Also, insert

an ammeter in series with the line and a voltmeter connected across the line terminals.

2. Have the instructor check the circuit before closing the main switch.

3. Close the main switch and read and record all instrument readings.

4. Open main switch, disconnect the ammeter in series with R_1 and insert in series with R_2; also, disconnect voltmeter across R_1 and connect across R_2.

5. Close the main switch and read and record all instrument readings. Be sure that line voltage remains the same for all readings.

6. Repeat Steps 4 and 5 for R_3.

OBSERVATIONS

Use a table similar to the one shown here to record your observations.

OBS. NO.	E LINE	E_1	E_2	E_3	I LINE	I_1	I_2	I_3
1								

QUESTIONS

1. What is the relation between line current and current through each resistance?

2. What is the relation between the line voltage and the voltage across each branch?

3. What is the relation between the combined resistance and each resistance?

4. Where are parallel circuits used?

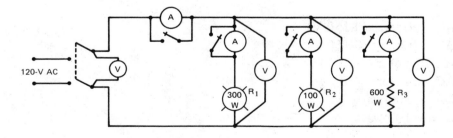

Fig. 3-29

Experiment 5 PARALLEL (OR SHUNT) MULTIPLE CIRCUIT

OBJECT

To study the path an electron may take through a divided-path circuit.

APPARATUS

1 - SPST switch
3 - AC ammeters (0-150)
2 - Lamp bases
2 - Lamps

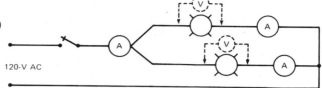

120-V AC

PROCEDURE

Fig. 3-30

1. Trace current from line through meters and lamps, and back to line as in figure 3-30. Observe that there is more than one path that an electron can take. Such a divided-path circuit is called a parallel (or shunt) multiple circuit.

2. Check and record the ammeter readings.

3. Check the voltage at each lamp and record findings.

4. Compare voltage with a similar series circuit.

5. Make a statement giving the relation between the individual lamp voltages in series with the line voltages.

QUESTIONS

1. Is the voltage at each lamp in a parallel circuit the same?

2. Does each lamp in a parallel circuit use the same amount of current?

3. Is there a relationship between the individual lamp currents and line current?

Experiment 6 SERIES-PARALLEL CIRCUITS

OBJECT

To check and verify the calculated resistance values of a series-parallel circuit with an ohmmeter.

APPARATUS

1- Ohmmeter
4 - 1-watt resistors of proper ohmic value and percent tolerance as indicated in the circuit diagram, figure 3-31.
4 - Mounting stands for resistors

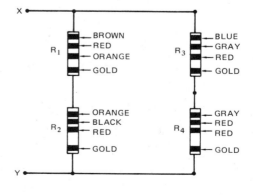

PROCEDURE

1. Connect resistors as shown in figure 3-31.

Fig. 3-31

Caution: Be sure that the ohmic values of the resistors used are the same as indicated in the circuit diagram and are placed in the same position as shown, figure 3-31.

2. Record the values of resistors R_1, R_2, R_3, and R_4.

OHMIC VALUE OF RESISTORS USING COLOR CODE

3. Using the resistor color code, determine the ohmic value of resistors R_1, R_2, R_3, and R_4, and record.

4. Short the test leads of the ohmmeter and adjust the potentiometer until the needle indicates zero.

5. Connect the test leads of the ohmmeter across terminals X and Y and record the reading in ohms in the proper column of your observation chart.

Caution: Be sure you are reading the instrument on the proper scale and are using the correct multiplier.

OBSERVATION AND CALCULATION DATA FOR CIRCUIT DIAGRAM 1

Use a table similar to the one shown here to record your observations.

OBS. NO.	OHMMETER READING IN OHMS	CALCULATED RESISTANCE IN OHMS
1		
2		
3		
4		
5		
6		

6. In Observation No. 2, connect the test leads of the ohmmeter directly across the resistor R_1 and record the reading.

7. For Observation No. 3, repeat the procedure in Step 6 with the ohmmeter leads connected directly across R_2.

8. In Observation No. 4, short circuit resistor R_1 and record the ohmmeter leads connected directly across resistor R_2.

9. In Observation No. 5, disconnect resistor R_1 and repeat the procedure used in Step 8.

10. In Observation No. 6, reconnect resistor R_1 back into the circuit and short circuit resistor R_2. Again connect the ohmmeter across R_2 and record the reading.

11. Calculate the resistance for each of the six circuit conditions. Use the resistor color code to determine the ohmic values of resistors R_1, R_2, R_3, and R_4. Show all calculations neatly arranged in your experiment report.

12. Reconnect resistors as shown in figure 3-32.

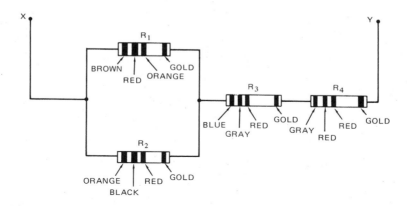

Fig. 3-32

Caution: Be sure that the ohmic values of the resistors used are the same as indicated in the diagram and are placed in the same positions as shown.

13. Short the test leads of the ohmmeter and move the zero positioning adjustment until the needle again indicates zero.

14. Measure the total resistance of the series-parallel circuit by placing the ohmmeter test leads across the terminals X and Y. Record the reading in ohms.

15. Calculate the total resistance of the series-parallel circuit and show all calculations neatly arranged. Use the resistor color code to determine the values of R_1, R_2, R_3, and R_4.

QUESTIONS

1. Explain how the resistor color-coding system is used to determine the ohmic value of resistors.

2. How is the percent tolerance of resistors determined by the resistor color-coding system?

3. In the second observation for figure 3-31, the ohmmeter was connected directly across resistor R_1. Explain why the ohmmeter indication was considerably less than the ohmic value of resistor R_1.

4. Why is the total resistance of the circuit shown in figure 3-31 considerably lower than the total resistance of the circuit shown in figure 3-32?

PROJECTS

Use the symbols listed here for Projects 1 through 9.

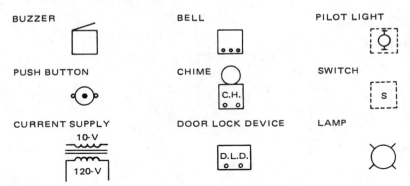

Project 1

Draw a wiring diagram showing a circuit containing a buzzer, a push button, and current supply. Wire and test.

Project 2

Draw a wiring diagram showing a series circuit containing a buzzer, two push buttons, and current supply. Wire and test.

Project 3

Draw a wiring diagram showing a parallel circuit containing a buzzer, two push buttons, and current supply. Wire and test.

Project 4

Draw a wiring diagram showing a series circuit containing two bells, a push button, and current supply. Wire and test.

Project 5

Draw a wiring diagram showing a parallel circuit containing two bells, a push button, and current supply. Wire and test.

Project 6

Draw a wiring diagram showing a circuit containing a bell and push button for the front door, a buzzer and push button for the rear door, and current supply. Wire and test.

Project 7

Draw a wiring diagram showing a circuit containing a chime, a push button for the front door, a push button for the rear door, and current supply. Wire and test.

Project 8

Draw a wiring diagram showing an apartment door-lock circuit containing a buzzer, door-lock device, two push buttons, and current supply. Wire and test.

Project 9

Draw a wiring diagram showing a circuit containing a switch, a lamp, and a pilot light. Wire and test.

Chapter 4
Electrical energy and power

Some electrons are of special interest and value because they can perform useful services. This ability to perform useful work is called energy. Energy appears in many forms, some of which will be discussed in detail because they contribute to an understanding of electrical energy.

ENERGY

Energy can be defined as the ability to do work or the ability to accomplish physical changes. The type of work or change referred to is one that involves force and motion. For example, the following changes all require energy: producing mechanical movement, producing heat or light, producing sound, changing one chemical compound into another, and producing radio waves. The amount of energy required for changes of these types, although it is nonmaterial and invisible, is readily measured. In common conversation, the words *work* and *energy* have broader usage. The physical work or energy discussed here does not include such things as the work done by someone sitting and counting cars passing a corner, the work done in making someone change their mind, or the energy with which one tackles an arithmetic problem.

Most daily activities involve the conversion and control of energy. One of the main functions of a house is to control heat (which is the energy of motion of molecules). The main function of an automobile is to convert the energy of gasoline into the energy of mechanical movement. Gasoline and other fuels are useful only because of the energy they possess. For farmers, the energy of the sun can be used to change water, soil, and air into sugar, vitamins, and other useful chemical compounds.

The energy of the sun also produces most electrical energy. Last year, the heat of the sun evaporated water from the oceans, which fell as rain and snow. This water, stored behind dams, supplies the energy to run electrical generators. A hundred million years ago some sun energy was stored in the dehydrated vegetation called coal; this year some of the coal will be burnt in boilers. This releases heat to form steam which, in turn, will run turbine-generators, figure 4-1, page 66.

Mechanical Energy

A few mechanical situations will be used to show the meaning of such terms as work, energy, and power. These terms are used in electrical measurements, but it may be easier to understand them from some mechanical examples.

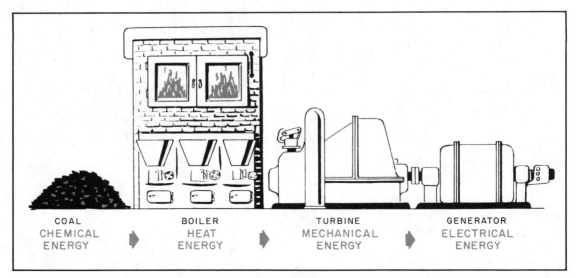

Fig. 4-1 Conversion of Energy

The lifting of a weight illustrates the meaning of one unit for measurement of energy. To lift a one-pound weight one foot is said to require one foot-pound of energy, or one *foot-pound* of work. (Work and energy are measured in the same way.)

A foot-pound is the energy used when a one-pound force moves something a distance of one foot, with the one-foot movement being in the same direction as the force, figure 4-2.

How much work is done in lifting a 20-pound weight five feet vertically?

> Work = Force x Distance

A 20-pound force traveling one foot accomplishes 20 foot-pounds of work; if it must travel five feet, it does 100 foot-pounds of work.

> Foot-pounds = Feet x Pounds

Potential Energy

What becomes of this 100 foot-pounds of energy? It is saved up; the lifted 20-pound weight has 100 extra foot-pounds of energy that it did not have when it was on the ground. This saved energy is called *potential*

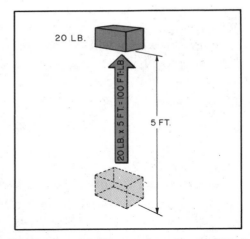

Fig. 4-2 Potential Energy from Vertical Work

energy. When the weight is allowed to fall back to earth, it will deliver 100 foot-pounds of energy to whatever it hits.

How much work is done in dragging a 200-pound box horizontally along the floor a distance of 6 feet? Stated this way, the question cannot be answered. The 200 pounds is a vertical force, not in the same direction as the motion. Suppose we find, with a spring scale, the additional information that the horizontal force needed to drag the box is 50 pounds. Now we can find how much work is done, 50 x 6 = 300 foot-pounds.

What becomes of this 300 foot-pounds of energy? It is wasted, converted into heat by the process of friction against the floor. The box has not gained potential energy. The energy has not vanished; it has uselessly warmed the floor and the bottom of the box.

Kinetic Energy

How much work is done when a 3/4-pound ball is thrown, if a force of 8 pounds is applied through a distance of 6 feet?

Work = Force x Distance = 8 x 6 = 48 ft.-lb.

What becomes of this energy? It exists as energy of motion of the ball, which is called *kinetic energy*. The ball in flight has 48 foot-pounds of energy, which it will deliver to whatever it hits.

Turning back to the example of potential energy, when the 20-pound weight which rests 5 feet above the floor is allowed to fall, its potential energy of 100 foot-pounds becomes kinetic energy, energy of motion, figure 4-3.

POTENTIAL ENERGY OF ELECTRONS

If an electron is forcibly taken away from one neutral object and put onto another, the electron has gained potential energy, figure 4-4. Force had to be used to pull the electron away from A, because the electron is attracted back to A by the positive charge that remains behind. If several electrons are transferred to B, they repel additional electrons. Force has to be used to push electrons onto B. Given the opportunity, electrons will fly back to A from B, producing heat as they return. Or, they could be permitted to return to A through an electric motor and convert their potential energy to mechanical energy as they pass through the motor.

The above behavior may be compared to the transfer of water from a pond to a high tank. Water can be taken from the pond

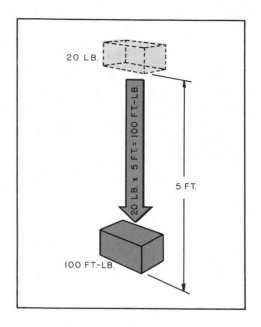

Fig. 4-3 Energy of Motion

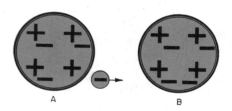

Fig. 4-4 Giving Electrons Potential Energy

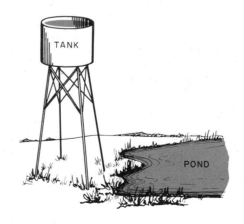

Fig. 4-5 Giving Water Potential Energy

and put into the tank, but it does not go there by itself. Energy must be used to carry the water up to the tank, figure 4-5. That energy is not wasted. It is the potential energy of the water at a higher level. If given the opportunity, water in the tank will run back down to where it came from, doing work on whatever it hits. The water could produce heat by friction. Or, if it passes over a water wheel, it could do useful work.

Each gallon of water in the tank has more potential energy than a gallon of water in the pond. There is a *potential energy difference* between water in the tank and water in the pond.

Likewise, there is a potential energy difference between electrons in B and electrons in A. This is called the *potential difference* between B and A.

Potential energy of water is measured in foot-pounds. The potential difference between tank and pond can be expressed as a number of foot-pounds of energy per gallon of water.

The potential energy of electrons is often measured in a metric-system unit called the *joule.* (A joule equals 0.738 foot-pound, which is a little less than a foot-pound.) The transfer of each coulomb of electrons from A to B requires the use of a certain number of joules of energy. The potential energy difference between A and B can be expressed as a number of joules of energy per coulomb of electrons. (Comparable to energy per gallon of water.) The cumbersome expression, "joules of energy per coulomb," is the accurate definition of a *volt.*

We may continue to think of a volt as a measure of electrical pressure or electromotive force because these terms may be easier to understand. But in case of any question about the exact meaning of the quantity measured in volts, it is *potential difference,* the potential energy difference between two points.

Volts = Joules/Coulomb

POWER

In everyday language, people use the word power to mean a variety of things. In technical language, *power* means how fast work is done, or how fast energy is transferred.

Power is the rate of doing work and power is the rate of energy conversion are two useful definitions of the term.

The fact that power is a rate deserves attention. Accurately, we do not buy or sell power; the commodity that we buy or sell is energy. Power tells how fast energy is used or produced.

For a mechanical example: If an elevator lifts 3 500 pounds a distance of 40 feet, and it takes 25 seconds to do it, 40 x 3 500 = 140 000 foot-pounds of work is done in 25 seconds. The rate of doing work can be stated in foot-pounds per second: The rate is $\frac{140\ 000\ \text{ft.-lb.}}{25\ \text{sec.}}$ = 5 600 ft.-lb./sec. (The division sign / is read as "per"; ft.-lb./sec reads as foot-pounds per second.)

Power can be expressed in foot-pounds per minute as well. If a pump needs 10 minutes to lift 5 000 pounds of water 60 feet, it is doing 300 000 ft.-lb. of work in 10 minutes, which is a rate of 30 000 ft.-lb./min.

Horsepower

When James Watt started selling steam engines, he had to rate his engines in comparison with the horses they were to replace. He found that an average horse, working at a steady rate, could do 550 foot-pounds of work per second, figure 4-6. This rate is the definition of one *horsepower* (hp).

1 horsepower = 550 ft.-lb./sec.

The elevator referred to above, doing work at the rate of 5 600 foot-pounds per second, is doing work at the rate of a little over 10 hp.

In the previous discussion of energy, a metric-system unit called a joule was mentioned. In metric units, power can be measured in joules per second, which corresponds to foot-pounds per second in British units. Joules per second is used so often that the word *watt* is used to replace the three words, "joules per second."

> *A watt is a measurement of power. One watt is a rate of one joule per second.*

Heating Rates

The rate of using or producing heat energy can be measured in Btu per second, or calories per second, or calories per minute. Household heating equipment is rated in Btu per hour (Btu stands for British thermal unit.)

For comparison of power units:

1 hp	= 746 watts
1 watt	= 3.42 Btu/hour
1 Btu/sec.	= 1 055 watts
1 cal./sec.	= 4.19 watts
1 ft.-lb./sec.	= 1.36 watts

Kilowatts and Kilowatt-hours

The statement "Power is rate of using energy" can be written as the formula

$$\text{Power} = \frac{\text{Energy}}{\text{Time}}$$

(If we had needed a formula, we could have used this one to find foot-pounds per second when we knew the foot-pounds and the time in seconds.)

This formula can be rearranged to say: Energy = Power x Time. This formula seems hardly necessary because one could do this problem without it:

"How much work (energy) is done by a 1-hp engine in 20 seconds?"

1 hp is 550 ft.-lb./sec. and we would

ONE HORSEPOWER = 746 WATTS

Fig. 4-6 One Horsepower Equals 550 ft.-lbs. per Second

multiply 550 by 20 seconds to find the work done. The formula, Energy = Power x Time, can serve to introduce another common unit of energy. By multiplying a power unit by a time unit, we get a unit of measurement of energy.

Use the metric system units that are used for electrical quantities for the following. If a device uses electrical energy at the rate of one watt and it uses energy at this rate for one hour, the energy used = one watt-hour.

Watt-hours	=	Watts	x	Hours
(energy)		*(power)*		*(time)*

Watt-hours are generally lumped together and sold by the thousand. (The prefix kilo means thousand.) One kilowatt-hour (kwh) is 1 000 watt-hours.

The *electric meter* in a house is a meter for recording the number of kilowatt-hours of energy that have passed through the meter. Correctly, it is called a kwh meter, figure 4-7, page 70.

Figure 4-7 represents the register of a kilowatt-hour meter with four dials. The dial readings represent units, tens, hundreds, and thousands. In reading the meter, starting with the number 1 dial, simply write down the last number the pointer has passed and get a registration of 4 294 kwh.

If we read the meter a month later, figure 4-8, page 70, and find the dials register

Fig. 4-7 Reading a Kilowatt-hour Meter

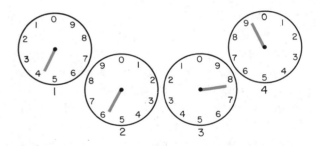

Fig. 4-8 The Reading a Month Later

4 579 kwh, by subtracting the 4 294 from 4 579 we find we have used 285 kwh during the month.

The reason for using kwh for measuring energy is that it is a unit of convenient size. If we calculated our ordinary energy requirements in foot-pounds or in joules, the numbers would be inconveniently large. One kwh = 2 655 000 ft.-lb. or 3 600 000 joules or 3 413 Btu.

THE ENERGY OF ELECTRONS

Electrical energy and power measurements are based on the metric unit of energy, the joule. Our more familiar units are derived from it.

What Does the Consumer Buy From the Power Company?

We buy electrons when we buy the light bulb. The power company is paid to move the electrons. Power is *rate*. Current is also a rate. *Energy* is what the consumer buys from a power company. Electrons do useful *work,* or produce useful *heat,* by giving up energy. We buy the energy that electrons bring to us.

CALCULATING POWER

There are two useful formulas for calculating watts of power:

Watts = Volts x Amps (or Power = EI)
and
Watts = $I^2 R$ (I is amps, R is ohms)

Why Volts x Amps Equals Watts

Volts tell the energy carried by each coulomb of electrons. *Amps* tell how many coulombs pass by each second. If the voltage is 120, that means that each coulomb will deliver 120 units of energy. If the current is 5 amps, that means that five coulombs are delivering energy each second. Five per second, each delivering 120 energy units, means that 600 energy units (joules) are delivered each second. 120 volts x 5 amps gives 600 joules each second, but we call joules per second, *watts.*

Using W = EI (Watts = Volts x Amps)

1. How many watts can we have on a house circuit if it is a 120-volt circuit, protected by a 15-amp fuse?

Watts = 120 x 15 = 1 800. This 1 800 watts is the maximum total of all appliances in use on the one circuit.

2. How much current (amps) is in the line when four 60-watt lamps are in use? Four lamps, of 60 watts each, equal 240 watts. If it is a 120-volt line:

W = V x A (E x I)
240 = 120 x X   X = 2 amps.

3. A 220-volt dc motor takes 10 amps. How many watts is that?

220 x 10 = 2 200 watts, or 2.2 kilowatts. W = EI is useful for all dc equipment, and also for finding power converted to heat in ac devices. For some ac circuits, this formula

does not give the true power requirement. Later, we will find out how W = EI is corrected, when needed, to apply to ac machinery.

Using Watts = $I^2 R$

This formula can be derived from W = EI used above. Since E in this formula is volts, and volts are found by the formula E = I x R, then the expression I x R can be substituted for E in any general formula.

W = EI becomes W = (I x R) x I. The formula can be written more simply as W = I^2 x R. This formula is often used for finding the heating rate of current in a resistor or long conductor in both ac and dc applications.

1. An iron of 11-ohms resistance is intended to operate on 10 amps. What is its rating in watts?

Watts = $I^2 R$

Watts = 10^2 x 11 = 100 x 11 = 1 100 watts

2. A 10 000-ohm resistor in a radio has a current of 15 milliamperes in it. Find watts.

15 milliamperes = 0.015 amps

Watts = $(0.015)^2$ x 10 000

= 0.000 225 x 10 000

= 2.25 watts

Which formula is better to use, Watts = EI or Watts = $I^2 R$? Both give right answers. If E and I are known at the start, use EI. If I and R are known use $I^2 R$. If E and R are known, but not I, find I from Ohm's Law (I = E/R); then use either formula. To review some relationships:

Coulombs = Quantity of electrons
Amperes = Flow rate of electrons, coulombs/sec.
Joules = Energy
Watts = Joules/sec. the energy rate
Volts = Joules/coulomb
Watts = Volts x Amps, because

Watts = Joules/sec.; Volts = Joules/coul.; Amps = Coul/sec.; and

$$\frac{Joules}{sec.} = \frac{Joules}{Coul.} \times \frac{Coul.}{sec.}$$

CALCULATING ENERGY AND COST

Electrical energy is usually calculated in kilowatt-hours (kwh). Watts x hours gives watt-hours. The answer is expressed in kwh by dividing watt-hours by 1 000.

1. A 500-watt heater is operated for 10 hours. How many kwh of energy are used?

500 x 10 = 5 000 watt-hours, which is 5 kilowatt-hours.

If kilowatt-hours of energy cost 3 cents each, the cost of this operation is 15 cents.

2. A 120-volt, 10-amp iron is operated for 3 hours. Find the cost at 2.5¢/kwh.

120 volts, 10 amps = 1 200 watts

1 200 watts x 3 hours = 3 600 watt-hours = 3.6 kwh

3.6 kwh x 2.5¢ per kwh = 9¢

3. What does it cost to leave two 60-watt lamps on all night (8 hours)?

120 watts x 8 hours = 960 watt-hours (call it about one kwh)

How much does one kwh cost?

Household energy rates are based only on energy used per month. Industrial energy rates are based on energy used, plus a charge based on maximum power required.

CONVERSION OF ENERGY

"Energy cannot be created or destroyed" is a way of stating a principle long known as the *Principle of Conversion of Energy*. Each kwh used in lamps or motors comes from the burning of coal or the release of stored water. If lamps were lit by batteries the energy of coal would still be responsible, since coal was oxidized to release the zinc or lead of the battery from other elements when the metal was refined. The electrical energy that is used daily is soon converted to heat by one process or another, lost to the air, and radiated out into space. Electric energy users store some energy when they charge batteries or pump water into a storage tank, but such examples of energy storage are few. We convert electrical energy from one form to another so that people can do different things, for example seeing at night, or listening to music. The efficiency of this energy conversion is a way of measuring how well the energy-converting device accomplishes its task.

$$\text{Efficiency} = \frac{\text{Useful energy obtained}}{\text{Total energy used}}$$
or
$$\text{Efficiency} = \frac{\text{Power output}}{\text{Power input}}$$

For example, a dc motor that takes 4.2 amps on a 120 volt line, produces one-half horsepower. What is its efficiency?

The power output is 0.5 hp, or 373 watts, because 1 hp = 746 watts.

The power input is 120 volts and 4.2 amp = 504 watts.

The efficiency of the motor is 373 ÷ 504 = 0.74 , or 74 percent.

The efficiency of any device can be no more than 100 percent, which is a way of saying that the device cannot give out more energy than it takes in. The efficiency of all electrical heating devices is 100 percent; electrical production of heat is easy. Heating devices may vary, however, in how effectively they deliver heat from the coils in which it is produced to the place where it is to be used.

Referring back to the motor with 74 percent efficiency, the other 26 percent of the energy used appears as heat. If the motor is stalled so it can produce no mechanical power, it becomes a 100 percent efficient heating device and burns up its coils.

Another example: Find the efficiency of an electrical generator, requiring 10 hp input, and producing 50 amps at 100 volts.

Power output: 50 amps x 100 v is 5 000 watts

Power input: 10 hp is 7 460 watts

Efficiency is $\frac{5\ 000}{7\ 460}$ = 0.67

An often-proposed scheme is, "How would it work to drive an electrical generator with an electric motor, and let the generated current run the motor?" It would not work very well according to the preceding discussion. Both the motor and the generator waste some of the energy applied, so one will not be able to produce enough energy to run the other.

POINTS TO REMEMBER

- Work = Force x Distance.
- Energy is ability to do work.
- Power is rate of using energy.
- Power x Time = Energy.
- Watts = Volts x Amps.
- Watts = $I^2 R$
- Watts and kilowatts measure power, which is a rate.
- Watt-hours and kilowatt-hours measure energy.
- Efficiency = Output ÷ Input.

REVIEW QUESTIONS

1. What is energy?
2. What is a foot-pound?
3. A crane lifted a 25-ton freight car a distance of 6 feet. How much work did it do? How much energy is needed to throw a 6-pound projectile 50 000 feet into the air?
4. What is potential energy?
5. What is kinetic energy?
6. How is the potential energy of electrons measured? What is it called and what portion of a foot-pound does it represent?

7. What is power?

8. Define horsepower.

9. If 300 000 foot-pounds of work has to be done in 5 seconds, how much horsepower is required? How many watts?

10. Define a watt.

11. What is a watt-hour? What is a kilowatt-hour (kwh)?

12. Calculate watts for each of these devices:

    a. A 60-volt, 10-ampere arc lamp.
    b. A 100-ohm resistor carrying 1/2 ampere.
    c. A 12-ampere, 110-volt heater.
    d. A 2-ohm resistor on a 6-volt line.

13. Calculate resistance of a 60-watt, 120-volt lamp when operating.

14. Find the operating current of an 800-watt, 115-volt toaster.

15. A 20 000-ohm resistor is rated 5 watts. What is the maximum current it can carry without exceeding its 5-watt rating?

16. Find the cost of operating the toaster of Question 14 for 5 hours per month, if energy costs 3 cents per kwh.

17. Would it be a good idea to arrange cost-free electrical energy distribution, like free use of public roads and schools?

RESEARCH AND DEVELOPMENT

Experiments and Projects on Electricity, Energy, and Power

INTRODUCTION

The experiments and projects listed for this unit will use the information presented in Chapter 4. The experimental generator suggested as the partially developed project will give you the opportunity to actually produce electric energy.

EXPERIMENTS

1. Measure the power consumed by some common electrical appliances. Determine the cost of operation per hour at the local power rate.

2. Measure the power consumed by a motor running free and under load. Determine its efficiency.

3. Calculate resistance by measuring volts and amperes in a circuit which uses a variable transformer.

PROJECTS

1. Examine electrical heating appliances at home (iron, toaster, heater). Find the rating in volts and watts or amperes on them. Look for the power rating on some other electrical devices.

2. Plan and construct a model substation complete with transmission towers and distribution transformers.

3. Plan and construct a model hydroelectric generating plant.

4. Plan and construct a model steam turbine generating plant.

5. Plan and construct a miniature water-wheel power plant complete with a generator that will produce a measurable amount of electricity.

6. Design and construct a simple experimental generator.

EXPERIMENTS

Experiment 1 ELECTRIC POWER AND ENERGY

OBJECT

To measure the power consumed by common household electrical appliances such as an electric iron, an electric toaster, and an electric heater. Also, to determine the cost of operation per hour at the local power rate.

APPARATUS

1 - AC voltmeter (0-150 V)
1 - AC ammeter (0-10 A)
1 - Wattmeter (0-150 V) (0-5, 0-150 A)
1 - DPST knife switch
1 - SPST knife switch
1 - Electric toaster, nameplate data
1 - Electric iron, nameplate data
1 - Electric hotplate, rating

> Lamps, heating elements, soldering irons, and other electrical devices should be allowed to cool before handling.

PROCEDURE

1. Connect apparatus, as shown in figure 4-9, using a 200-watt, 120-volt lamp as a load.

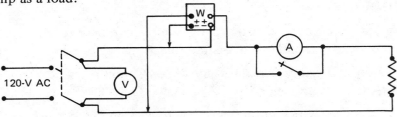

Fig. 4-9

2. Have the instructor check the circuit before closing the main switch.

3. Close main switch and read and record simultaneous readings of all instruments.

4. Repeat Step 3, using a toaster as the load.

5. Repeat Step 3, using an electric iron as the load. *Caution:* If an electric iron is used, care should be taken to prevent it getting too hot.

6. Repeat Step 3, using an electric hotplate.

7. Calculate the resistance and power in watts required by each type of load.

8. Calculate the kilowatt-hours used by each type of load for a 10-hour period, and also the cost of operation at $.03 per kilowatt-hour.

OBSERVATIONS

Use a table similar to the one shown here to record your observations.

OBS. NO.	LINE VOLTS	LINE CURRENT	WATTS OBSERVED	TYPE OF LOAD
1				200 WATT, 115 VOLT LAMP
2				500 WATT, 115 VOLT TOASTER
3				550 WATT, 115 VOLT ELECTRIC IRON
4				1 200 WATT, 120 VOLT HOTPLATE

CALCULATIONS

Use a table similar to the one shown here to record your calculations.

OBS. NO.	CALCULATED RESISTANCE	CALCULATED WATTS	KWH	COST FOR 10 HR. OPERATION
1				
2				
3				
4				

Sample Calculations:

$$R = \frac{E}{I}$$

$$W = EI$$

$$kwh = \frac{W \times hours}{1\,000}$$

$$Cost = kwh \times rate$$

QUESTIONS

1. Is electrical power the same as electrical energy?

2. Is it possible to purchase a watt or kilowatt of power?

3. Name several electrical appliances found at home.

4. What is the power input in watts to a 5-horsepower motor at rated load if the operating efficiency is 80 percent?

Experiment 2 ELECTRIC MOTOR POWER AND EFFICIENCY

OBJECT

To measure the power consumed by the motor running free and under load, and to determine the efficiency of the motor.

APPARATUS

1 - Fractional horsepower motor (1/4 - 1/3 hp)
1 - 4- or 5-inch (100- or 125-mm) pulley
1 - Pc. heavy sash cord
2 - Spring scales (0-25 lb.)
1 - Wattmeter (0-300 V)
1 - SPST knife switch

PROCEDURE

1. Connect apparatus as shown in figure 4-10.

2. Read data on motor nameplate. Note rating: horsepower (hp), amperes (amps), and revolutions per minute (rpm).

3. Calculate the available torque of the motor when it is operating at rated horsepower, using the formula:

$$\text{Torque (lb.-ft.)} = \frac{\text{hp}}{0.000\ 19 \times \text{rpm}} \quad \text{or} \quad \frac{\text{hp} \times 5\ 252}{\text{rpm}}$$

4. From this amount of torque, find how much turning force on the pulley slot is produced by this torque, using the definition:

$$\text{Torque (lb.-ft.)} = \text{Turning force (lb.)} \times \text{Radius (ft.)}$$

5. (a) With the motor running, adjust the tension on the spring scales so that the *difference* in their readings equals the *pound-force* found in Step 4. The motor should now be operating at full-rated horsepower.
    (b) While this force is maintained, read the wattmeter.

6. Find efficiency of motor. Convert hp output to watts, using: 746 watts = hp. "Power In" is wattmeter reading. Efficiency = Power Out ÷ Power In.

OBSERVATIONS   CALCULATIONS

Use a table similar to the one shown here to record your observations.

OBS. NO.	LINE VOLTS	LINE CURRENT	WATTS OBSERVED	AMOUNT OF LOAD
1				
2				
3				

QUESTIONS

1. Does the motor consume more power under load than when running free?

2. What happens to the current consumption when the motor is stalled?

3. Was the starting current and running current the same?

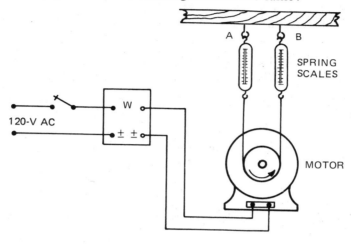

Fig. 4-10

Experiment 3 OHM'S LAW AND ELECTRIC POWER

OBJECT

To make resistance measurement from volts and amperes.

APPARATUS

1 - Variable transformer
1 - AC ammeter (0-1 A)
1 - AC ammeter (0-10 A)
1 - AC voltmeter (0-150 V)
1 - Lamp base
1 - Lamp
1 - Switch
1 - Glocoil

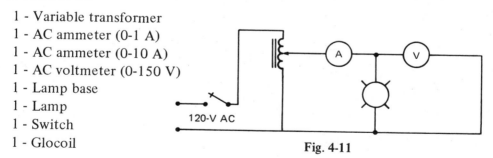

Fig. 4-11

PROCEDURE

1. Connect 0-1 ammeter in a circuit as shown in figure 4-11.

2. Start with variable transformer set at zero and turn on the current.

3. Increase the voltage in steps of 20 volts at a time, as read on the voltmeter.

4. For each value of voltage, record current.

5. For each set of voltage and current readings, calculate resistance of lamp and power. As the current in the lamp is increased, you can see that the temperature of the tungsten filament rises. This rise in temperature causes an increase or decrease in the resistance of the lamp. Practically all metals behave this way, changing resistance with changing temperature.

Use a table similar to the one shown to record your calculations.

VOLTS	AMPS	RESISTANCE	WATTS
20			
40			
60			
80			
100			
120			
140			

6. Turn off the current and set variable transformer at 0.

7. Replace the 0-1-ampere meter of the above circuit with a 0-10-ampere meter.

Use a table similar to the one shown to record your calculations.

VOLTS	AMPS	RESISTANCE	WATTS
40			
60			
80			
100			
120			

8. Remove the lamp, and place the glocoil in the receptacle. *Note:* The glocoil consists of nichrome (nickel-chromium alloy) wire wound on a porcelain spool. This wire is the same type as used in the heating elements in toasters, flatirons, and similar heating devices.

9. Turn on the current and record current in the circuit as the emf is increased in steps from 40 to 120 volts.

10. Calculate ohms and watts as before.

QUESTIONS

1. When both are at 120 volts, which produces the most heat — the lamp or the glocoil?

2. At 120 volts, which has more resistance — the lamp or the glocoil?

3. What produces heat in an electrical circuit?

PROJECTS

DESIGN AND BUILD AN EXPERIMENTAL GENERATOR

This simple generator can be a valuable aid in the study of how a generator produces electrical energy. A miniature power plant could be constructed with a water wheel, turbine, or windmill used to turn the generator.

The experimental generator can be constructed in the shop or at home to produce a small amount of direct or alternating current. To produce dc, a segmented commutator is employed. To produce ac, substitute slip rings for the commutator.

MATERIALS

1 - Large horseshoe magnet
1 - Shaft, 1/8″ welding rod
1 - Wooden base, size to fit individual plan
2 - Binding posts
2 - Shaft-bearing brackets, 24-gauge sheet metal
2 - Retaining straps, 24-gauge sheet metal
2 - Brushes, thin spring brass
1 - Wooden armature coil support, size optional
1 - Commutator, wooden dowel and copper tubing, size optional
12 - Wood screws, #5 x 1/2″ RH
1 - Pulley
    Quantity of #28-30 AWG magnet wire

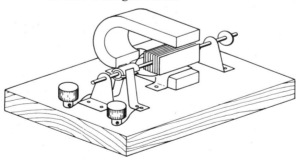

Fig. 4-12

PROCEDURE

Procure a large horseshoe magnet.

Make an assembly drawing of the model and a working drawing, with dimensions of each part.

Make the parts.

Assemble and test.

Evaluate.

Suggestions for Construction of Parts

Armature: Make the coil support of soft wood. Drill a hole through the center with a drill slightly smaller than the shaft. The coil support must be a tight fit on the shaft.

Wind the same number of turns of magnet wire on each side of the support and in the same direction. Cross on one side. Secure the wire ends with string looped over wire, laid under several turns, and pulled tight. Coat with shellac or varnish.

Commutator: Select a piece of cooper tubing, turn a piece of wooden dowel to fit the inside of the tube, and drill a hole through the center the same size as for the armature, figure 4-13. Next, cut a piece of tubing the required length, finish the ends, and cut it in half from end to end. Both pieces must be the same size. The halves should be cemented or pinned to the wooden dowel.

An ac slip-ring arrangement may be constructed by pressing two copper rings on a longer piece of dowel.

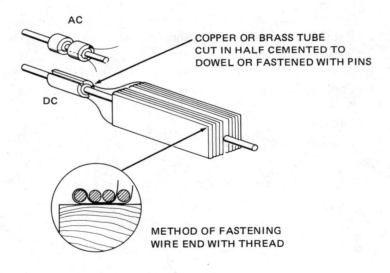

Fig. 4-13

Chapter 5
Electric heating and lighting

When electrons flow through a material, some of their potential energy is used in jiggling the atoms and molecules of the material. The energy of motion of atoms and molecules is called *heat*. In many devices heat production is undesirable, and some care has to be taken to prevent excessive heat production. When heat is wanted, we need only to supply a current through a resistance and the equation watts = I^2R tells the rate at which electrical energy is converted to heat. This is a complete conversion of energy: all electrical heating devices are 100 percent efficient in this energy conversion, though they may differ in how effectively the heat is directed to some useful task.

HEAT FROM CURRENT IN A SOLID RESISTOR

Most common heating devices, such as toasters, irons, and electric ranges, are heated by current in a coil of nickel-chromium alloy wire or ribbon. For localized production of small amounts of heat, the ease of control of electrical devices is sufficient reason for their wide use. Nickel-chromium (nichrome, chromel, etc.) alloys provide a good combination of features: small space requirement and a reasonable relationship of cost, life, and freedom from breakage.

To illustrate factors to be considered in building a heating element, assume we want to design a heavy-duty resistance-type soldering iron, which will develop heat at a rate of about 180 watts. According to the Table 5-1, No. 26 gauge wire might be used.

The current in the wire, found from Watts = Volts x Amps, 180 = 120 X, will be 1.5 amps. The resistance of the wire is found next from Ohm's Law: E = IR, 120 = 1.5 X, X = 80 ohms. The length of the No. 26 nichrome wire is found from R = KL/C.M. For nichrome wire, K is about 600, figure 2-13, and the circular mils for No. 26 wire is 254 (appendix), R is 80 ohms; calculating L, we find that about 33 feet (10 meters) of wire is needed.

Perhaps winding and insulating a coil containing 33 feet (10 meters) of wire will result in a bulky coil. What would happen

Heater Watts (115 volts)	Wire Gauge No.
100-200	26-30
200-350	24-28
350-400	22-26
450-500	20-24
550-650	19-23
700-800	18-22
850-950	17-21
1 000-1 150	16-20
1 200-1 350	14-18
1 400-1 500	12-16
2 000	10-14

Fig. 5-1 Nichrome Wire Recommendations

if No. 28 wire were used? No. 28 has 160 C.M. instead of 254, and R = KL/C.M. shows that a length of 21 feet (6.4 meters) will do. Therefore, use No. 28 because the coil is smaller, lighter weight, and cheaper to build. The 180 watts is produced in a smaller volume of wire, resulting in a higher surface temperature of the wire, therefore, the No. 28 wire will oxidize faster and have a shorter life.

Electric stoves often use a heater element consisting of a coiled resistance wire enclosed in a steel tube. The space around the wire, inside the tube, is packed with magnesium oxide or a similar filler to electrically insulate the wire from the protecting tube. *Three-heat* hot plates have two heater elements controlled by a four-position switch. On the high heat setting, both elements are in use, connected in parallel to the 120-volt line. On the medium and low positions, either the larger or smaller of the elements is in use individually or the coils are connected in series with each other.

For producing larger amounts of heat in industrial kilns and furnaces, solid rods of silicon carbide are used as heating elements. These so-called "Glo-bar" heaters are made in lengths from 4 inches to 6 feet (102 mm to 1 850 mm). Their resistances range from 0.4 to 5 ohms. These resistors are used for temperatures up to about 2 800°F. (1 538°C).

The trade name, Kanthal, is applied to a group of alloys containing iron, chromium, aluminum, and cobalt. These alloys are useful for heater elements with a high-temperature requirement. Such applications include resistor-type soldering irons and cigarette lighters, involving a concentrated heat source, and also high-temperature kilns and furnaces. Kanthal builds up a very adherent aluminum-oxide coating, which resists further oxidation. Various alloys are intended for operation at temperatures from 2 100°F. to 2 460°F. (1 150 to 1 350°C). A powder-metallurgy

product, containing molybdenum, silicon, and other metals and ceramics, operates at 2 900° F. (1 590°C).

TEMPERATURE CONTROL

For stepless regulation of the average temperature of a heating element, a *compound bar* (also called a bimetal strip) can be used to close and open a pair of contacts, as shown in figure 5-2. A compound bar is a strip of brass welded to a strip of invar. When warmed, the brass expands but the invar alloy does not, causing the bar to curve. As shown in figure 5-2 (A), contacts X and Y are closed, the load resistor (heating element) is warming up, and a coil of thin resistance wire around the compound bar is warming the bar also. When the bar is heated, it bends and disconnects from the supply. The bar soon cools enough to bring contact Y up to contact X again, restoring power to the heating element. Control of temperature is accomplished by a movable cam or similar mechanical arrangement, not shown in the sketch, for adjusting the position of the contacts. For example, if contact X is pushed down farther toward Y, then the compound bar will have to be heated to a higher temperature before Y is pulled away from X. Incidentally, the load resistor will be heated to a

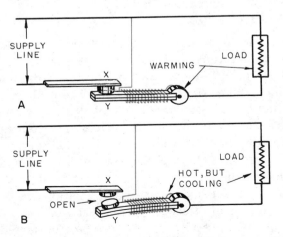

Fig. 5-2 A Bimetal Thermostat

higher temperature also, because the contacts remain closed for longer time intervals.

Resistance Welding

Welding of sheet steel is often done by heat developed by current in the steel itself. Copper electrodes are pressed against the steel, the electrodes are then connected to a low-voltage, high-current supply for a short time, figure 5-3. The resistance of the steel is small, but I^2R is large enough to weld the metal. After the current is turned off, electrode pressure is maintained briefly, then the work is released. In production machines, the "squeeze-weld-hold-off" cycle and the control of current are accomplished by specialized electronic equipment.

Heat From Arcs

Electric arcs are a common source of heat. Motor driven dc generators supply the power for most industrial arc welding. Before the arc is struck, voltage between the rod and the work is about 60 volts. When the arc is struck, voltage is reduced to approximately 25 volts. During welding the current should remain quite constant, even though the resistance of the arc varies.

Arc Furnaces

A large-scale example of heat production by carbon arcs is the use of electric furnaces in the production of alloy steels. In the laboratory, a small, high-temperature furnace may be constructed by surrounding a carbon arc with refractory material.

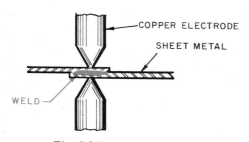

Fig. 5-3 Resistance Welding

Induction Heating

If a piece of metal is placed inside a coil which is supplied with high-frequency ac, the rapidly changing magnetic field induces eddy currents in the surface of the piece of metal, thereby heating it. This method is often used in surface-hardening heat treatment for gears and similar machine parts.

DIELECTRIC HEATING

Dielectric is another name for insulator. Heat can be produced throughout the body of an insulating material by putting the material between two metal plates and applying a high-frequency alternating voltage to the plates. For example, in the making of plywood the wood and glue are heated to dry and bond the glue to the wood. Wood is a poor conductor of heat, and gluing would be time-consuming if the wood had to be heated by direct contact with hot metal. When high-frequency (such as 20 million hertz, or cycles per second) voltage is applied to metal plates on each side of the wood, electrons in the wood vibrate back and forth in the molecules in accordance with the rapid positive to negative changes of the metal plates. This electron vibration heats the molecules of wood quickly and internally. Dielectric heating may be also used with rubber, plastic, or ceramic materials. Microwave ovens cook by this process. In medicine, it is called a diathermy unit.

ELECTRIC LIGHTING

The first practical electric light to be developed was the carbon arc, followed by the carbon filament incandescent lamp. *Incandescent* means white-hot and glowing due to heat.

HOW LIGHT IS MEASURED

The term measurement of light may refer to either of two quantities: (1) the intensity of the source, or (2) the illumination on a surface.

The intensity of a light source can be measured by candlepower. For many years this measuring unit was the light of a *standard candle,* made of sperm wax of specified size and shape, burning at the rate of 120 grains per hour. This standard was unsatisfactory on several counts. The present standard, adopted in 1948, is based on the light produced by white-hot powdered thorium oxide in a furnace held at 3 216°F. The light coming from a 1/60-square centimeter hole in this furnace is defined as having a source intensity of one candle, or one *candela,* figure 5-4. This is far more convenient and more accurately reproducible than any candle flame.

The unit of measure for surface illumination was the *foot-candle,* defined as the illumination on a surface at a distance of one foot from a standard candle. It became advisable to introduce another unit, called the *lumen.* One lumen is the rate of flow of visible light energy through a one-square foot hole at a distance of one foot from a one-candle light source. If a one square foot surface were placed at a distance of one foot from a standard candle, its illumination would be one lumen per square foot. One lumen per square foot is the same as a foot-candle. Since the surface area of a sphere = $4\pi r^2$, a one-candle source produces 4π lumens.

If a 100 percent efficient light source could be produced, converting all of its electrical energy into visible light, it would produce 621 lumens per watt. An ordinary 100-watt tungsten lamp has an efficiency of 15 1/2 lumens per watt; fluorescent tube lamps give about 50 lumens per watt.

The eye is highly adjustable to changes in illumination. Outdoor illumination may vary from 8 000 or 10 000 lumens/sq. ft. in clear, bright sunlight to 100 lumens/sq. ft. on a dull, dark day, to 0.03 lumens/sq. ft. on a night lighted by a full moon. Recommended illuminations for artificial lighting vary from 0.5 lumens/sq. ft. for sidewalks to 10 or 12 for classrooms and offices to 25 lumens/sq. ft. for drafting rooms. For ordinary work, illumination is measured by light meters which consist of a photoelectric cell and a microammeter which is scaled to read lumens/sq. ft., figure 5-5.

Source, Distance, and Illumination

Observe the geometry of the distribution of light in figure 5-6.

A. Light from a single point source spreads as it radiates, so that the surface illumination is inversely proportional to the square of the distance of the surface from the source. In (a), the light which would fall

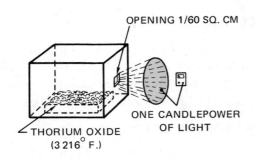

OPENING 1/60 SQ. CM

ONE CANDLEPOWER OF LIGHT

THORIUM OXIDE (3 216° F.)

Fig. 5-4 One Candela Standard

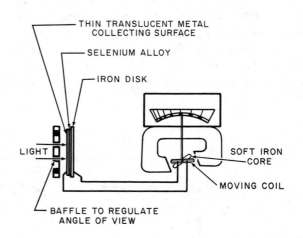

THIN TRANSLUCENT METAL COLLECTING SURFACE

SELENIUM ALLOY

IRON DISK

LIGHT

SOFT IRON CORE

MOVING COIL

BAFFLE TO REGULATE ANGLE OF VIEW

Fig. 5-5 Diagram of a Light Meter

on a one-foot square surface held 2 feet below the lamp would cover 4 square feet if allowed to travel twice as far from the source.

B. Light from a line source, such as a long string of fluorescent tubes, spreads out so that the surface illumination is proportional to the distance from the source. Illumination of surfaces below these two light sources is improved by placing reflecting surfaces above the source.

C. Light from a source which is a flat surface indefinitely wide in extent gives an illumination which is independent of distance from the source, figure 5-6.

LIGHT IS ENERGY

Light is energy radiated by electronic disturbances in atoms. Electrons in an atom can accumulate energy in many ways: from heat, as in a red-hot object; or by being hit by other electrons, as in a gas-conduction tube; or by absorbing energy radiated by other materials. Sooner or later, this absorbed energy is given out. The amount of energy an electron can emit in one burst depends on where the electron is, that is, what kind of atom it is in and its location in the atom.

The energy is radiated as a wavelike pulse of electric lines of force and magnetic lines of force, which is called an *electromagnetic wave*. The vibration frequency of these traveling lines of force is proportional to the amount of energy the electron gives off. Frequencies in the range from 4.3×10^{14} to 7.5×10^{14} vibrations per second affect electrons in the eyes. Electromagnetic waves in this frequency range are called *light*.

Waves of a frequency slightly higher than 7.5×10^{14} are called *ultraviolet;* waves in a lower frequency range, from 10^{11} to 10^{14} vibrations per second are called *infrared* and heat radiation. The term *black light* may mean either ultraviolet or infrared. Ultraviolet and infrared differ greatly in their effects and uses.

A listing of the applications of electromagnetic waves of various frequencies is given in figure 5-7, page 86.

The speed of travel of all of these electromagnetic waves is the *speed of light,* 186 000 miles per second (299 500 km/second), provided there is nothing in the way.

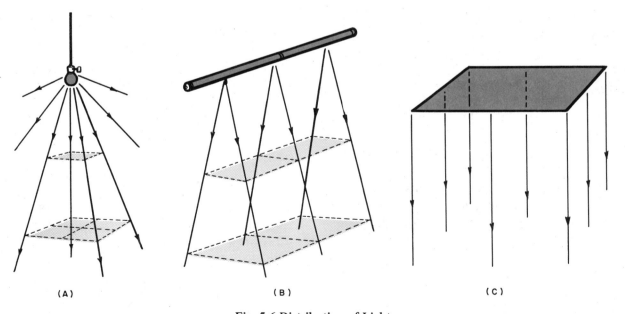

(A) (B) (C)

Fig. 5-6 Distribution of Light

COMMERCIAL LIGHT SOURCES

Carbon Arc

The one occasionally desirable feature of a carbon-arc lamp is the extremely concentrated brilliance of the source of light. Direct current carbon arcs are still preferred for 35 mm movie projectors and for searchlights. This is because the brilliant small-area light source works well with a simple system of mirrors and lenses for projection of the light.

For comparison purposes, here are a few surface-brightness figures (in candles per square centimeter):

The sun, 160 000; ordinary carbon arc, 13 000; high-intensity arc, 80 000 plus; melted tungsten (6 120°F. or 3 382°C), 5 740; interior of furnace at (3 216°F. or 1 770°C), 60; frosted-glass surface of 40-watt lamp, 2.5; filament of tungsten projection lamp, 2 100.

Despite the high surface brilliance of the arc, so much energy is converted to heat that its efficiency is low enough (10 to 20 lumens per watt) to help make it unsuitable for general lighting. At the high temperature of the arc (about 10 000°F. or 5 000°C), it radiates considerable amounts of ultraviolet light. Eye irritation noticed after exposure to arcs is caused by ultraviolet radiation.

Incandescent Lamps

Considering all the specialized types of lamps manufactured, there are hundreds of sizes and types of tungsten incandescent lamps. The readers may provide their own cross-section diagram of construction by smashing a few old lamp bulbs. Be careful to protect eyes from flying glass and hands from sharp edges.

Filament temperature in ordinary 25- to 300-watt lamps ranges from 4 200°F. to 4 800°F. (2 318°C to 2 649°C); average lamp life is 750 to 1 000 hours. Efficiencies

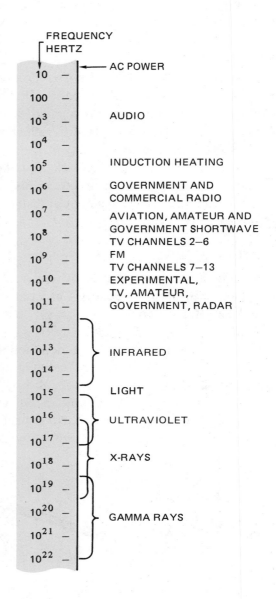

Fig. 5-7

range from 10.5 lumens per watt for the 25-watt lamp to 20 lumens per watt for the 300-watt lamp.

Which gives more light — ten, 40-watt lamps or four, 100-watt lamps? Both sets total 400 watts. Because the efficiency of the 100-watt lamp is higher (16.8 lumens per watt as compared with 11.7 for the 40-watt), four 100-watt lamps give about 42 percent more light than ten 40-watt lamps. Both sets produce the same total energy; most of it is heat.

How long a lamp lasts before it burns out is conditional. Both the life and the brightness of an incandescent lamp depend on the temperature of the filament, which is controlled by filament dimensions and current. The higher the temperature, the more light the bulb produces, and the shorter its life. "Extra long-life" lamps do not produce as much light as a standard bulb. 100-watt, 120-volt incandescent lamps can be built for any predetermined life; ordinary bulbs average 1 000 hours and produce 1 675 lumens. A 100-watt lamp can be built to last 300 hours and put out 2 285 lumens, almost as much as a standard 150 watt; or to last 3 000 hours, but giving only 1 260 lumens, because its filament temperature must be lower to give the longer life. The main purpose of a lamp is to produce light. The energy the lamp uses during its life costs more than the lamp. Of the three possibilities described, the 300-hour lamp produces the most total light per dollar.

Originally, incandescent lamp filaments operated in a vacuum to prevent oxidation of the filament. Forty-watt and larger lamps are now filled with nitrogen at near-atmospheric pressure. The gas pressure retards evaporation of the filament, permitting the filament to be operated at higher temperature and higher efficiency.

Infrared Lamps

If you hold your hand below an operating 100-watt lamp bulb, you can feel heat radiated from the lamp. Actually, about 70 percent of the energy given to the heated filament is radiated away in vibration frequencies too low to be visible as light. This energy is called infrared. When infrared radiation strikes and is absorbed in a material, the material is warmed. These lamps are often called *heat lamps,* and the sizes are 250 to 5 000 watts. They are used industrially for drying and baking, at home for treating sore muscles, and commercially for heating open areas such as grandstands, warehouse platforms, or sidewalks in front of store windows. The filaments in infrared lamps operate at lower temperatures than in lamps intended for lighting. Therefore, their efficiency as visible light sources is appropriately low, and their life is long, about 5 000 hours.

High-Pressure Mercury Lamps

The high efficiency of mercury lamps (about 50 lumens per watt) is responsible for their wide use in streetlights and industrial lighting. Light is produced by current through mercury vapor, rather than through a solid wire. Conduction through a gas or metallic vapor is called an *arc;* there is more information about electric arcs in Chapter 8. The light produced by the glowing gas is not uniformly white, its energy is concentrated at a few wavelengths, so colors may not appear normal when viewed by mercury light. The lamps are made in sizes from 100 to 3 000 watts and require special equipment in the fixture to regulate current through the lamp. Life of these lamps averages over 16 000 hours.

Fluorescent Lamps

Fluorescent lamps are also high-efficiency lamps, about 50 lumens per watt even after 6 000 hours of operation. Mercury vapor at very low pressure is the conductor; current through it produces a small amount of blue and violet light, but most of the energy given off by the mercury atoms is invisible ultraviolet. This ultraviolet radiation is absorbed by the coating on the inside of the glass tube,

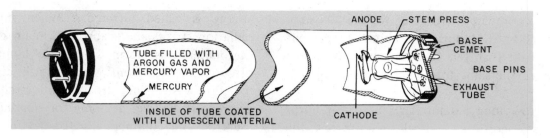

Fig. 5-8 A Fluorescent Light Tube

figure 5-8. When struck by ultraviolet radiation the coating glows, reradiating some of the energy at lower frequencies which are visible light. The color of light radiated by the coating is characteristic of the coating material itself. Substances which have this ability to absorb radiation and immediately reradiate visible energy are called *fluorescent materials* or *phosphors*.

The special circuitry used with fluorescent lamps is described in Chaper 14.

Ultraviolet Lamps

Hundreds of specialized uses for ultraviolet light are found in various industries working with foods, minerals, textiles, metal parts inspections, dyes, coatings, oils, and decorative effects. Most of these uses depend on the ability of many materials to glow visibly when illuminated by invisible ultraviolet light. Other uses depend on the high ability of ultraviolet light to initiate chemical changes. For example, it is used in photographic copying processes.

Low-pressure mercury arcs are sources of ultraviolet light, and an incandescent lamp operated at a higher than normal filament temperature produces a little ultraviolet light. The most used ultraviolet sources, however, are fluorescent lamps. In principle they operate like ordinary fluorescents, but with a coating that radiates almost all of the energy as ultraviolet. When desirable, ultra-violet lamps are covered or enclosed by a special glass filter that stops visible light, but lets the invisible ultraviolet pass through.

OTHER LIGHT SOURCES

Sodium vapor lamps were introduced in a few highway lighting systems approximately forty years ago. They had long life and high efficiency (55 lumens per watt) but produced only yellow light. Recent improvements have put a high-pressure sodium arc in a translucent ceramic tube, giving a whiter light at about 100 lumens per watt.

Small neon-glow lamps (0.04 watt to 3 watts) giving an orange red light are often used as indicator lamps. Their life is over 3 000 hours, and they are unusually reliable in that they fail gradually, having no filament to burn out. The 2-watt NE-34 lamp deserves consideration as a night light for home use.

Another good type of night light for the home is the luminescent panel, in which light is produced by excited electrons in a phosphor sandwiched between thin conducting films. A three-inch panel takes 0.02 watts and is very long-lived.

THE HOME WIRING SYSTEM

Most homes are supplied with electrical energy from a commercial power company, figure 5-9. The company generates ac power in a central location, sends it to community distribution centers at extremely

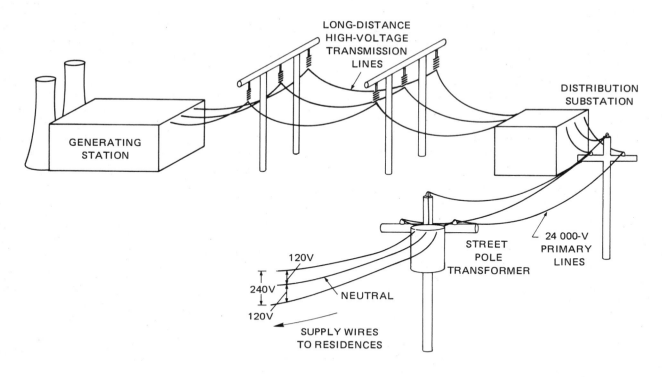

Fig. 5-9 Commercial AC Power

high voltage (150 000 volts or higher), and then uses transformers to lower the voltage for residential use. Chapters 9 and 10 discuss transformer and generator operation. Usually, three conductors supply energy to homes. One of the conductors is at earth ground potential and is called the *neutral* lead. It may be a bare cable, because there is no more danger of receiving a shock by touching it than there is danger from touching the earth. The other two wires are called *hot leads* because they have approximately 120-volts potential with respect to ground, and 240-volts potential with respect to each other. In house wiring, a neutral or ground wire is colored green, white, or left bare. A hot wire is usually either black or red.

The capacity of the service entrance panel, figure 5-10, determines the total amount of power available to circuits in the house. Each circuit line leading from the panel is really a multiwire cable containing a hot wire (probably black) taking power to

an outlet or device; a neutral wire (white) providing a complete circuit back to the source; and a bare ground wire (bonding wire) which grounds the frame of each circuit box or device. This bonding wire insures that the frame of an appliance is at earth potential, thus causing no shock hazard to the user.

> Fuses and circuit breakers are safety devices. When a circuit breaker trips or a fuse blows, locate the cause and provide a remedy before resetting the circuit breaker or replacing the fuse.

Current overload protection is provided by fuses or circuit breakers, or a combination of both, in the service panel. The entire house can be disconnected from the power line by removal of a main fuse holder or the opening of a main circuit breaker. In a fuse, a lead-alloy wire melts when overheated by excessive current, disconnecting the circuit as a result. Bimetal strips and magnetic effects cause an overloaded circuit breaker to open. After the overload is remedied, the breaker can be reset easily. Circuits in damp locations (bathrooms or outdoors) should be protected by ground-fault circuit breakers. The *ground-fault breaker* is intended as a lifesaver. It disconnects the circuit in about 0.03 seconds if someone standing in water takes only a few thousandths of an ampere from a hot wire. Convenience outlets and lighting circuits are protected by small fuses or breakers in the *hot* side of their supply. Any 240-volt device, such as a water heater, must have *both* of its hot leads protected. Removal of fuses or opening of switches in the hot lead is necessary to remove the possibility of shock when working on a device. Connecting a switch in the neutral line is a very dangerous and illegal performance; it would turn the device on or off, but would leave a dangerous shock or fire hazard if the device malfunctions.

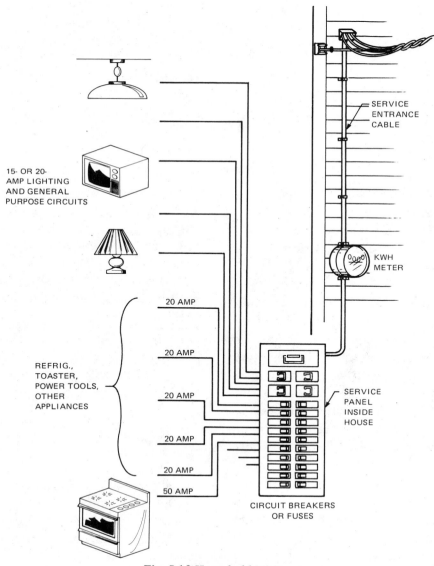

15- OR 20- AMP LIGHTING AND GENERAL PURPOSE CIRCUITS

SERVICE ENTRANCE CABLE

KWH METER

20 AMP

REFRIG., TOASTER, POWER TOOLS, OTHER APPLIANCES

20 AMP

20 AMP

SERVICE PANEL INSIDE HOUSE

20 AMP

20 AMP

50 AMP

CIRCUIT BREAKERS OR FUSES

Fig. 5-10 Household Wiring

Figure 5-11 omits the kwh meter and other details, since it intended only to show some relationships of voltages and currents in the household three-wire circuit. (This arrangement is called three-wire single-phase, not three-phase, which is an entirely different arrangement.) The two 120-volt transformer coils at the left of figure 5-11 are in series, so their voltages add to produce 240 volts. Yet, there is no more than a 120-volt potential difference between any wire and the earth in the residence circuits. Notice also that the neutral conductor in the middle of the diagram carries only the difference in the currents carried by the two outside wires. Circuit breakers or fuses act as on-off switches in the hot wires, the neutral wire is solid throughout.

HOME HEATING

In an increasing number of new homes under construction, electricity is used to heat the house. The two most used types of resistance heating devices are: (1) flexible cable installed in the ceiling before plastering (the warm ceiling heats the room by radiation); and (2) baseboard units, which heat the room by convection. Radiant wall units and radiant baseboard units are also available. The above heating methods are noiseless, safe, and should be long-lived if well built and installed. In new construction, too, the expense of a chimney is avoided. Figure 5-12, page 92, gives examples of energy costs.

The Heat Pump

Figure 5-13, page 92, shows the essentials of a system used in refrigeration, air conditioning, and house heating. The freezer section of a household electric refrigerator can properly be called the evaporator. Inside its double-walled chambers is a liquid that evaporates very readily. As you may have noticed, hands wet with gasoline, cleaning solvents, or water are cooled as the liquid evaporates. The faster the liquid evaporates, the more cooling effect, that is, more heat is taken from the hands. In the refrigerator, when the interior warms up a little, a thermostat starts the electric motor that runs a pump, pumping vapor away from the liquid in the evaporator and giving more liquid a chance to evaporate. As it does so, it takes heat from the metal evaporator itself and everything nearby. As the vapor goes

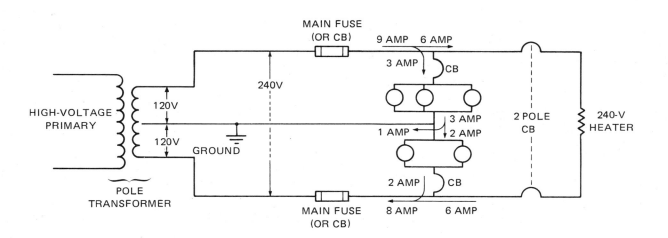

Fig. 5-11   3-Wire Domestic Wiring

COMPARATIVE COST OF ENERGY

	Heat Content	Assumed Efficiency	Price	B.t.u. per dollar	
				Purchased	Used
Electricity (resistors)	3 413 Btu./KWH	100%	3¢/K.W.H.	113 750	113 750
LP Gas (propane, butane)	92 000 Btu./gal.	83%	40¢/gal.	230 000	191 000
Oil	138 000 Btu./gal.	80%	40¢/gal.	345 000	276 000
Natural Gas	1 000 Btu./cu. ft.	83%	25¢/100 cu. ft.	400 000	332 000
Coal	12 500 Btu./lb.	76%	$25/ton	1 000 000	760 000

Fig. 5-12

through the pump, it is warmed by compression. The hot vapor, under pressure, is cooled enough in the condenser so that it changes back to liquid, giving off the heat that it absorbed back in the evaporator when it became vapor. The air of the room cools the condenser, that is, it takes heat from the condenser.

To make this device serve as a house heater, put the evaporator outdoors where it will cool the outside air and put the condenser in the house, so it can warm the inside air. In summer, the valves can be reversed so that condensation takes place outdoors, and evaporation takes place in the pipes inside the house so the house will be cooled. As a heating device, this system is especially appropriate in mild climates, providing winter

heating and summer air conditioning. The transfer of heat requires less expenditure of energy than the production of heat. Hence, the heat pump conserves more energy than resistance heating.

SWITCHING CIRCUITS

The control of one lamp, or one group of lamps, from two locations is accomplished by the use of two *three-way* switches.

As indicated in figure 5-14, a three-way switch is a type of single-pole, double-throw switch. The movable blade is always in connection with the common terminal. There is no on or off position marked on the switch, because the common terminal is always connected to one or the other traveler terminal.

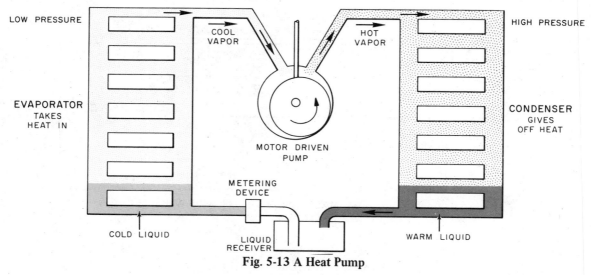

LOW PRESSURE
COOL VAPOR
HOT VAPOR
HIGH PRESSURE
EVAPORATOR TAKES HEAT IN
MOTOR DRIVEN PUMP
CONDENSER GIVES OFF HEAT
METERING DEVICE
COLD LIQUID
LIQUID RECEIVER
WARM LIQUID

Fig. 5-13 A Heat Pump

Figure 5-15B shows two three-way switches in a lamp circuit. As shown, the lamp is on, and it can be turned off by either switch. When the lamp is off, operation of either switch can again close a circuit to the lamp, using one or the other of the two traveler wires between the switches.

For control of a lamp from three or more locations, another type of switch, called a *four-way,* has to be put into the circuit between the two three-way switches, figure 5-16.

A four-way switch has the same effect in a circuit as a reversing switch. The schematic diagram, figure 5-17, shows lamps controlled from four locations. (A) and (D) are three-way switches. (B) and (C) are two four-way switches. Operation of switch (B) converts its internal circuit to that shown in (C); moving the handle of (C) makes its connections like those drawn for (B). Tracing the circuit through the switches in the diagram, one finds that the circuit is open. Operating any one of the four switches will close the circuit.

Remote Control Systems

If a system were desired to control several outlets in a building from each of several different locations, the 120-volt outlets could be turned on or off by low-voltage operated relays. A less expensive

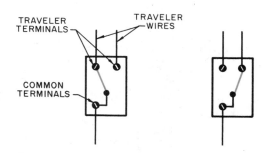

Fig. 5-14 Two Positions of a Three-Way Switch

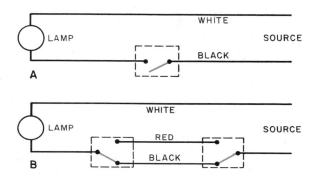

Fig. 5-15   (A) Circuit with Single-Pole Switch Control; (B) Circuit with Three-way Switch Control

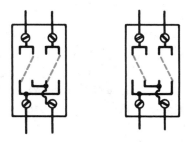

Fig. 5-16 Two Positions of a Four-Way Switch

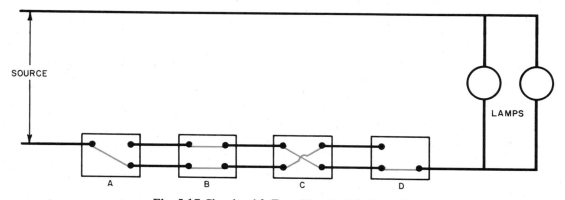

Fig. 5-17 Circuit with Four-Way Switch Control

Fig. 5-18 Low-Voltage Relay and Wiring Diagram

wiring installation is obtained with the low-voltage system, since there is no need to run 120-volt cables to the switches. Low-cost, easily installed #18 or #20 wire is used.

At each controlled outlet, one 24-volt relay is mounted on the outlet box. The relay contains two coils. A momentary current in the on coil closes the 120-volt contacts, which remain closed until a momentary current in the off coil opens them.

The relay coils are operated from a 24-volt transformer by normally open, momentary-contact switches, several of which may be connected in parallel to control one relay. The connection of such a switch in the low-voltage circuit is shown in figure 5-19.

Low-voltage systems can be expanded to include a variety of automatic lighting controls, timers, and security systems.

POINTS TO REMEMBER

- Resistors are 100 percent efficient as heat producers.

- Heating rate, in watts = $I^2 R$.

- Resistance welding uses large currents at low voltages; arc welding uses moderate current and voltage.

- An induction furnace generates heat-producing ac in the metal to be heated.

- Intensity of a source of light is measured in lumens. 12.57 lumens equals one candlepower.

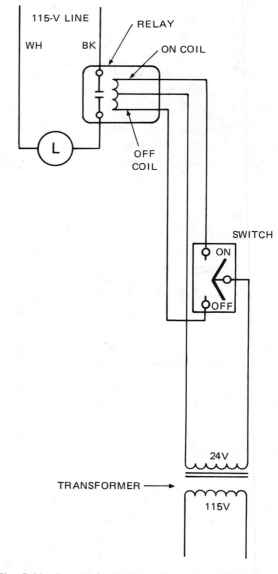

Fig. 5-19 One Light Control from One-Switch Point

- Intensity of illumination of a surface is measured in lumens per square foot.

- Surface illumination is increased not only by increasing power of source, but also by reducing distance between source and surface and placing reflecting surfaces behind the light source.

- Light is produced by electron disturbances in atoms. Light is one narrow group of vibration frequencies of the entire electromagnetic spectrum.

- Larger incandescent lamps are generally more efficient than small ones. Fluorescent lamps and mercury arcs are 2 1/2 to 3 times as efficient as incandescent lamps.

- Hot wires in a house are red or black; white is used for the current-carrying neutral.

- Grounding wires are green or bare.

- 120 V appears between any hot wire and a neutral or ground.

REVIEW QUESTIONS

1. Assuming that 10 000 Btu per hour are required to keep a room warm (70°F./21°C) when the outdoor temperature is 25°F. (–4°C), calculate the necessary equivalent rate of electrical heating, in watts.

2. Resistors are connected to a three-wire system as shown. Calculate the current at each of the lettered points (A, B, C, and D), figure 5-20.

3. Calculate appropriate amount of nichrome wire for a 440-watt heating element.

4. A certain heater element is rated "1 600 watts, 220 volts". Calculate its resistance. What will be its watts if operated at 110 volts?

5. Name two units for measuring intensity of a light source. State the numerical relation between them.

6. Name two units for measure of surface illumination.

7. A 25-watt lamp has an efficiency equal to 10.4 lumens per watt. How much light does it produce?

8. A work surface is illuminated by a single incandescent lamp with no reflector placed 5 ft. (1 500 mm) above the surface. If the lamp is lowered so it is 4 ft. (1 200 mm) from the surface, the illumination on the work directly below the lamp has been increased how many times?

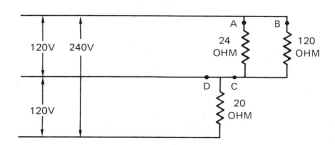

Fig. 5-20

9. State two uses for carbon-arc lights.

10. State uses for mercury arcs, sodium arcs, and neon-glow lamps.

11. What is a phosphor?

12. Diagram a circuit for operating a lamp by switches at two different locations. Indicate the color of the wires and the current source.

13. Diagram a circuit for operating a lamp by switches at three different locations. Indicate the color of the wires and the current source.

14. What are the advantages of remote control relay systems?

15. Under what circumstances are remote control relays appropriate?

16. Give several reasons why you believe the electric power company uses high voltages in their distribution lines.

17. Explain how the current is divided at the distribution panel. What voltages are available? Name some advantages and uses of each.

18. What color wires are used in wiring a house? Name those used on the service drop, for the electric stove, and for convenience outlet and light fixtures.

19. What are the colors of the wires in a 120-volt extension cord? In a 240-volt extension cord? How many wires should each have to provide a ground?

RESEARCH AND DEVELOPMENT

Experiments and Projects on Electric Heating and Lighting

INTRODUCTION

The energy of motion of atoms and molecules produces heat. Light is produced by electron disturbances in atoms. Chapter 5 explains the use, measurement, and control of electron motion to produce heat and light for various purposes. The experiments and projects which follow provide an opportunity for you to control electron flow as heat or light is produced, test and repair appliances, and construct experimental devices.

EXPERIMENTS

1. To observe operation of a bimetal thermostat in controlling electron flow in a circuit.

2. To compare the lighting effect and efficiency of an incandescent lamp and a fluorescent lamp of equal power.

3. To test heating effects of electricity in different metals.

PROBLEMS

Information and Problems in circuit wiring are provided as follows:

Problem 1. Circuit with single pole switch: feed at switch.

Problem 2. Circuit with single pole switch: feed at light.

Problem 3.   Lamp controlled by a single-pole switch with live convenience receptacle: feed at switch.

Problem 4.   Circuits controlled with three-way switches:
A. Feed at switch
B. Feed at light
C. Feed at light between switches

Problem 5.   Circuit with control at three different locations: feed through switches.

Problem 6.   Circuit with a pilot light and single-pole switch.

> Replace all worn or deteriorated cords on lamps, appliances, and power tools.   Use good quality cord large enough to carry the load.   Frayed cords with deteriorated rubber are a fire hazard.

PROJECTS

1.   Design and build an adjustable bimetal thermostat.

2.   Design a heating element such as an electric hot plate. Calculate the amount of resistance wire of the proper size and kind to produce a 600-watt heating element.

3.   Wire a lamp.

4.   Repair a heating appliance cord.

5.   Make a 120-volt, three-wire extension cord.

6.   Make a three-wire replacement cord for an electric power tool, such as a hand drill, with a ground connection.

7.   Make a 240-volt extension cord.

8.   Plan a circuit and wire a remote control device.

9.   Design a heating element which will develop heat at a rate of about 220 watts. Make the element and test for current consumed and find the resistance in ohms.

EXPERIMENTS

Experiment 1 CONTROLLING A CIRCUIT WITH A THERMOSTAT

OBJECT

To observe the operation of a bimetal thermostat in controlling the flow of electrons in a circuit.

APPARATUS

1 - Bimetal thermostat
1 - SPST switch
1 - Lamp base

1 - Lamp, 100-watt

1 - Electric soldering copper or other suitable source of heat

PROCEDURE

1. Assemble the apparatus according to the circuit diagram in figure 5-21.

2. Adjust thermostat so that contacts are just closed.

3. Close switch; lamp should light.

4. Hold a hot soldering copper or other heat source near the bimetal element in the thermostat until the lamp goes out.

5. Observe the action of the mechanism in the thermostat as you heat it.

6. Allow the thermostat to become cool. Watch the bimetal element as it loses heat.

7. Use the thermostat-controlled lamp as a source of heat. Arrange the lamp and thermostat so that the light turns off and on automatically.

QUESTIONS

1. What happened to the lamp when the thermostat became hot?

2. Why did the contact points close when the bimetal element became cool?

3. Where is this device generally used in a home? Name three other uses.

4. Could a device operating on this same principle be used to protect a stalled electric motor from overheating? Explain your answer.

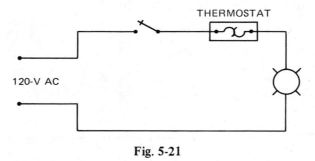

Fig. 5-21

Experiment 2 MEASURING ILLUMINATION

OBJECT

To compare the lighting effect and efficiency of an incandescent lamp and a fluorescent lamp of equal power.

APPARATUS

1 - Lamp receptacle

1 - Incandescent lamp

1 - Fluorescent lamp (15 or 25 W)
1 - Fluorescent tube to fit fixture
1 - Light meter
2 - SPST switches

OBSERVATIONS

Use a table similar to the one shown here to record your observations.

LAMPS	WATTS	LUMENS/SQ. FT.
FLUORESCENT		
INCANDESCENT		

PROCEDURE

1. Connect a fluorescent lamp in a circuit according to the diagram in figure 5-22.

2. From a given point 8 to 10 feet (2.5 to 3 meters) or more, measure the intensity of the light it produces. *Note:* Lamps should not have reflectors if a true comparison is to be made.

3. Record the candlepower or the lumens per square foot.

4. Connect an incandescent lamp in a circuit according to the diagram.

5. Measure the intensity of light from an incandescent lamp of the same wattage as the fluorescent lamp. *Note:* The lamps should be placed in the same position and the light reading taken from the same point.

6. Measure the intensity of light from several incandescent lamps of different power until you find one about equal to the fluorescent lamp. Record the results of your measurement for each lamp.

QUESTIONS

1. What was the difference in power (watts) between the fluorescent lamp and the incandescent lamp producing approximately the same amount of light?

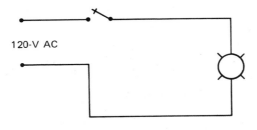

120-V AC

Fig. 5-22

2. What is the ratio of illumination produced by the fluorescent lamp to that of the incandescent in terms of foot-candles or lumens per square foot?

3. Which type of illumination is most efficient?

Experiment 3 HEATING EFFECTS OF ELECTRICITY IN DIFFERENT METALS

OBJECT

To test the heating effect of a length of copper wire and a length of nichrome wire connected in series and then in parallel.

APPARATUS

1 - Push-button switch
1 - Battery or dc power source, (1.5 - 3 and 6 - 12 volts)
1 - Board, 5 1/2" wide 1' long (140 mm wide x 300 mm long)
    12" (300 mm) #30 Nichrome wire
    12" (300 mm) #30 Copper wire
2 - Strips thin copper about 3/8" wide 4" long (9 mm wide x 100 mm long)
4 - Fahnestock clips
4 - Wood screws, #8 x 3/4" RH

PROCEDURE

1. Fasten the copper strips, Fahnestock clips, and wire on the board according to figure 5-23.

2. Connect the switch and 1.5-volt power supply. to circuit.

3. Close the switch. After a few moments, test the heat in the copper and nichrome wire and record the results. Repeat the experiment, using a 3-volt power supply.

4. Record your observations.

5. Disassemble the right side of the circuit and arrange the wires according to figure 5-24.

6. Connect the switch and 6-volt supply to the circuit.

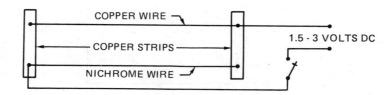

Fig. 5-23 Wires Connected in Parallel

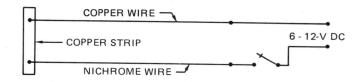

Fig. 5-24 Wires Connected in Series

7. Close the switch and test the heat in each wire as in figure 5-23 and record. Repeat the experiment, using a 12-volt power supply.

8. Record your observations.

OBSERVATIONS

Use a table similar to the one shown here to record your observations.

CIRCUIT	VOLTS	EFFECT ON NICHROME	EFFECT ON COPPER
PARALLEL	6		
SERIES	6		
PARALLEL	12		
SERIES	12		

QUESTIONS

1. In figure 5-23, which kind of wire gained the most heat? Give a reason for your answer.

2. Did the parallel arrangement cause one kind of wire to gain considerably more heat than the other?

3. If both wires were made of the same material, would one heat more than the other?

4. In figure 5-24, which kind of wire gained the most heat? Explain.

5. If both wires were made of the same material, would one gain more heat than the other in the series circuit? Why or why not?

6. Did you notice any difference in the amount of heat produced using 12 volts in the circuit as compared to 6 volts? Give a reason for your answer.

PROBLEMS

Since it is important for you to know how the electric light switches in your home operate, and how connections should be made, several problems are suggested together with appropriate information about each type of switch. Indicate the color of each numbered wire in each pictorial. Then solve Problems 1 through 6.

Problem 1 Circuit With Single-Pole Switch: Feed at Switch

A single-pole switch is used where it is desired to control a light or group of lights, or other load, from one switching point. This type of switch is used in series with the ungrounded "hot" wire feeding the load. The diagrams given in figure 5-25 show a single-pole switch controlling a light from one switching point where the 120-volt source feeds directly through the switch. Note that the identified white wire goes directly to the load. The unidentified wire or black wire is broken at the single-pole switch. The cable is nonmetallic, with conductors and a separate wire for grounding.

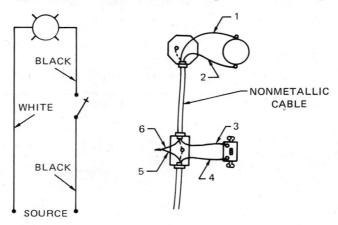

Fig. 5-25

Problem 2 Circuit With Single-Pole Switch: Feed at Light

The 120-volt source feeds directly to the light outlet in figure 5-26. This results in a two-wire cable with black and white wires being used as a switch loop between the light outlet and the single-pole switch. The National Electrical Code permits the use of a white wire in a single-pole switch loop in

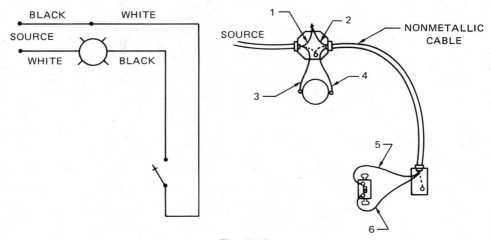

Fig. 5-26

Section 200-7. However, the unidentified or black conductor must connect between the switch and the load. This requirement is observed in figure 5-26.

Problem 3   Lamp Controlled by a Single-Pole Switch With Live Convenience Receptacle: Feed at Switch

Figure 5-27 gives another application of a single-pole switch control. The feed is at the switch which is to control the light outlet, with the convenience outlet independent of the switch.

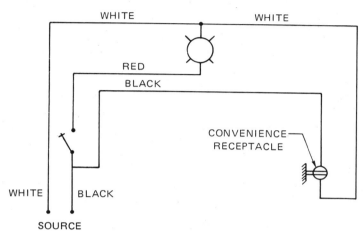

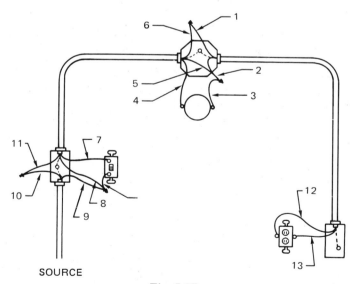

Fig. 5-27

THREE-WAY SWITCH CONTROL

A three-way switch may be compared with a single-pole, double-throw switch. There is one terminal called the common terminal to which the switch blade is always connected. The other two terminals are called the traveler wire terminals.   In one position the switch blade is connected between the common terminal and one of the traveler terminals.   In the other position the switch blade is connected between the common terminal and the other traveler terminal.   Figure 5-28 illustrates the operation of the three-way switch.

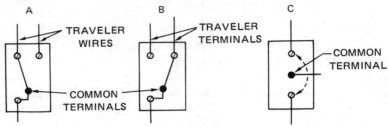

A AND B: TWO POSITIONS OF THE THREE-WAY SWITCH   C IS A S.P.D.T. SWITCH

Fig. 5-28

The three-way switch has no on or off position. As a result, there is no on or off position marked on the switch handle. The three-way switch can be further identified by its three terminals. One of these terminals, called the common terminal, is a darker color than the other two traveler wire terminals which are a natural brass color.

Two three-way switches are used where it is desired to control a light or group of lights, or other load, from two different switching points. Therefore, they are used in any circuit where it is desired to control the load from either of two switching points.

Problem 4 Circuits Controlled With Three-Way Switches

Figure 5-29 represents a circuit where one light is to be controlled from either of two switching points. Many times it is convenient to be able

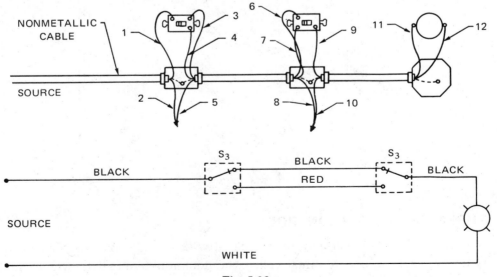

Fig. 5-29

to control a hall light from either upstairs or downstairs or possibly a garage light from either the house or the garage. Note that the feed in this circuit is at the first switch-control point.

Figure 5-30 represents a different arrangement using three-way switch control where the feed is at the light. This circuit arrangement makes it necessary to use the white wire in the cable as part of the three-way switch loop. In compliance with Section 200-7, Exception No. 2, National Electrical Code, the unidentified or black wire is used as the return wire to the light outlet.

Exception No. 2. Cable containing an identified conductor may be used for single-pole, three-way or four-way switch loops where the connections

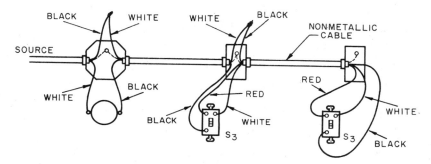

Fig. 5-30

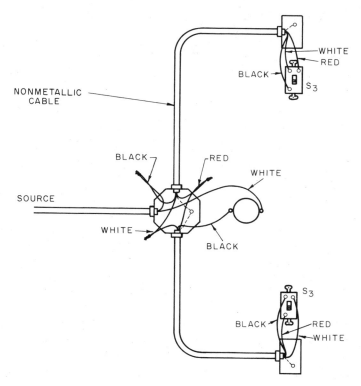

Fig. 5-31

are so made that the unidentified conductor is the return conductor from the switch to the outlet. This exception makes it unnecessary to paint the terminal of the identified conductor at the switch outlet.

Figure 5-31, page 105, represents another arrangement for three-way switch control. The feed is at the light with cable runs from the ceiling outlet to each of the three-way switch control points which are located on each side of the light outlet.

FOUR-WAY SWITCH CONTROL

A four-way switch, figure 5-32, is like a double-pole, double-throw switch. Like the three-way switch, it has two positions and neither of these positions is on or off. Thus, the four-way switch has no on or off marked on the switch handle.

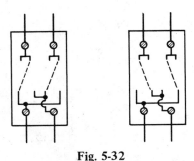

Fig. 5-32

The four-way switch is used where it is desired to control a light or group of lights, or other load, from more than two switching points. The switch that is connected to the source and the switch that is connected to the load will be three-way switches. However, at all other control points, four-way switches are used.

Problem 5   Circuit With Switch Control at Three Different Locations: Feed Through Switches

Figure 5-33 illustrates control of a lamp from any one of three switching points. Care must be used in connecting the traveler wires to the proper terminals of the four-way switch. Always make sure that the two traveler wires from one three-way switch are connected to the two terminals on one side of the four-way switch while the two traveler wires from the other three-way switch connect to the two terminals on the other side of the four-way switch.

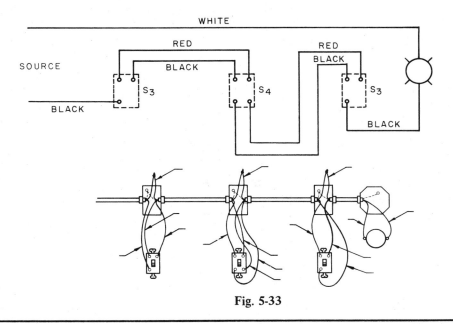

Fig. 5-33

Problem 6 Circuit With a Pilot Light and Single-Pole Switch: Feed at Switch

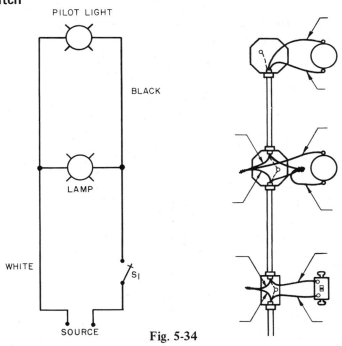

Fig. 5-34

| PROJECTS |

DESIGN AND BUILD A BIMETAL THERMOSTAT

The bimetal or compound-bar thermostat is actually a switch, figure 5-35, page 108. It operates on the same basic principle as the control devices used in many heating appliances.

The brass expands when heated, breaking the contact. As it cools, it contracts and makes contact again.

MATERIALS

4 - Binding posts or Fahnestock clips
1 - Strip brass, #22 gauge, 3/8″ x 6″ (9 mm x 150 mm)
1 - Strip iron, #24 gauge, 3/8″ x 6″ (9 mm x 150 mm)
1 - Pc. band iron, 1/8″ x 3/8″ x 3 1/2″ (3 mm x 9 mm x 90 mm)
1 - Base, wood, 3/4″ x 4 1/2″ x 7 1/2″ (20 mm x 100 mm x 200 mm)
1 - Pc. asbestos paper, 2″ x 3″ (50 mm x 75 mm)
1 - Pc. nichrome wire, #28 or #30
1 - Brass machine screw, #8-32 x 3/4″ RH
1 - Brass nut, #8-32
4 - Wood screws, #6 x 1/2″ RH

PROCEDURE

Cut and shape the various pieces. Braze or solder the brass and iron bimetal strips together. Remove the excess fastening material with a file and slightly round all edges. Lay out and drill the holes. Moisten the asbestos paper and carefully wrap it around the bimetal strip as shown in the drawing. Allow the paper to dry.

Calculate the amount of nichrome wire required and wind it over the asbestos paper. The wires should be about 1/16″ (2 mm) apart. Allow enough to fasten under the binding post on the bent end of bimetal strip and to wrap tightly, several turns around the bare metal on the other end.

Assemble the parts.

The contact screw should be adjusted to just make contact with the bimetal strip.

Connect several lamps to the load binding posts. About 100 watts is sufficient.

Connect the other binding posts to 115-volt current supply and test the unit. The frequency of the flash may be controlled by adjusting the gap between the bimetal strip and the contact screw.

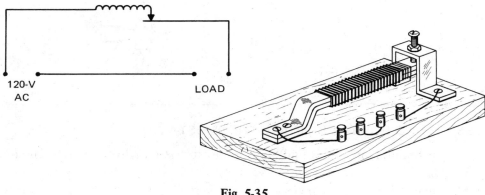

Fig. 5-35

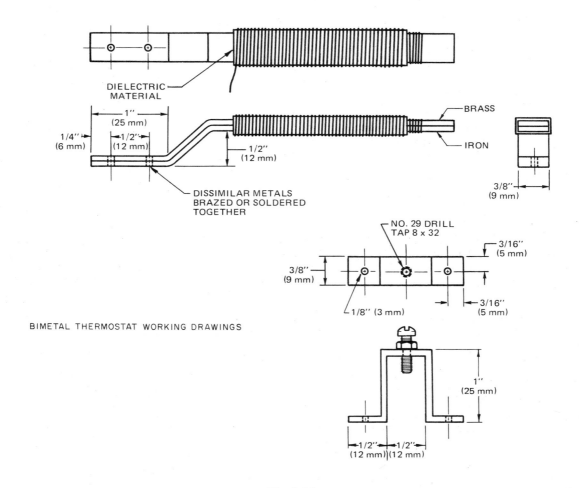

BIMETAL THERMOSTAT WORKING DRAWINGS

Fig. 5-36

DESIGN AND CONSTRUCT AN ELECTRIC STOVE

MATERIALS

Sheet metal, aluminum or steel, #22 gauge Transite board, 1/8" or 3/16"
(3 mm or 5 mm) thick (a good quality asbestos shingle is satisfactory)
Steel welding rod, 1/8"

16 - Steel machine screws, #6-32 x 1/4" RH

16 - Steel machine screw nuts, #6-32

 2 - Brass machine screws, #8-32 x 1/4" RH

 4 - Brass machine screw nuts, #8-32

 1 - Brass rod, 3/16" x 3" (5 mm x 75 mm)

 4 - Brass nuts, #5-40

 6 - Brass washers, 1/8"

 4 - Brass washers, #8

    Nichrome resistance wire, #19-23 AWG

    Sheet mica or dielectric bushings

PROCEDURE

Appraise the pictorial sketch, figure 5-37, and determine whether you wish to make a stove like it or whether you wish to modify the design.

If you decide to change the design to one which will be more appropriate for your use, make some idea sketches to present to your instructor for approval. Make working drawings for your design.

Analyze your job and list the major operations you will have to perform in making each part.

Calculate the amount of resistance wire you will need to make a heating element of

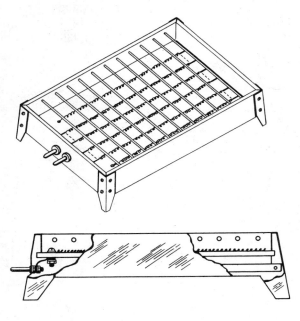

Fig. 5-37

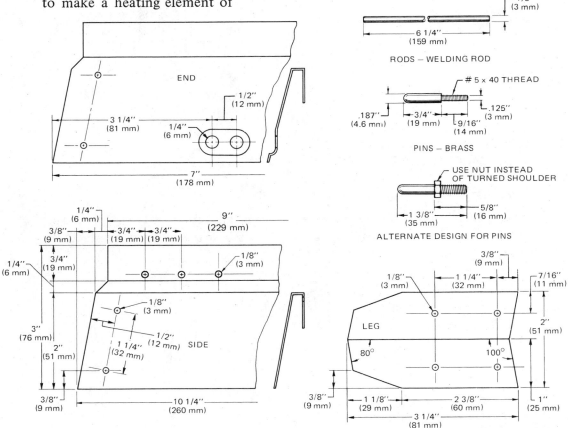

Fig. 5-38

about 650 watts. No. 22 gauge resistance wire is an appropriate size. Wind the element on a 1/8" (3 mm) rod.

To make a winding rod, saw a slot 3/16" deep in the center of one end of a length of 1/8" (3 mm) drill rod. Use a thin hacksaw blade.

To wind the element, make a 1/8" (3 mm) right angle bend on one end of the wire. Place the angle bend in the saw slot of the rod and wind the wire tightly around the rod in a clockwise direction. Keep a space between the turns about the same or slightly less than the thickness of the wire.

Place this end of the rod between leather and clamp it in a vise. This will form a guide to insure that all the turns of wire will be the same distance apart. Turn the crank slowly until all the element wire is wound on the rod, figure 5-39.

Fasten the element wire to the transite base, figure 5-40. Assemble the terminal pins on the end piece. After the shell is assembled, place the heating unit in position and fasten the connecting wires from the element screws to the terminal pins. Next, fasten the support pan in place and test the stove on a 115-volt power source.

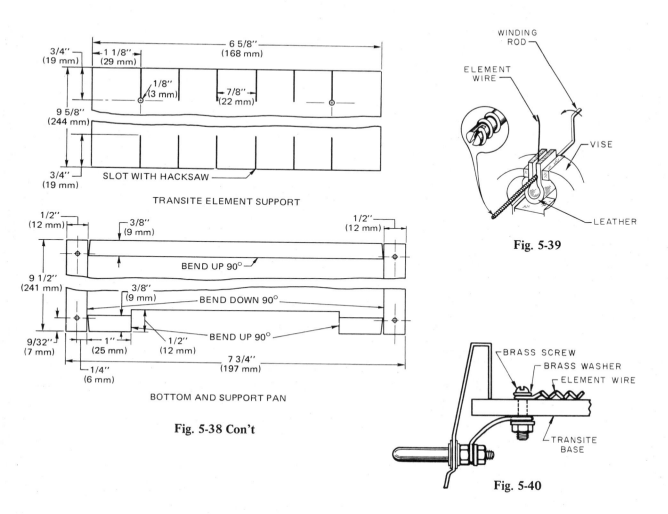

TRANSITE ELEMENT SUPPORT

BOTTOM AND SUPPORT PAN

Fig. 5-38 Con't

Fig. 5-39

Fig. 5-40

WIRING A LAMP

MATERIALS

Lamp socket
Attachment plug
Wire, lamp cord

PROCEDURE

Select an appropriate lamp cord and cut to desired length. Obtain a dead front attachment plug. Disassemble the plug and open the prongs. See figure 5-41(A). Cut the end of the cord square and thread it through the cap as in figure 5-41(B). Next, insert the wire into the plug as far as it will go and close the prongs, figure 5-41(C). Hold the prongs together and carefully slide the cap in place, figure 5-41(D).

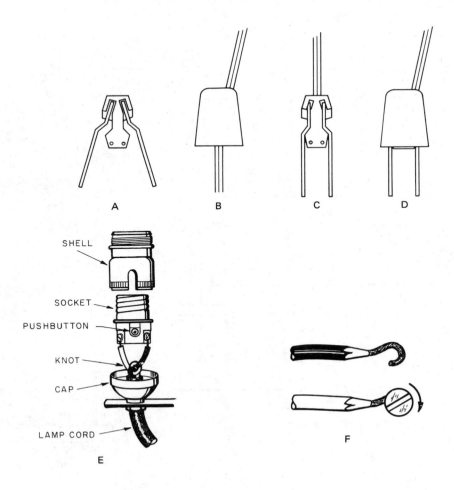

Fig. 5-41

Part the wires on the other end of cord about 2 3/4 inches (70 mm) and remove 5/8 of an inch (160 mm) of insulation from the end of each wire. Be careful not to cut into the wire. Damage to the fine wires can be avoided by holding the knife at an angle and removing a small amount of insulation at one time. If a wire stripper is available, use it instead of a knife. Twist the strands of wire firmly together and tin.

Disassemble the socket and mount the cap on the lamp, figure 5-41(E). Thread the wire through the lamp socket and tie an underwriters' knot as shown in figure 5-42.

Attach the wires as in figure 5-41 (E). The white wire should be attached to the silver-colored screw and the black wire to the brass-colored screw. Wires should always be wound around a screw in a clockwise direction as shown in figure 5-41 (F).

After you have securely fastened the wires to the socket, fit the insulating sleeve and socket shell in place. Now, carefully pull the cord through the lamp until the socket fits into the base and the shell can be snapped into place.

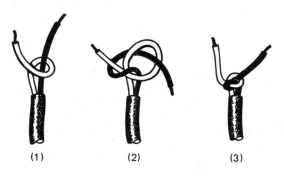

(1)   (2)   (3)

MAKING AN UNDERWRITERS' KNOT

Fig. 5-42

REPAIRING A HEATING APPLIANCE CORD

MATERIALS

Appliance plug
Attachment plug
Wire, Type HPN heater cord, 300-volt or equivalent asbestos-insulated
    heater cord

PROCEDURE

Insulation should be removed from the cord. Asbestos-insulated heater cord requires special attention since the asbestos frays easily and the insulating value will be lost if it is not held in place. The asbestos can be secured by winding a few turns of thread around it after the two wires are separated.

Attach the cord to the male plug as for a lamp cord. The correct method of fastening the cord to an appliance plug is shown in figure 5-43.

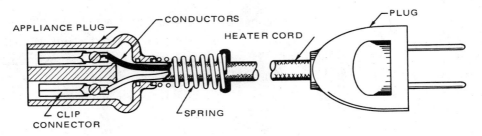

Fig. 5-43

MAKING AN EXTENSION WITH A GROUND WIRE

MATERIALS

Attachment plug, parallel-ground
Connector, parallel ground
Wire, Type SJ, 300-volt, 16/3 rubber-covered cord

PROCEDURE

Thread cord through the plug, remove insulation, twist strands together, tin, and fasten them around the screws in a clockwise direction as shown in figure 5-44 (A). The black wire must be fastened to the brass-colored screw, the white wire to the silver-colored screw, and the ground wire to the green, hexagonal-shaped screw. Press the wire firmly around the pin and prongs, press the fiber insulating cap on, and tighten the clamp on the reverse side. *Note:* When making a three-wire extension cord, always match the cord end colors on the male and female plugs.

Next, thread the cap of the connector on the other end of the cord as in figure 5-44(C). Prepare the wires and fasten them as indicated in figure 5-44(B) and (C). Again, the black wire must be fastened to the brass-colored screw, and the ground wire to the green, hexagonal-shaped screw. Fit the pieces together, being careful to put the key in the right place, tighten the screws until they are snug (not too tight), tighten the clamp on the reverse side and the extension cord is ready to test.

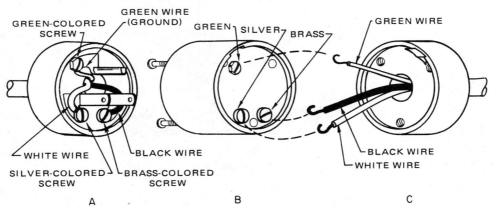

Fig. 5-44

MAKING A REPLACEMENT CORD FOR AN ELECTRIC POWER TOOL

MATERIALS

Attachment plug, parallel-ground
Wire, Type SJ, 300-volt, 16/3 rubber-covered cord

PROCEDURE

Prepare attachment plug end of cord the same as for an extension cord with a ground wire. The wires on the other end of the cord should be fastened to the marked screws or wires in the tool, figure 5-45. It is good practice, when disassembling a power tool, to observe how the wires were fastened and how the cord was secured to the frame.

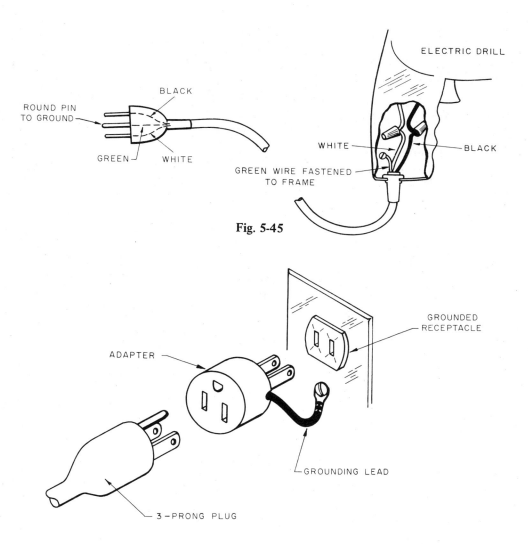

Fig. 5-45

Fig. 5-46

When it is necessary to use an electric power tool with a convenience outlet that does not have provision for 3-prong plugs, an adapter should be used, figure 5-46. Attach the grounding lead on the adapter to the center screw of the receptacle. Check the system for ground. If it is not grounded, attach the lead to a water pipe.

MAKING A 240-VOLT EXTENSION CORD

MATERIALS

Attachment plug, 3-wire twist-lock, 20-amp, 250-volt
Connector, 3-wire twist-lock, 20-amp, 250-volt
Wire, type S, 600-volt, 12/3 rubber-covered cord

PROCEDURE

Remove the required amount of rubber cover from both ends of the cord. Free the wires and remove one-half inch of insulation. Twist the strands in each wire together, trim ends, and fit the copper sleeves over the wires as shown in figure 5-47.

Next, fit the wires through the proper holes in the attachment plug according to the color or witness marks, figure 5-48. The black wire should be connected to the terminal marked B, the white wire to the terminal marked C, and the green wire to the terminal marked GR.

Adjust the cord so the wires protrude about 7/16" (11 mm) above the base. Hold the cord in this position and tighten the clamp on the reverse side. Check the distance the wires extend and whether they are in the right holes; then bend them toward the outer shell of the plug. Press them firmly into place, fit the cap on, press it into place, and tighten the screws until the connection is firm. Too much pressure will strip screws.

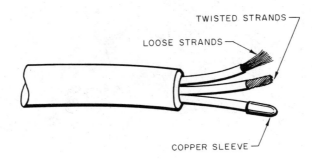

TWISTED STRANDS
LOOSE STRANDS
COPPER SLEEVE

Fig. 5-47

The same procedure should be followed in fastening the other end of the cord to the conductor.

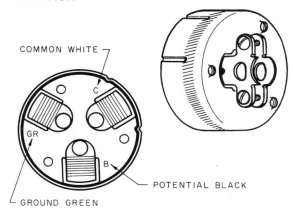

Fig. 5-48

Chapter 6
Electromagnetism

One of the most familiar and most used effects of electric current is the ability to produce the force called magnetism. This magnetic force is responsible for the operation of motors, generators, electrical measuring instruments, communication equipment, transformers, and a great variety of electrical control devices.

All magnetism is electromagnetism, an effect of the energy of electron motion. A great deal was learned about magnets before this basic fact was discovered: currents traveling in the same direction attract each other; currents traveling in opposite directions repel each other. First, some of the earliest known facts of magnetism will be discussed. How they are explained by action of electrons will be discussed later.

Magnetism is a different force than the attraction and repulsion forces due to static electric charges. Both kinds of force have been recognized separately for centuries. The first hint that magnetism was in some way connected with electrical behavior appeared in 1819, when Hans Oersted, a physics professor in Denmark, noticed that a magnetic compass needle was affected by a wire that he had connected to a battery.

If a wire is punched through this black spot on the page, figure 6-1, connected to the battery, and held perpendicular to the paper so that electrons come from below the page toward the reader, compasses placed on the paper near the wire will point in the directions shown in the diagram. The north ends of the compasses are pushed in the direction shown by the clockwise arrows around the wire. Some already known facts that must be reviewed about magnets are used as a help in picturing the reason for this compass-affecting force.

SIMPLE MAGNETS

A magnet attracts iron, steel, and a few other materials, such as nickel, cobalt, and a

COLOR IS NORTH

Fig. 6-1 How Current in a Wire Affects Magnets

few minerals and alloys. Magnets do not attract copper, aluminum, wood, or paper. They do not attract most substances. This is unlike electrical attraction, which works on everything.

The force of the magnet is strongest at two spots on the magnet, called *poles.* It was found that if a magnet is supported by a string or on a pivot, one of these poles turns toward the north, the other toward the south. The end of the magnet toward the north is called its north end, or north pole; the other, the south pole. A compass needle is a pivoted lightweight strip of magnetized steel.

By bringing compasses near each other, it was discovered that the north end of a compass (or any magnet) repels the north end of another magnet and attracts the south. The south pole of a magnet repels the south pole of another magnet, and attracts the north, figure 6-2. This is summarized in the magnetic attraction and repulsion law: like poles repel, unlike poles attract. Although this sounds like the electrical attraction and repulsion law, remember that magnets are one thing, electrical charges are another.

The term poles means "opposite parts," as used in "positive and negative poles of a battery" or "north and south geographic poles" of the earth. It might have been better if magnet poles had been names, instead of north and south, some other pair of names, such as black and white, or right and left. One must realize that the geographic poles of the earth are the ends of the axis on which the earth turns, and they are not magnetic

poles. There is a place in northern Canada that has the same kind of magnetic force as the south pole of a steel magnet. There is a place in the Antarctic that has the same kind of magnetic force as the north pole of a steel magnet.

THE MAGNETIC FIELD

The force in the space around a magnet can be pictured by examining the pattern made by iron filings sprinkled on a card placed over the magnet. Each little splinter of iron acts like a compass needle, attracting other filings at its ends and repelling those lying parallel to it. These chains of filings led to the assumption that the region (field) around a magnet contains invisible *lines of force,* figure 6-3.

English scientist Michael Faraday suggested in 1840 that the behavior of magnets

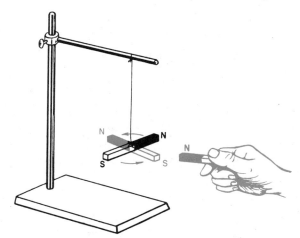

Fig. 6-2 Magnetic Attraction and Repulsion

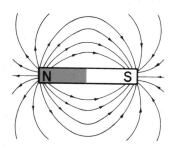 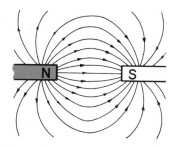 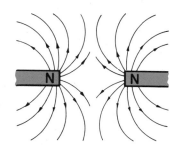

Fig. 6-3 Magnetic Lines of Force

could be explained by the interaction of these lines of force. The lines act like stretched rubber bands in attempting to contract lengthwise, thus pulling north and south together, and at the same time pushing against each other sidewise as they try to contract. They are closed curves that could apparently be continued through the inside of the magnet. *Magnetic flux* is another term for magnetic lines of force.

Magnetic lines of force have a direction that is defined as the direction in which the north pole of a compass tries to point when it is put in a magnetic field. The density, or concentration, of the lines represents the amount of force. The lines cannot cross, by definition, because the compass does not point two ways at once.

Like electric lines of force, these magnetic imaginary lines represent a real force.

THE MAGNETIZING PROCESS

Iron, nickel, and some metal oxides and alloys are called magnetic materials because they can be magnetized. A *magnet* is a piece of magnetic material that has magnetic poles developed on it. The magnetizing, or pole-forming, process is accomplished either by putting the material inside a coil of wire that has current in it, or by placing the material near another magnet.

Long ago, it was found that heating or hammering a magnet would cause it to lose some of its strength. Both heating and hammering tend to shake and stir up the atoms of the metal. Furthermore, if an ordinary steel bar magnet (or any magnet) with only two poles is cut into fragments, each fragment has two poles, north (N) and south (S). Theoretically, this cutting process could go on until we came to the smallest possible fragment of iron, which is an atom. Thus arose the idea that all atoms of magnetic materials are themselves permanent magnets, figure 6-4.

In an unmagnetized piece of iron, the atoms of iron are arranged haphazardly, with the atomic north and south poles pointing in all directions. The magnetizing process rotates these atoms so that the north pole sides of atoms are facing in the same direction.

When a magnet is cut, without disturbing the atom arrangement, the atomic south poles are exposed on one side of the break, and a group of north poles are exposed on the other, figure 6-5. Before the cutting, these atomic poles exerted their attraction force of north and south pole on each other, so that there was no force reaching out into space around them.

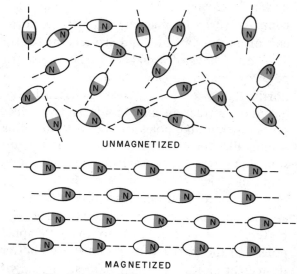

Fig. 6-4 Atomic Arrangement of Iron

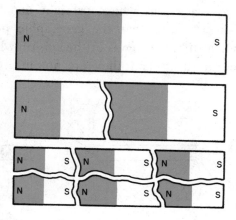

Fig. 6-5 A Magnet Before and After Cutting into Fragments

Recently, the above ideas have been changed slightly by the discovery of a degree of order in the magnetic disorder of an unmagnetized piece of iron. Within a crystal grain of iron, a few thousand atoms form a group called a *magnetic domain.* Within one domain, the atoms are lined up with north poles all facing in one direction. This group of atoms acts like a little permanent magnet but produces no external effect because of nearby oppositely magnetized domains.

The strength of a given permanent magnet is limited. When all of the atoms are perfectly in order facing in the same direction, the magnet is at its maximum strength, or *saturation.*

Types of Magnetic Materials

For a rough classification, magnets may be divided into two groups: permanent and temporary. A permanent magnet is intended to keep its atomic arrangement steady after the magnetizing force is removed. Permanent magnets are used in telephone receivers, door latches, small dc motors, electrical measuring instruments, magnetos, speedometers, and a variety of gadgets. For years, high carbon tool steel, and a few alloy steels (cobalt, molybdenum, chrome-tungsten) were the only useful permanent-magnet materials. Later, various alloys were developed, such as alnico (aluminum, nickel, cobalt, iron), which is widely used as a high strength permanent magnet. Flexible rubbery magnets are made by mixing magnetic oxide powders with vinyl and other plastics.

Temporary-magnet materials are easily and strongly magnetized but lose most of their strength when the magnetizing force is removed. They are of more importance than permanent-magnet materials, both in total amount of use and in a variety of applications. The material first used for temporary magnets was the purest iron obtainable, softened by

annealing (slow cooling). A soft metal is one in which atoms slide around readily, permitting atoms to be easily disarranged magnetically. The most used material is silicon iron (2-4% Si), a soft alloy used in transformers, most motors and generators, relays, and other magnetic equipment built in large quantities.

THE RELATION OF MAGNETIC FIELD TO CURRENT DIRECTION

For a Single Wire

Information similar to that shown in figure 6-1 can also be obtained with the help of coarse iron filings, which make the circular pattern of the magnetic field evident.

If the direction of electron flow is known, the direction of the magnetic field may be found as shown in figure 6-6. Imagine grasping the wire with the left hand, with the thumb pointing in the direction of the electron current; the fingers then encircle the wire in the same direction as the magnetic

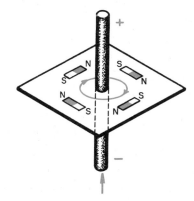

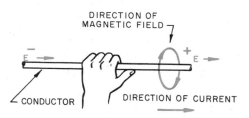

Fig. 6-6 Left-Hand Rule for Magnetic Field Around a Wire

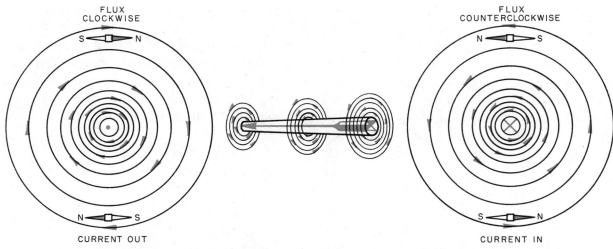

Fig. 6-7 Relation of Magnetic Flux According to Direction of Current in a Wire

lines of force. (Direction of the field means the direction which the north pole of the compass takes; the thumb gives the direction of electron flow.)

The pattern of the magnetic field may also be shown by figure 6-7. The dot in the center of the left-hand wire indicates the point of the current-direction arrow coming toward the observer. The X at the right represents the tail of the current arrow pointing away from the observer.

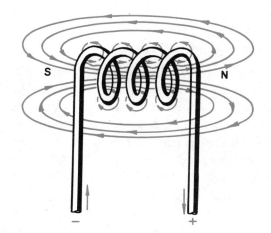

Fig. 6-8 Magnetic Polarity of a Coil

For a Coil

When wire is wound into a coil, as in figure 6-8, each turn of wire is surrounded by its own circular magnetic field. These little whirls of force combine to make the one large field shown surrounding the entire coil.

A magnetic coil of this shape is termed a solenoid. Figure 6-9 shows a way to remember the relation between current direction and field direction for a coil.

The ends of the coil are, in effect, magnetic poles, regardless of whether there is any iron core in the coil.

If the coil is grasped with the left hand so that the fingers point in the same direction as the electron current in the wires, the

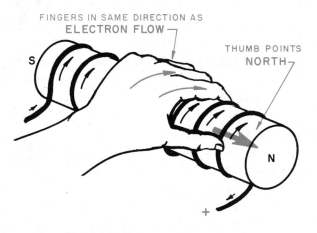

Fig. 6-9 Left-Hand Rule for a Coil

extended thumb points in the direction of the magnetic field inside the coil and points at the north end of the coil.

If the current direction is unknown, but the field of the coil is known or can be found with a compass, then the current direction can be found by using this rule.

Forces Between Parallel Currents

It has been pointed out that all magnetism is an effect of electron motion. Currents in the same direction attract each other; currents traveling in opposite directions repel each other, figure 6-10.

Andre Ampere first reported in 1822: "I observe that when I passed a current of electricity in both of these wires at once, they attracted each other when the two currents were in the same direction and re-pelled each other when they were in opposite directions ... the attractions and repulsions ... are facts given by an experiment which is easy to repeat ... We now turn to the exam-ination of this action and of the action of two magnets on each other and we shall see that they both come under the law of the mutual action of two electric currents." Ampere's idea that there are currents inside a permanent magnet was not justified until 100 years later by the discovery that atoms contain spinning electrons.

If the attraction and repulsion shown in figure 6-11 are not immediately reasonable,

one might apply the left-hand rule for magnet poles and locate north and south on each coil.

THE MAGNETIC CORE IN THE COIL

The magnetizing ability of a coil may be described and measured conveniently in either of two ways:

1. A system of formulas has been set up so that a certain magnetizing ability, or mag-netic strength, can be represented by a certain number of lines of force in each square inch of sectional area of the coil. The number of lines of force per square inch is called the *flux density*. *Flux* means the total number of lines.

2. The magnetizing ability can also be represented by the *ampere-turns* of the coil. Ampere-turns are the result of multiplying the

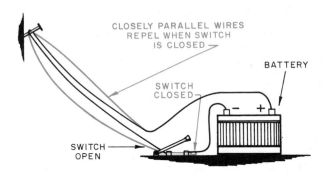

Fig. 6-10 Magnetic Effect of Current in Parallel Wires

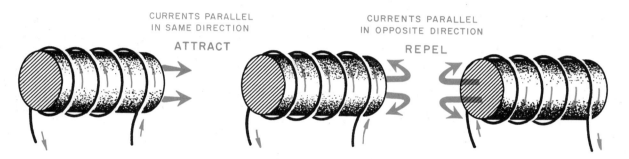

Fig. 6-11 Magnetic Effect of Direction of Current in a Coil

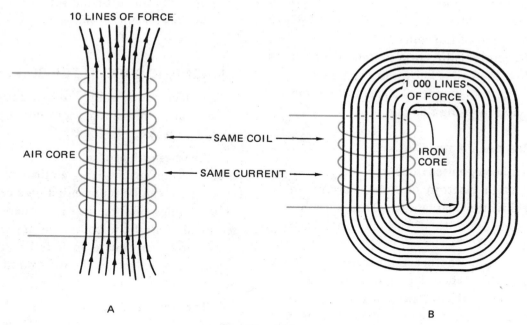

Fig. 6-12

number of turns of wire by the current (amps) in the coil. Two amps in 20 turns has the same magnetic effect as 4 amps in ten turns, or 1/2 amp in 80 turns. Two amps in a 100-turn coil has five times as much magnetizing force as 2 amps in a 20-turn coil.

The presence of nonmagnetic materials in the dc magnet coil has no appreciable effect on the coil magnetism. Insertion of any magnetic material results in a great increase in the total number of lines of force. Assume we have a long coil of wire, figure 6-12(A), with enough current in it to produce a magnetic field whose strength we shall indicate by drawing 10 lines of force. When a bar of magnetic material is placed in and around the coil, figure 6-12(B), it may be found that, due to the magnetization of the material, there are now 1 000 lines of force instead of 10. Inserting the bar of material has increased the magnetic field 100 times.

The ability of a magnetic material to increase field strength in this way is called permeability. *Permeability* is a measure of the willingness of the material to become

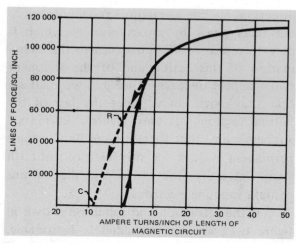

Fig. 6-13 Graph Illustrating Permeability of a Magnetic Material

magnetized. More accurately, it is the number of times that the flux density is increased by adding the material. In figure 6-12, the permeability of the iron core is 100.

A manufacturer of magnetic devices also needs to know exactly how the magnetic flux density builds up when the current in the coil increases. This is best shown by a graph, such as figure 6-13. The makers of magnetic alloys can provide such charts for their materials. The expression *ampere-turns per inch of length* on the graph means the ampere-turns of the coil divided by the total inches of length of the path of the lines of force.

Two other properties of special interest can also be indicated on the graph, figure 6-13. The dotted line shows the magnetic behavior of the material as the current in the magnetizing coil is reduced. The height of point R above the zero level represents *residual magnetism,* which is the amount of magnetism remaining in the core after the magnetizing force (current in coil) is removed. Residual magnetism should be high in permanent magnets — 50 000 lines per square inch or more. In good temporary-magnet materials, the residual magnetism is very low.

The distance on the ampere-turns scale from zero to the point C is a measure of *coercive force.* The measurement to the left on the scale indicates current in the reverse direction that has to be put through the coil in order to remove the residual magnetism, bringing the magnetization of the core down to zero level. If the coercive force is a large amount, that means that the magnet is difficult to demagnetize, which is a desirable property for permanent magnets. The best temporary-magnet materials have a coercive force very close to zero.

ENERGY IN THE MAGNETIC FIELD

When a battery circuit is closed and a current starts, electrons are set in motion by potential energy stored in the battery. This potential energy becomes a sort of kinetic energy, but the moving electrons, unlike flying baseballs, have neither enough weight nor speed to account for the energy that the current possesses. This energy of the current lies in the magnetic field around it. The calculation of this energy is of value in the design of induction coils and magnetos, in which the energy of a magnetic field is delivered usefully to a circuit when a current is suddenly stopped. This is discussed further in Chapter 9.

Around a coil carrying a constant dc, the magnetic field does not have to be continually supplied with more energy to maintain itself. Once the current is established, and the field set up, all of the energy of the current is used to produce heat, overcoming the resistance of the coil of wire ($W = I^2 R$).

Interaction of Magnetic and Electric Fields

The connection of electric and magnetic forces is often summarized this way: "A magnetic field is produced by a moving electric field; and an electric field is produced by a moving magnetic field." In other words, a magnetic field is produced by electric charges in motion; and the motion of magnets can produce electric force. The last part of this statement, the production of an electron moving force from the motion of magnets, will be discussed later in connection with generators.

MAGNETIC DEVICES

A listing of all of the applications of magnetism would be extremely long. Measuring instruments, motors and generators are important enough to merit special discussion. The few devices described in this section are chosen for their own importance, and also to illustrate principles that are used in many devices.

Permanent Magnets

Commonly used permanent magnets are often of the horeshoe shape, figure 6-14, page 126. Bringing poles close together in this manner creates a much stronger magnetic field than would be obtained from the same amount of steel in a straight bar.

If a permanent magnet is to keep its strength for a long period of time:

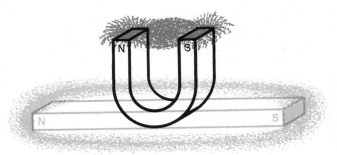

Fig. 6-14 A Horseshoe Magnet

1. It should be made of a properly hardened alloy of high-coercive force. (Alnico is preferable to steel.)

2. Excessive heat, mechanical shock, and nearness to the magnetic field of ac machinery should be avoided.

3. The presence of other iron on or between the magnet poles is desirable, in that it tends to prevent self-demagnetizing, figure 6-15.

A small permanent magnet may be remagnetized by placing its poles against the poles of an electromagnet, figure 6-16, or against a strong large permanent magnet. While in this position, a strong applied field will remagnetize the permanent magnet quickly. If the poles of the permanent magnet are to be kept the same as originally, the north pole of one magnet must be placed in contact with the south pole of the other.

Lifting Magnets

Lifting magnets were one of the first applications of magnetic force developed by electric current. As now constructed, the coil is nearly surrounded by iron, with one pole formed on the core inside of the coil, the other on the shell that surrounds the coil. This circular horseshoe is a way of producing a strong, concentrated magnetic field.

Magnetic Separators

Hundreds of industries use various magnetic separation devices to sort magnetic ore from nonmagnetic rock, scrap iron from coal, scrap iron from other metals, iron particles from sand and ceramic clays, and

LIKE POLES OF ATOMS AT THE POLE-FACE OF A MAGNET REPEL EACH OTHER

AND SOON SCATTER THEMSELVES INTO THIS SORT OF AN ARRANGEMENT

 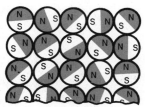

A SOFT IRON "KEEPER" BECOMES MAGNETIZED AND ITS POLES HOLD THE ATOMS OF THE PERMANENT MAGNET IN ALIGNMENT

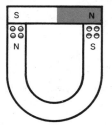

Fig. 6-15

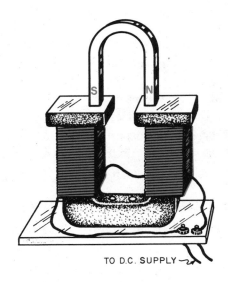

TO D.C. SUPPLY

Fig. 6-16 Device for Magnetizing Magnets

remove nails and baling wire from cattle feed. In the arrangement shown, figure 6-18, magnetic materials are attracted to the pulley and carried around until they fall into hopper No. 2 as they move beyond the magnetic field and the belt finally separates them from the pulley.

Magnetic Control at a Distance

In 1826, Joseph Henry found out how to build stronger lifting magnets than had been made before. He also found that by using a high-voltage battery to force a small current through poorly conducting wire then available, and by using a magnet with a large number of turns, he could get a magnet to work a sounder at a long distance from the battery and switch. He published a full description of his magnetic signaling device in a little periodical, a copy of which came into the hands of Samuel F. B. Morse, a portrait painter. Morse secured a patent on the electromagnetic telegraph, got a subsidy from Congress, and began stringing wire from Washington to Baltimore.

The Electromagnet Pulls a Switch: The Relay

A *relay* is a switch which is closed and opened by the operation of an electromagnet. It was originally devised by Henry to solve a problem met in the early telegraph circuits. As first used, the small current at the left, figure 6-19, page 128, was the feeble current in a long telegraph line, too weak to operate a sounder or recording mechanism. This small current could cause the electromagnet (M) to tip the pivoted iron armature (A) over toward the magnet. C is a stationary contact. When A touches C the circuit at the right is completed, and the voltage source can cause a large current in the nearby controlled device. As used in the first telegraph systems, the controlled device was the recording mechanism, which required a larger current to operate it than could be supplied from a distant battery.

Relays are now used in hundreds of control devices. The small current is often the current in a transistor circuit, which is

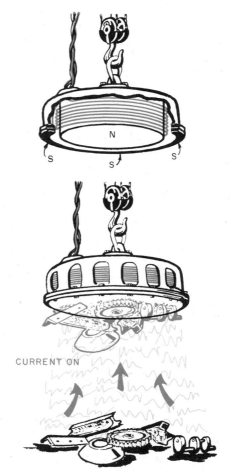

Fig. 6-17 Lifting Magnet

CURRENT ON

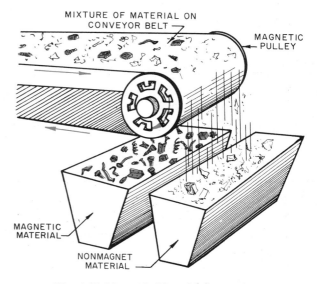

MIXTURE OF MATERIAL ON CONVEYOR BELT

MAGNETIC PULLEY

MAGNETIC MATERIAL

NONMAGNET MATERIAL

Fig. 6-18 Magnetic Material Separator

made to respond to change in temperature, light, sound, or position of a machine part. The voltage source is the power line, and the controlled device may be a motor, a lighting circuit, or anything imaginable. Relays are also used to control switches in remote locations, or locations that are inaccessible due to space, temperature, radiation, or some high-voltage hazard.

Note from the circuit, figure 6-19, that the magnetic current is isolated from the controlled circuit. The magnet merely closes a switch in the controlled circuit. If the stationary contact C is placed to the right of A in figure 6-19, the magnetic pull will open the controlled circuit, rather than close it. Often, a relay may have multiple contacts on both sides of the movable armature so that several contacts are opened and several are closed by one movement of the iron armature.

Magnetic Vibrators — The Electric Bell or Buzzer

The flat spring and iron armature is a movable assembly, pivoted at the left end of the spring in figure 6-20, where the spring is held. When the bell is idle, the loose end of the spring touches the stationary contact. When an external switch (pushbutton) is closed connecting the bell to a battery, the current path is that shown in the diagram. When the iron horseshoe is magnetized, the armature is attracted, pulling the spring away from the stationary contact and breaking the circuit at this point. When the spring leaves the contact, current in the circuit stops so the magnet loses its magnetism, no longer holding the armature. With the magnet turned off, the elasticity of the spring brings armature and spring away from the magnet, the spring touches the contact again, and the whole process repeats. Removal of the gong converts the bell to a buzzer.

This arrangement is used to produce vibratory motion, or it may be used for the main purpose of turning the current on and off rapidly. There is a high-voltage-producing device called the induction coil, also developed by Joseph Henry, in which a direct current has to be started and stopped rapidly. This is accomplished by placing a buzzer in the circuit. The Model T Ford spark coil is an example of an induction coil with a buzzer.

The Solenoid-and-Plunger Magnet

This arrangement, no different in principle from the armature type magnets shown previously, is used where more distance of motion is wanted. Magnetizing the core causes the plunger to be drawn into the coil. The magnet may be arranged so that the plunger falls out of the magnet because of its own weight, or it may be pulled out by a spring when the magnetic force is released, figure 6-21. These solenoids are used for such things as operating switches, opening or closing valves, operating magnetic brakes, and operating circuit breakers. Automation is used in industry to control many operations. The solenoid is found in many of these control devices.

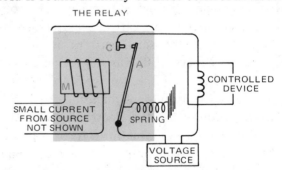

Fig. 6-19 Magnetic Relay

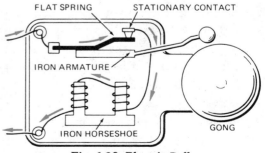

Fig. 6-20 Electric Bell

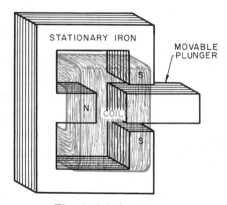

Fig. 6-21 Solenoid

Magnetic Production of Sound

Mechanical vibration produces sound. Vibrations in the range of 20 to 18 000 vibrations per second can be heard by human ears. The higher the frequency of vibration, the higher the pitch. The greater the extent of the back-and-forth movement, the greater the loudness.

Figure 6-22 represents a coil of wire wound on a paper sleeve and suspended so that it can move freely near the pole of a permanent magnet. If an alternating current is put through the coil, the coil is alternately attracted (A) and repelled (B), and vibrates at the same frequency as the frequency of the electron vibration of the alternating current.

Most radio and hi-fi loudspeakers are built similar to the one shown in figure 6-23. To obtain a uniform magnetic field in which the moving coil can vibrate, one pole is just inside the moving coil, and the other is brought around to surround the moving coil. The moving coil is attached to a paper-composition cone, vibration of which produces sound when alternating currents are fed into the movable voice coil from an amplifier.

Before alnico was developed, most speakers used a dc electromagnet, instead of a permanent one, to provide the field for the voice coil to work in.

The phone receiver, figure 6-24, uses a stationary coil of many turns of fine wire

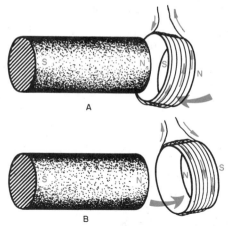

Fig. 6-22 Alternating Current in a Coil Produces Vibrations

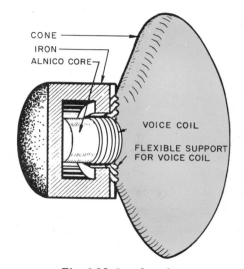

Fig. 6-23 Loudspeaker

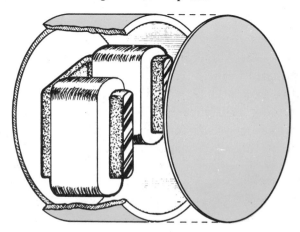

Fig. 6-24 Phone Receiver

wrapped around the poles of a permanent horseshoe magnet. This receiver is intended to operate on a much smaller current than that which is needed to operate a loudspeaker, hence the large number of turns of wire. The alternating current in the coils alternately strengthens and weakens the pull of the magnet on a flexible iron disk (diaphragm) which is supported at its edges. This varying pull causes the disk to vibrate in accordance with the alternations of the incoming current.

PRINCIPLES COMMON TO ALL MAGNETS

A general law applies to electrical and mechanical processes:

Accomplishment is directly proportional to effort, and inversely proportional to the hindrances. One example of this general statement is Ohm's Law, I = E/R, stating that the current produced is directly proportional to the electromotive force and inversely proportional to the resistance. In the attempt to produce magnetism, the accomplishment is the production of magnetic lines of force, or flux. The applied effort is the ampere-turns of the magnetizing coil, or, more accurately, the magnetomotive force is the ampere-turns per inch. The hindrances to magnetism are called *reluctance*. Air is a high-reluctance material; iron is a low-reluctance material. This is another way of saying that air is nonmagnetic and iron magnetizes easily. Reluctivity is the opposite of permeability.

The similarity of this relation, "flux equals magnetomotive force ÷ reluctance" to Ohm's Law, "current equals electric force ÷ resistance," leads to using the name *magnetic circuit* for the path of the lines of force through a magnetic device, even though there is no motion along the lines of force. All magnetic devices have a magnetic circuit which can be traced by following the path of lines of force from any point, around through iron and air, and back to the starting point.

Since it is easy to magnetize iron, but it is relatively difficult to produce lines of force through air, one aim in all magnetic devices is to make the path of the lines through air or nonmagnetic solids as short as possible.

Figure 6-25 represents two arrangements in which a pivoted iron bar is to be attracted to an electromagnet. If the ampere-turns are the same in each coil, the pulling force in the first one will be only a small fraction of the pulling force in the second. This is because the reluctance is so large in the first magnetic circuit. Or, in order to achieve the same flux in each, many more ampere-turns would be required in the first coil.

In many dc electromagnets, figure 6-26, the parts are assembled from solid bars or rods. In magnets of dc vibrators, motors and generators, and in ac equipment, the magnetic parts are an assembly of thin sheets of iron called *laminations*. In the last applications named, magnetic fields are in motion; the lines of force are bouncing around as current is turned on or off, or reversed.

Whenever lines of force sweep across a piece of any kind of metal, they tend to

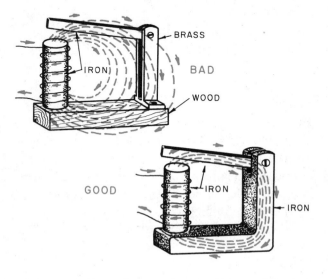

Fig. 6-25 Effect of Reluctance and Permeability of Materials on a Magnetic Circuit.

VARIOUS ARRANGEMENTS FOR ELECTROMAGNETS

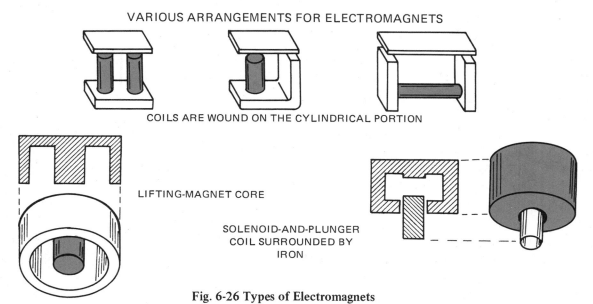

COILS ARE WOUND ON THE CYLINDRICAL PORTION

LIFTING-MAGNET CORE

SOLENOID-AND-PLUNGER
COIL SURROUNDED BY
IRON

Fig. 6-26 Types of Electromagnets

generate a current in the metal. In figure 6-27 (A), the colored line is the direction of current that would be generated inside the iron core of the electromagnet any time that the current in the coil is changed in amount or direction. This current, called an *eddy current* is a useless nuisance because it takes energy from the coil circuit and heats the iron core.

In figure 6-27 (B) the core is a stack of laminations (thin sheets of iron). Not much current can circulate around in that pile of iron sheets because of poor electrical contact between the sheets. Lacquer and iron oxide on the sheets can make this contact still less.

By preventing the formation of an eddy current, the loss of energy and production of heat is prevented. Iron containing a few percent of silicon has high permeability, which is favorable, and high electrical resistance, which is also good. High electrical resistance in a core material hinders the setting-up of these unwanted eddy currents inside the core material.

Demagnetizing

The removal of magnetic poles on a piece of steel is accomplished by completely

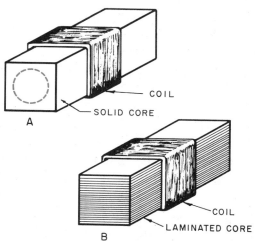

Fig. 6-27 Solid and Laminated Iron Cores

disarranging the atoms in the steel. The stator windings of a discarded 1/4-hp (187-W) motor, connected to an ac power line through a current-limiting resistor, such as a 1 000-watt heating element, make a convenient demagnetizer for small objects.

A piece of steel in position (1), figure 6-28, page 132, will be magnetized, its upper end changing from north to south and back again 60 times per second on an ordinary ac line.

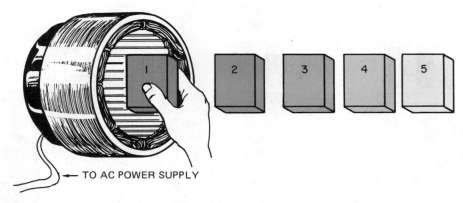

← TO AC POWER SUPPLY

Fig. 6-28 A Demagnetizer

As it is removed from the coil at position (2), it is still being magnetized and remagnetized 60 times per second, but not as strongly as at (1).

At (3), the same effect is still weaker.

At (4), the upper end is alternately very weak north or very weak south.

By the time it gets to (5), the directions it has been receiving from the coil are so faint that the atoms are left in general disarrangement.

The steel hairspring of a watch, if magnetized by approach to magnets or dc machinery, is put out of order because the coils of the spiral spring stick together when magnetized. A watch can be demagnetized in a few seconds by the above process. Most good watches use nonmagnetic springs, so there is no worry about magnetism.

Steel plates often need to be demagnetized before they are arc-welded. A magnetic field pushes the arc current sidewise, making it difficult to maintain the arc and do a smooth welding job. The steel plates are passed through large coils carrying alternating current to accomplish the demagnetizing.

POINTS TO REMEMBER

- All magnetism is due to electron motion; either the movement of electrons around and around as they pass through a coil, or the spinning-like-a-top motion of electrons in atoms.

- Like poles repel; unlike poles attract.

- The strength of a magnetic field is represented by the density of lines of force. Direction of the field means the direction in which the north pole of a compass points.

- Most materials are nonmagnetic. The internal electron spins in atoms of iron, nickel, and some alloys, and oxides make them magnetic.

- Atoms of iron are little permanent magnets. The magnetizing of a piece of iron is a matter of arranging these atoms so that like poles face in the same direction.

- A left-hand rule for a single wire: thumb in direction of electron flow, fingers in direction of field.

- A left-hand rule for a coil: thumb at north pole, fingers in direction of electron current.

- Parallel currents in the same direction attract; in the opposite direction, they repel.

- Permeability is the ability of a material to become magnetized. Residual magnetism is flux density that it retains after the magnetizing force is removed. Coercive force is oppositely directed magnetizing force applied to make the material lose its magnetism.

- A magnetic field contains useful energy.
- Moving electrons exert force on magnets; moving magnets exert force on electrons.
- Maintenance of strength of a permanent magnet depends on: (1) type of alloy, (2) avoidance of excessive heat, shock, and ac magnetic fields, and (3) keeping iron between the poles.
- The magnetizing force of a coil is the number of ampere-turns per inch of the magnetic circuit. The ampere-turns may be produced by large current and few turns or small current and many turns.
- A relay is a switch operated by a magnet, so that one current can turn a larger current on or off.

- Sounds are produced in speakers and phones by using the changing magnetic field of an alternating current to produce vibration of mechanical parts.
- The complete path of lines of force through a magnet, air, and the iron that the magnet pulls on, is called the magnetic circuit.
- The unwillingness of a material to be magnetized is called reluctance.
- Total number of lines of force = $\dfrac{\text{Magnetizing force in ampere-turns/inch}}{\text{reluctance of the magnetic circuit}}$
- The use of laminated cores reduces energy loss in eddy currents.

REVIEW QUESTIONS

1. Tell why all magnetism is considered electromagnetism.
2. What is the polarity of the end of a compass needle that points northward?
3. State the magnetic attraction and repulsion law.
4. What is meant by the term *magnetic field?*
5. What happens to the atoms in a piece of iron or steel when it is magnetized? Why does a hardened piece of steel retain magnetism? Why does a soft piece of iron or steel lose magnetism rapidly?
6. Explain the rule-of-thumb method of determining the direction of the magnetic field. Also, explain the method of finding the direction of the electron current.
7. Why do electric currents traveling in the same direction attract each other, and electric currents traveling in the opposite direction repel each other?
8. A permanent magnet hung on a post is given the job of holding up ten pounds of iron for six months. Does this weaken the magnet?
9. Define coercive force, residual magnetism, and permeability.
10. A certain air-core coil has a flux density of 60 lines per square inch (9.3 lines per square centimeter). The insertion of an iron core raises the flux density to 60 000 lines per square inch (9 300 lines per square centimeter). Calculate the permeability of the iron.

RESEARCH AND DEVELOPMENT

Experiments and Projects on Magnetism

INTRODUCTION

All magnetism is due to electron motion. The discovery of magnetism was one of the most important in the history of mankind. After studying Chapter 6, you will understand why magnets and magnetic fields have had a tremendous impact on industrial technology and our way of life.

EXPERIMENTS

1. To investigate magnets and magnetic fields.

2. To verify the principles of electromagnetism.

3. To find out why a metal is magnetized under one current condition and demagnetized under another.

4. To observe the heat effect in a coil:
   a. With an air core
   b. With a laminated iron core
   c. With a solid iron core

PROJECTS

1. Design and construct a lifting magnet, bar magnet, or solenoid.

2. Design and construct a device using a solenoid and plunger to move something such as a model railroad crossing signal, switch, or gate.

3. Design and construct a demagnetizing-magnetizing coil.

4. Design and build a device for magnetizing horseshoe and bar magnets.

5. Design, build, and operate a simple relay or assemble a relay from a kit and operate it in a circuit.

EXPERIMENTS

Experiment 1 MAGNETS AND MAGNETIC FIELDS

OBJECT

To investigate magnets and magnetic fields.

APPARATUS

2 - Bar magnets
1 - Horseshoe magnet
2 - Pocket compasses
1 - Box of iron filings
1 - Sheet of cardboard

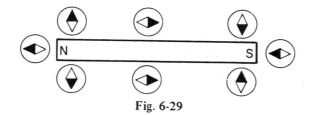

Fig. 6-29

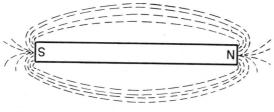

Fig. 6-30

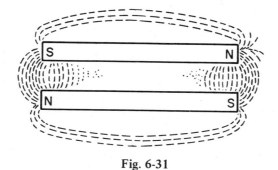

Fig. 6-31

PROCEDURE

1. Place the pocket compasses in the various positions indicated in figure 6-29 and explore the magnetic field around the bar magnet.

2. Lay a piece of cardboard over a bar magnet and sprinkle it with iron filings and note the pattern of the iron filings. See if this pattern is practically the same as indicated in figure 6-30.

3. Using two bar magnets, place unlike poles adjacent and cover with cardboard, sprinkle iron filings on top, and note the pattern of the filings. This pattern of filings should be practically the same as that shown in figure 6-31.

4. Repeat Step 3, but with like poles adjacent. The pattern you obtain will be somewhat the same as shown in figure 6-32, page 136.

5. Repeat Step 3, using a horseshoe magnet. The pattern of iron filings you obtain should resemble that shown in figure 6-33, page 136.

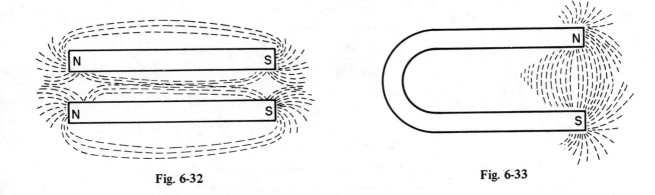

Fig. 6-32   Fig. 6-33

QUESTIONS

1. Where are the poles on a magnet located?

2. Which pole of the magnet points northward?

3. How would you magnetize a piece of hard steel?

4. What is a line of force and what is a magnetic field?

5. What is the law of repulsion and attraction?

Experiment 2 ELECTROMAGNETISM

OBJECT

To verify the principles of electromagnetism.

APPARATUS

1 - Storage battery
2 - Pocket compasses
1 - Soft iron bar
1 - Protective resistance
1 - Box of iron filings
1 - Sheet of cardboard
1 - SPDT switch
1 - DPDT switch
4' (1.2 m) #20 AWG solid wire
#18 AWG magnet wire

PROCEDURE

1. Pass a direct current through the four-foot (1.2 m) wire with the protective resistance in series so as not to overheat the wire as shown in
figure 6-34.

2. Place a compass first above and then below the wire and note the position of the compass needle in each position.

3. Place a strong direct current upward through a vertical wire passing through a piece of cardboard, figure 6-35 (A). Note the position of the pocket compasses.

4. Sprinkle iron filings on the cardboard and note the pattern of the field.

5. Repeat Steps 3 and 4 with the current downward and note the position of the pocket compasses.

6. Wind a small coil about 3″ (75 mm) long (#18 or #20 AWG magnet wire) on a soft iron bar.

7. Connect the ends of this electromagnet coil to a dry cell and check the polarity of the soft iron core with a pocket compass, figure 6-36(A), page 138.

8. Check the polarity of this electromagnet with the right-hand thumb rule.

9. Repeat Steps 7 and 8 with the current reversed in the turns of the coil. This is shown in figure 6-36 (B).

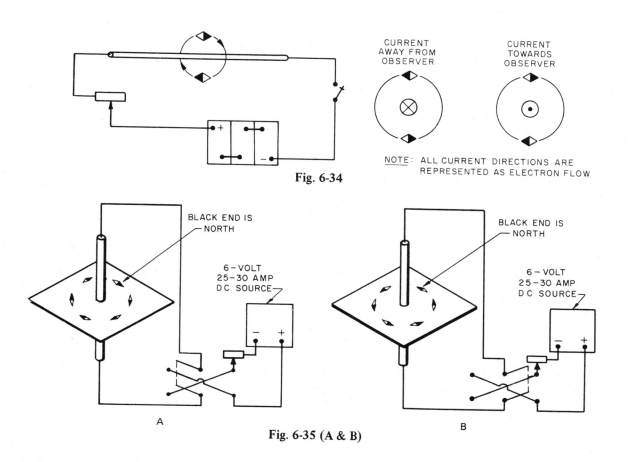

CURRENT AWAY FROM OBSERVER

CURRENT TOWARDS OBSERVER

NOTE: ALL CURRENT DIRECTIONS ARE REPRESENTED AS ELECTRON FLOW

Fig. 6-34

BLACK END IS NORTH

6−VOLT 25−30 AMP DC SOURCE

BLACK END IS NORTH

6−VOLT 25−30 AMP DC SOURCE

A

B

Fig. 6-35 (A & B)

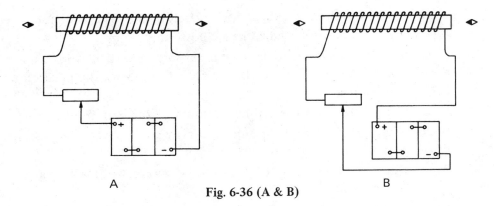

A   Fig. 6-36 (A & B)   B

QUESTIONS

1. What shape is a magnetic field around a conductor carrying a current?

2. What effect has the reversal of current on the polarity of a coil?

3. What effect does an iron core have on the strength of an electromagnet?

4. What is the relation between the direction of the current in a coil and the polarity of a coil?

5. Name some applications of the electromagnet.

Experiment 3 MAGNETIZING AND DEMAGNETIZING FERROUS METAL WITH ALTERNATING CURRENT

OBJECT

To find out why the metal is magnetized under one current condition and demagnetized under another.

APPARATUS

1 - Coil-type magnetizing device, ac (Page 146)
1 - Pc. hardened steel

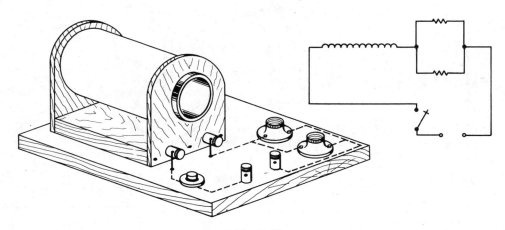

Fig. 6-37

1 - Pc. soft steel
1 - Pc. nonferrous metal (brass, copper, aluminum)
2 - Glocoils, 660-watt
    Small steel nails, tacks
1 - SPST switch

PROCEDURE

1. Place the piece of hardened steel in the coil and connect to 120-volts ac. Remove the steel from the coil and try its effect on lifting the nails.

2. Place the steel in the coil again. Turn off the current, remove the steel, and try its effect on lifting the nails.

3. Repeat Steps 1 and 2 with a piece of soft steel or iron and with a piece of nonferrous metal.

QUESTIONS

1. What happened to the hardened steel when you removed it from the coil with the current turned on? Why?

2. What happened when you removed it with the current turned off? Why?

3. Did the soft steel retain magnetism as well as the hardened steel? Give a reason for the answer.

4. Were you able to magnetize the nonferrous metal? Explain your answer.

5. Give several reasons why it is necessary to demagnetize metal.

Experiment 4 HEAT EFFECTS OF CURRENT PASSING THROUGH A COIL OF WIRE

OBJECT

To observe the heat effect in a coil:
a. With an air core
b. With a laminated iron core
c. With a solid iron core

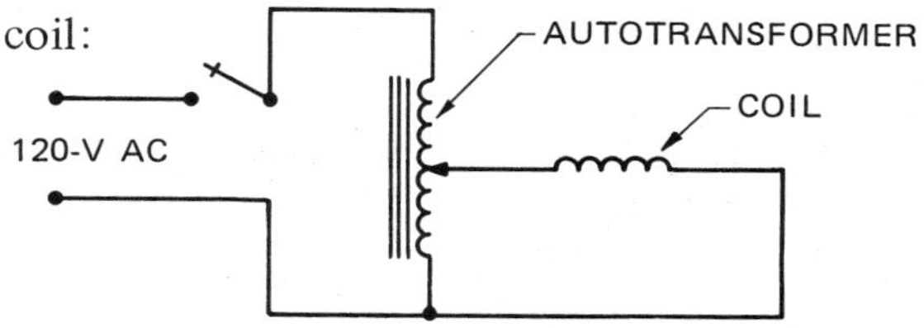

APPARATUS

1 - Coil
1 - Autotransformer or lamp bank
1 - Laminated iron core
1 - Solid iron core
1 - SPST switch

Fig. 6-38

PROCEDURE

1. Connect the apparatus as shown in figure 6-38.

2. Set the autotransformer at low current or set up a lamp bank with all lamps in place.

3. Turn on the current and observe the temperature of the coil. It should be cold. If it is not, reduce current.

4. Place the laminated core into the coil and observe temperature. If the coil is cold, increase the current until the coil and core become slightly warm and remain that way.

5. Remove the laminated core and insert the solid core, figure 6-39.

6. Observe the rise in temperature of the coil and core.

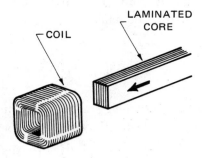

Fig. 6-39

QUESTIONS

1. What effect does the presence of nonmagnetic material (air) in a coil have on the temperature? Why?

2. What caused the solid iron core to get hot?

3. Explain why the laminated iron core did not get hot.

PROJECTS

LIFTING MAGNET

Electromagnets designed to lift articles can be made in a variety of styles and with a degree of strength to suit the purpose for which they will be used, figure 6-40.

This magnet is powerful enough to lift considerable weight when connected to a 3- or 6-volt battery or other dc source.

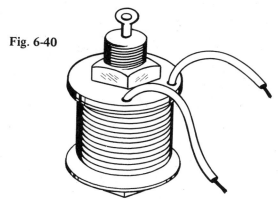

Fig. 6-40

MATERIALS

1 - Machine bolt, 1/2" x 2 1/2" (M12 x 60L)
1 - Machine bolt, nut - 1/2" (M12)
3 - Fiber disks, 1/16" x 1 3/8" (2 mm x 35 mm)
1 - Pc. insulating paper
1 - Eye bolt, #10 x 24 (M5)
Magnet wire, #18 Formvar
Plastic tape
Cotton tape

Note: Plastic laminate 1/16" (2 mm) thick, or hardboard 1/8" thick, is a good substitute for fiber.

PROCEDURE

1. Determine which kind of magnet you wish to make.
2. Analyze the problem.
3. Design a magnet suited to your needs.
4. Make working drawings of each part, figure 6-41, page 142.
5. Plan your procedure in making the parts.
6. Calculate the amount of wire needed for the coils. (See sample problem.)
7. Make the parts.
8. Wind the coil. See figure 6-42 (A, B, and C), page 142.
9. After the coil is wound, cover it with cotton tape and coat with insulating varnish.

SAMPLE PROBLEM

Find out how much wire is needed to wind a coil (1 1/2" (38 mm) long) on a 1/2" (M12) bolt, using #20 AWG enameled magnet wire, when 10 layers of wire are required on the coil. (For data, see Appendix.)

Formula: Circumference of center layer of wire x number of turns in a layer x number of layers = length of wire.

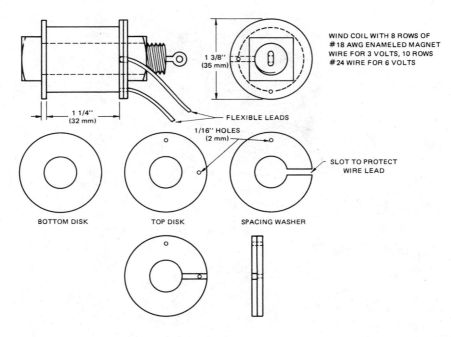

WIND COIL WITH 8 ROWS OF
#18 AWG ENAMELED MAGNET
WIRE FOR 3 VOLTS, 10 ROWS
#24 WIRE FOR 6 VOLTS

1 3/8"
(35 mm)

FLEXIBLE LEADS

1 1/4"
(32 mm)

1/16" HOLES
(2 mm)

SLOT TO PROTECT
WIRE LEAD

BOTTOM DISK   TOP DISK   SPACING WASHER

Fig. 6-41

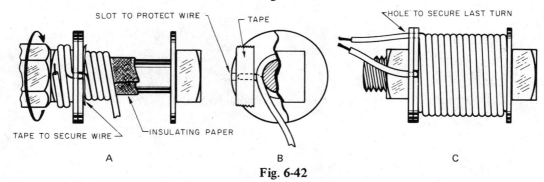

SLOT TO PROTECT WIRE   TAPE   HOLE TO SECURE LAST TURN

TAPE TO SECURE WIRE   INSULATING PAPER

A   B   C

Fig. 6-42

INCHES

Diameter of center layer = 1/2" + 10 x 0.032
= 0.500 + 0.320
= 0.820

Circumference = πD
C = 3.14 x 0.820
C = 2.574

Number of turns = 1.5" ÷ 0.032 = 46.8 (use 47)

Length of wire = Number of turns x
Number of layers x
C
L = 47 x 10 x 2.574
L = 1 209.78 inches
L = 100.8 feet

METRIC

Diameter of center layer = 12 mm + 10 x 0.8 mm

= 12 + 8

= 20

Circumference = πD

= 3.14 x 20

= 63 mm

Number of turns = 38 mm ÷ 0.8 = 47.5 (use 47)

Length of wire = Number of turns x

Number of layers x

C

= 47 x 10 x 63

= 29 610 mm

= 29.6 meters

Suggestions for Winding Coil

Place the coil end disks and nut on bolt and adjust the distance between the disks with the nut. Cut a piece of insulating paper the proper width and carefully wind it around the bolt. If heavy paper is used, one layer is sufficient; for thin paper, use two layers. Secure paper with plastic tape. Fit the bolt into a winding jig. Attach the wire, figure 6-42(A), and wind coil. Finish as shown in figure 6-42(C).

Alternate Designs

Alternate design, figure 6-43(A), page 144, is a horseshoe magnet composed of two coils fastened to an iron yoke. The coils may be wound on bolts or they may be two solenoid coils bolted to a yoke.

Figure 6-43(B) is a lifting magnet coil with 1/8" x 1" (3 mm x 25 mm) band iron shields placed on four sides of the coil.

Figure 6-43(C) is a more sophisticated and stronger magnet. It is a single coil wound on a nonferrous metal spool which is placed over a large iron core within a metal shield. An appropriate metal shield can be made from a large iron pipe cap. The threads should be bored out and the bottom machined smooth. After the core is pressed into place, the bottom of the core and pipe cap should be faced on the engine lathe so that they will both be in the same plane. The coil may be wound on a tube without ends, as illustrated on page 151 or in a plastic bobbin.

Lifting magnets fabricated as suggested in these alternate designs will vary in strength according to the size and style of the coil, number of turns, and gauge of wire used, size and kind of core, method of shielding coil, and amount and kind of current utilized.

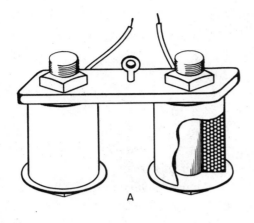

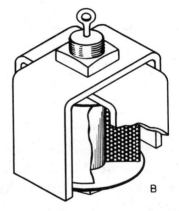

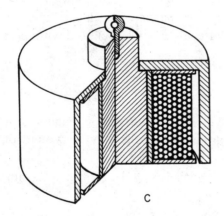

Fig. 6-43

SOLENOID COIL

A solenoid coil is an interesting device, figure 6-43. The ends of the coil are, in effect, magnetic poles, whether there is an iron core in the coil or not. When an iron core or plunger is placed in one end of a solenoid tube and the coil energized, the magnetic flux will pull the plunger into the coil until the center of the plunger is at the center of the coil.

Solenoids are extensively used to convert electrical energy into mechanical motion. A good example is the action of the control mechanism of the water valves on automatic washing machines and dishwashers or the hammer on an electric door chime.

MATERIALS

2 - Plywood disks, 1/8" x 1 3/8" (3 mm x 35 mm)
1 - Fiber, plastic, or nonferrous metal tube, 3/8" ID x 2" (9 mm x 50 mm)
2 - Pcs. flexible wire; Magnet wire, #22 AWG, enameled
   Insulating paper and cotton tape

PROCEDURE

Study the working drawings, figure 6-45, and make the parts.

Fit the coil end disks on to the fiber tube.

Cement disks to tube. (If a metal tube is used, expand the tube slightly on each end with a tapered punch.)

Calculate the amount of wire needed for 14 layers on the tube.

Wind coil, figure 6-46.

Cover coil with cotton tape and coat with insulating varnish.

Test and evaluate.

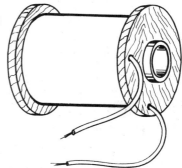

Fig. 6-44

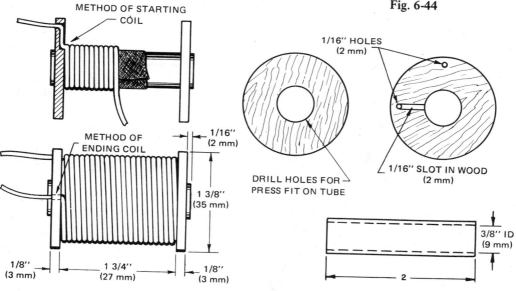

Fig. 6-45

USING A BOLT TO SUPPORT SOLENOID
COIL PARTS DURING WINDING PROCESS

Fig. 6-46

DEMAGNETIZING—MAGNETIZING COIL

This demagnetizing-magnetizing coil, figure 6-47, is a device which can be used to demagnetize tools such as punches, screwdrivers, and rules, when operated on 120-volt ac in series with two glocoils or in series with a lamp bank.

For magnetizing tools and other equipment, it can be operated on 6 or 12-volt dc with fuses in the lamp bases or on 120-volt ac connected as shown. See Experiment 3.

A coil wound with #16 AWG magnet wire will operate on direct current as follows: 6 volts, 5 amps, 30 watts; 12 volts, 10 amps, 120 watts.

A coil wound with #14 AWG magnet wire will operate as follows: 6 volts, 8 amps, 48 watts; 12 volts, 16 amps, 192 watts.

MATERIALS

1 - Fiber tube, 1 1/2 ID x 5 1/2" (38 mm ID x 140 mm) (Brass, copper or aluminum tubing may be used if insulated with paper or tape.)

2 - Pcs. tempered hardboard, 1/4" x 3 1/4" x 5" (6 mm x 80 mm x 125 mm)

1 - Pc. wood, 3/4" x 3 1/4" x 4 3/4" (20 mm x 80 mm x 120 mm)

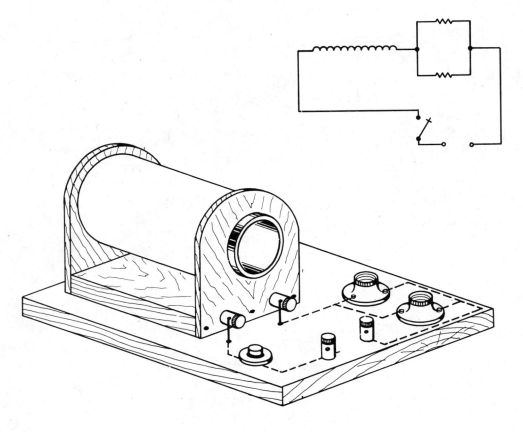

Fig. 6-47

1 - Pc. wood, 3/4″ x 5 1/2″ x 11″ (20 mm x 140 mm x 280 mm)

2 - Porcelain lamp receptacles

1 - Switch, pushbutton type

2 - Binding posts or Fahnestock clips

   Magnet wire, #14 AWG about 5 lbs. (2.5 kg) or #16 AWG about 2 1/2 lbs. (1.2 kg)

   Cotton tape

   Tubing, cotton or plastic

PROCEDURE

Study the drawings, figure 6-48, and list of materials. Make a working drawing for each part. Calculate the finished diameter of the coil, using about 5 lbs. (2.5 kg) of #14 or about 2 1/2 lbs. (1.2 kg) of #16 AWG enameled magnet wire. The finished coil should have an even number of layers of wire. The first and last layer should terminate at the same end of the coil. Adjust the dimensions on your working drawings according to the results of your calculations. Use data found in the Appendix for decimal equivalents of wire, wires per inch, etc., and formula, page 141, to determine amount of wire.

Make the parts.

Plan procedure for winding the coil.

Assemble tube and end pieces. Secure end pieces to the tube so they cannot move while winding the coil.

Thread a short piece of cotton or plastic tubing over the end of the wire and fasten it to the fiber end piece with plastic tape. Allow enough wire to connect to a binding post, figure 6-49(A), page 148.

Wind coil and secure end of wire with cotton tape, figure 6-49(B), page 148.

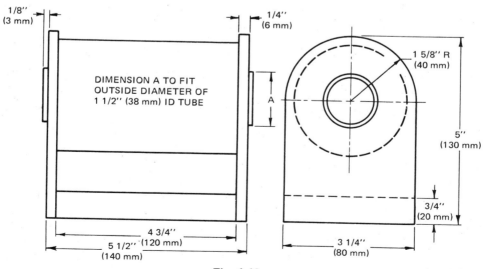

Fig. 6-48

Thread a piece of tubing over wire. Fasten wires to the binding posts. Assemble remaining parts and finish with shellac or varnish. Test and evaluate.

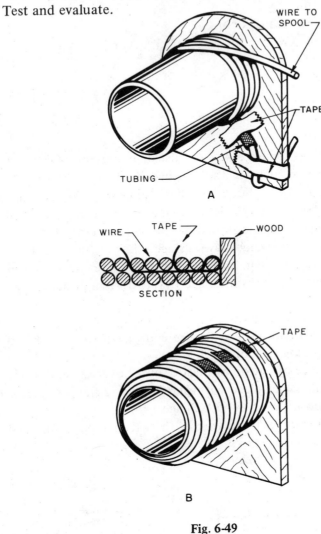

Fig. 6-49

MAGNETIZING DEVICE

This device, figure 6-50, was designed for use in magnetizing horseshoe magnets. It will work equally well for bar magnets and for magnetizing tools. It is constructed with a solid core to be operated on 6- to 12-volt dc. With glocoils in the circuit, it can be operated on ac for short intervals. If it is to be extensively used on ac, the core should be laminated.

MATERIALS

3 - Pcs. iron or soft steel, 1″ x 1″ x 3 1/4″ (25 mm x 25 mm x 80 mm)
1 - Pc. wood, 1 1/2″ (38 mm) thick (Other dimensions according to individual design.)

2 - Binding posts or Fahnestock clips
2 - Lamp receptacles
1 - Push-button switch, 15-amp
   Magnet wire, #18 enameled
   Fiber and fish paper or tag board
   Cotton tape

PROCEDURE

Examine the pictorial drawing, figure 6-50, schematic diagrams, figure 6-51(A & B), and suggestions for making cores and coils, figure 6-52, page 150.

The pieces for a solid core may be welded together or fastened together with countersunk screws.

The coils may be wound in a form on a mandrel, figure 6-53, page 151, or with tapered ends as in figure 6-54, page 151.

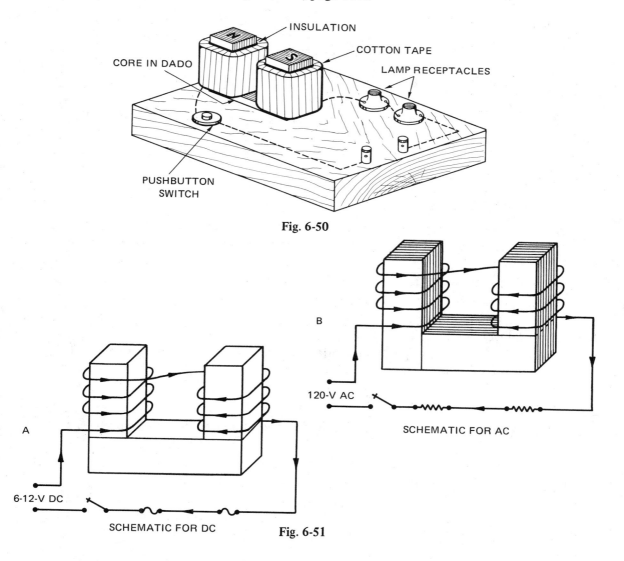

Fig. 6-50

SCHEMATIC FOR AC

SCHEMATIC FOR DC

Fig. 6-51

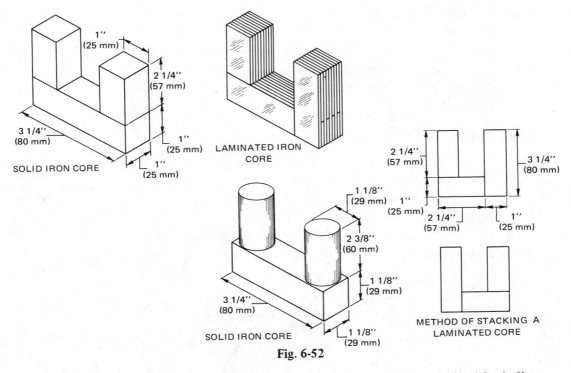

Fig. 6-52

Each coil should be bound with cotton tape and coated with shellac or varnish.

The lamp receptacles may be omitted if the magnetizer is intended for use only on low-voltage direct current.

Each coil will require between 185 and 200 ft. (56 m and 61 m) of wire. Calculate the number of layers of wire required for a coil made as illustrated in figure 6-53. If the alternate method of winding the coil is used, as shown in figure 6-54, more layers of wire will be required because of the tapered ends. Calculate the number and estimate the width of a finished coil. Will two such coils fit between the cores which are 1 1/4'' (32 mm) apart?

It is good practice to run the connecting wires in grooves on the bottom of the wooden base.

When all of the parts are made, give them a coat of finishing material.

Assemble the unit, test, and evaluate.

Suggestions for Winding Coils on a Mandrel

When a coil is wound on a mandrel the numerous turns of wire have a tendency to tighten the coil form, or other insulating material, on the mandrel. In order to remove the finished coil without damage, it is necessary to provide a means of relieving the pressure on the mandrel. Several methods of accomplishing this are illustrated in figure 6-55. The corners on a mandrel should be slightly rounded to avoid binding. Mandrels coated with a thin film of paraffin are usually easier to remove.

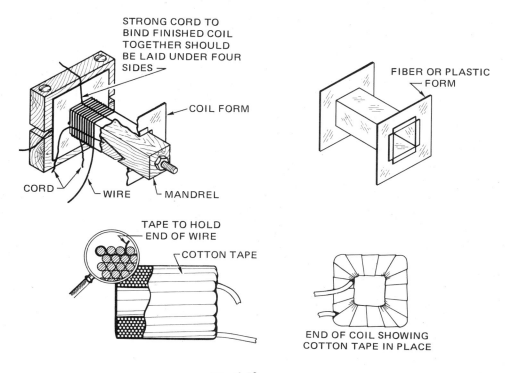

STRONG CORD TO BIND FINISHED COIL TOGETHER SHOULD BE LAID UNDER FOUR SIDES

COIL FORM

CORD

WIRE

MANDREL

FIBER OR PLASTIC FORM

TAPE TO HOLD END OF WIRE

COTTON TAPE

END OF COIL SHOWING COTTON TAPE IN PLACE

Fig. 6-53

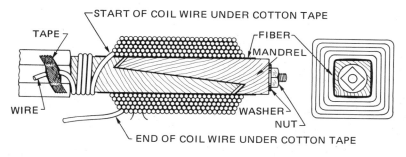

START OF COIL WIRE UNDER COTTON TAPE

TAPE

FIBER

MANDREL

WIRE

WASHER

NUT

END OF COIL WIRE UNDER COTTON TAPE

Fig. 6-54

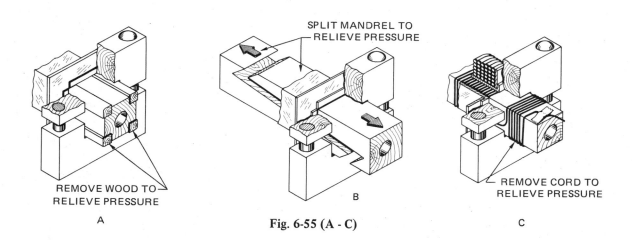

REMOVE WOOD TO RELIEVE PRESSURE

A

SPLIT MANDREL TO RELIEVE PRESSURE

B

REMOVE CORD TO RELIEVE PRESSURE

C

Fig. 6-55 (A - C)

Chapter 7
Measuring instruments

Previous discussion of current, potential difference, resistance, and power indicates the need for measuring these quantities to learn what is going on in an electrical circuit. Such measurements must be made to experiment with and test new devices, to repair equipment, to locate troubles, and to find whether portions of a circuit are functioning properly.

DC METERS

The most commonly used instruments are ammeters, voltmeters, and ohmmeters. All of these are similar in construction, and modifications of a simple basic instrument called a *galvanometer*. Like most measuring instruments, the galvanometer action depends on the magnetic effect of a small current.

DC Galvanometer

The essential parts of a dc galvanometer are shown in figure 7-1. North (N) and south (S) are the poles of a permanent magnet. The stationary cylindrical iron core between the poles makes an evenly distributed, uniformly strong magnetic field in the space where the moving coil operates. This uniform magnetic field makes it possible to have uniformly spaced numbers on the scale of the meter. The movable coil is generally supported by jewelled pivots similar to those that support the balance wheel of a watch.

What makes the pointer move? The coil, to which the pointer is fastened, becomes an electromagnet when it contains a current. In figure 7-2, if there is a current in a clockwise direction as we look at the coil, the coil acts as a magnet, with its north pole near the viewer and its south pole on the farther side of the coil.

Due to magnetic attraction and repulsion forces, the coil will try to turn so that unlike poles will be as close together as possible. The amount of turning force will depend on the strength of the permanent magnet, and the ampere-turns of the movable coil.

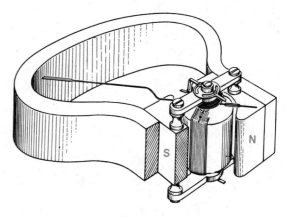

Fig. 7-1 Galvanometer

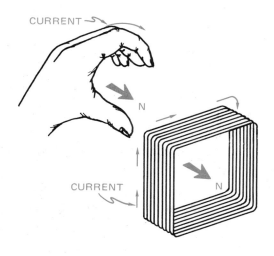

Fig. 7-2 Left-Hand Rule

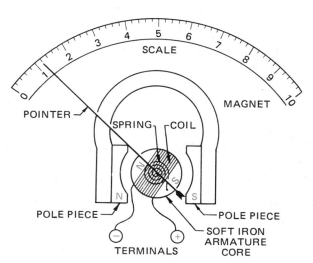

Fig. 7-3 DC Milliammeter

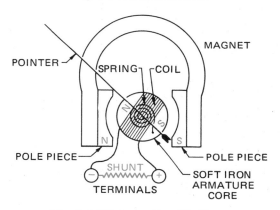

Fig. 7-4 DC Ammeter with Shunt

The motion of the coil is mechanically resisted by hairsprings. If the current in the moving coil is increased, the magnetic effect of the coil is stronger, so the coil turns farther, thus indicating the increased current on the scale. When the current is stopped, the spring returns the moving coil and pointer to the zero mark. Since one end of the hairspring is fastened to the moving coil and the other end of the spring is stationary, it is convenient to let these springs also serve as conductors to connect the movable coil to the stationary wiring in the meter.

This type of meter, and other meters based on this construction, operate only on direct current, figure 7-3. If an alternating current is put through this meter, the magnetic poles of the coil reverse rapidly. The coil is too large to swing back and forth 60 times per second (a common ac frequency); therefore, it does not turn at all. A meter intended for one ampere dc would not be damaged by one ampere ac but the meter would read zero on the ac.

The ordinary simple galvanometer, by itself, is of very limited use. As a current indicator, only a very small current can be allowed in the fine wire of the moving coil. Since this coil has a low resistance, only a very small voltage can be applied to the moving coil. The most useful galvanometers are scaled as milliammeters or microammeters, telling how many thousandths or millionths of an ampere pass through the meter.

DC AMMETERS

To make it possible to measure large currents with the galvanometer, a known large fraction of the large current is bypassed through a parallel low resistance called a *shunt*. Only a small fraction of the total current passes through the moving coil, figure 7-4. The scale is marked to tell the total current through the entire ammeter (galvanometer plus shunt combined).

Ammeter Shunts

Assume we have a one-milliampere meter movement (a galvanometer in which one milliampere in the moving coil is enough to move the pointer from the zero to the end of the scale). This meter has 50 ohms resistance, which is a commonly used meter movement, figure 7-5.

Problem: Convert this milliammeter to a 5-amp ammeter; that is, arrange a shunt circuit so that a current of 5 amps through the entire meter will cause a full-scale movement of the pointer. The scale will be renumbered to indicate 5 amps.

Notice that the shunt is a parallel resistance. When there is a 5-amp current through the meter, only 0.001 amp can pass through the moving coil, and the rest, 4.999 amps, must go through the shunt. Looking at this as an Ohm's Law problem, we could find the "X" ohms if we knew the potential difference between A and B. This voltage between A and B can be found by using the fact that there is a current of 0.001 amp in the 50-ohm coil. E = IR = 0.001 x 50 = 0.050 volt. This voltage is the same for the two parallel parts of the circuit, so we can use the 0.050 volt for the shunt-resistance problem. E = IR, 0.050 = 4.999 X, X = 0.01 ohm.

So, by combining a resistance of 0.01 ohm in parallel with the one-milliampere meter, the combination becomes a 5-amp ammeter.

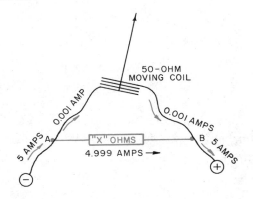

Fig. 7-5 Shunt Resistance

An experimenter planning to make this meter conversion need not look for a 0.01 ohm resistor in a supply catalog. A shunt can be made of copper wire. This shunt would be soldered or firmly attached in place to avoid introducing resistance due to poor contact. The procedure is:

1. Estimate the smallest wire size (gauge no.) which can carry the necessary shunt current without overheating. In this example the shunt current is about 5 amps. A reasonable conductor for 5 amps is no 18 copper.

2. Determine the length of wire needed to have the calculated resistance. In the example, .01 ohms is needed. No. 18 wire at 68°F. (20°C) has 6.39 ohms per thousand ft. or 0.006 39 ohms per foot (from the wire table in the Appendix). Dividing the ohms per foot figure into the total ohms needed will provide the number of feet needed.

$$\frac{.01 \text{ ohms}}{0.006\ 39 \text{ ohms/foot}} = 1.56 \text{ foot} = 18\ 3/4 \text{ in.}$$

3. Cut the wire slightly long to allow for making connections on the ends. For convenience, the shunt should be wound on a form. Using insulated wire reduces the chances of shorting out part of the length.

4. Test the meter with shunt for accuracy in a circuit with a 5-amp current. If the shunt feels hot to the fingers, recalculate the shunt using a larger diameter wire.

In manufacturing practice, meter shunts are made of manganin (a copper-nickel-manganese alloy) rather than copper. Its advantage is that its resistance does not change much with temperature changes. Furthermore, since its resistivity is greater than that of copper, a short strip of the alloy makes possible a sturdy assembly in a small space.

Always check the kind and rating of a meter before installing it in a circuit.

Using an Ammeter

In order to measure a current correctly, the meter should not interfere with the current being measured. The ammeter has a low enough resistance so that we may assume that putting the meter into the circuit will not reduce the current.

The purpose of an ammeter is to find out how much current there is in a piece of electrical equipment. The current through the equipment must also pass through the meter; hence, the ammeter must be in series with the equipment that is being tested, figure 7-6.

To measure current, connect ammeter in SERIES with the device.

Occasionally, someone will use an ammeter to find out how much current can come from a power line or car battery. There is enough current to burn out the ammeter. The ammeter, being of very low resistance, needs something in series with it to limit the current to a safe value. An ammeter connected directly to a current source, or in parallel with a device, acts as a short circuit.

However, an ammeter of sufficient range may be used as a short-circuit test of small dry cells. The internal resistance of the dry cell limits the current; contact is made only momentarily, otherwise the cell is damaged. When connecting an ammeter, the electrons should enter the negative terminal and leave the positive terminal. Failure to observe this polarity connection will result in the needle attempting to go backwards and may bend the needle.

Multirange DC Ammeters

Figure 7-7 represents a preferred arrangement of shunts for an ammeter with two scales. The circles marked 2 and 10 can represent either binding posts or selector switch contacts. A possible set of values of resistance is shown.

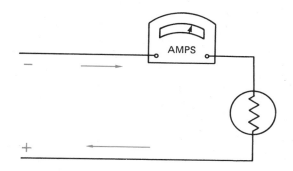

Fig. 7-6 Ammeter Connected in a Circuit

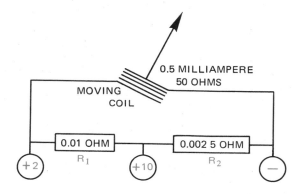

Fig. 7-7 Arrangement for Multiscale Ammeter

When the 2-amp contact is used, the shunt consists of R_1 and R_2 in series.

When the 10-amp contact is used, R_2 acts as the shunt; R_1 is in series with the moving coil.

This type of arrangement is called an *Ayrton shunt.* A three-scale ammeter contains a three-section shunt.

DC VOLTMETER

To make a *voltmeter*, a high resistance is connected in series with the galvanometer movement, figure 7-8, page 156. Unlike an ammeter, a voltmeter is intended to be connected directly across the source of energy, that is, in parallel with any other device to which the measured voltage is supplied.

The complete voltmeter must be a high-resistance instrument for two reasons:

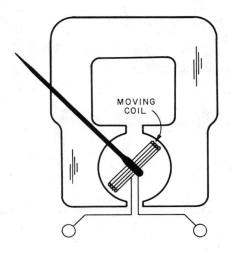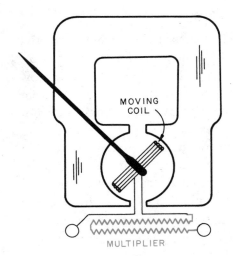

Fig. 7-8 A Resistor Converts the Galvanometer to a Voltmeter

1) Only a tiny current can be permitted through the moving coil.

2) The insertion of the voltmeter in the circuit should not cause a drain on the source which would reduce the electrical pressure being measured.

To measure emf, connect voltmeter in parallel or across the circuit.

The conversion of a galvanometer (milliammeter or microammeter) to a voltmeter is easy, both in calculation and construction. It is only necessary to calculate the series resistor which will limit the current to the full-scale galvanometer amount when the intended full-scale voltage is applied (Ohm's Law).

For example, there may be a meter scaled 0 to 200 microamperes (200 microamperes = 200 millionths = 0.000 2 amp). Assume a voltmeter reading up to 200 volts is wanted. The completed voltmeter must have a resistance found from:

$$R = \frac{E}{I} = \frac{200}{0.000\ 2} = 1\ 000\ 000\ \text{ohms}$$

This value is the total resistance of the voltmeter, the moving coil plus the series resistor. Ordinarily, resistance of the moving coil is so small (50-100 ohms) that this last figure is disregarded. In this case, a one-megohm resistor connected in series with the galvanometer coil makes the assembly a 200-volt voltmeter.

Anyone concerned about the inaccuracy introduced by disregarding 100-ohms coil resistance should consider these questions: If someone offered you a 999 900-ohm resistor, how would you tell whether it was 999 900 ohms or 1 000 000 ohms? How accurate is the microammeter movement that we started with? How accurately can a voltmeter be read? To answer one of these questions: inexpensive meters usually are accurate to 2 percent, more expensive meters to 1 percent.

Multirange voltmeters have several resistors. Which one is to be used is determined by choice of binding posts, or selector-switch setting.

Figure 7-9 shows theoretical values for the resistors. Actual values can vary from 1 to 2 percent.

Using the 2.5-volt connection, R = 2.5 ÷ 0.001 = 2 500 ohms total. (30 in meter plus 2 470 in series resistor.)

On 25 volts, $R = \frac{25}{0.001} = 25\ 000$ ohms

For 250 volts, R = 250 000 ohms

Sensitivity of Voltmeters

The sensitivity of voltmeters is stated in ohms per volt. In the first example, the 200-volt meter had 1 000 000 ohms resistance.

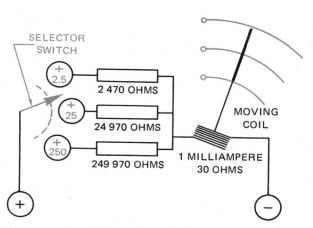

Fig. 7-9 A Multirange Voltmeter

$$\frac{1\ 000\ 000}{200} = 5\ 000 \text{ ohms per volt of full-scale reading.}$$

The multirange voltmeter shown in figure 7-9 has a sensitivity of 1 000 ohms per volt, on all scales. A large number of ohms per volt is desirable, for a high-resistance voltmeter takes a very small current to operate the meter movement. A meter rated 1 000 ohms per volt takes 1 milliampere for a full-scale reading; a 20 000 ohms/volt meter operates on 50 microamperes at full-scale. For checking electron-tube circuits, the 1 000 ohms/volt meter would be undesirable because at some points in the circuit there is not enough current available to operate the meter, or the current drain of the meter would cause serious error in the measurement taken.

OHMMETER

An *ohmmeter* contains a battery, series resistors, and a galvanometer (microammeter) movement. Batteries used may range from 1.5 to 45 volts. The moving coil assembly of this meter is the same type that has been used in all of the meters discussed so far. Increasing the current in the meter moves the pointer to the right. The meter is scaled, however, so that it indicates the amount of ohms resistance placed between the tips of the external test leads, figure 7-10.

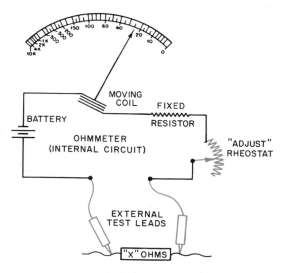

Fig. 7-10 Ohmmeter Circuit

To use this instrument, the tips of the test leads are first held together (short-circuited) and the rheostat adjusted so that the pointer of the meter moves over to the right-hand end of the scale where the zero-ohms mark is placed. The meter now indicates something already known. There is zero resistance between the test leads. The adjustment compensates for changes in the resistance of the battery as it ages. If this adjustment is not made, the readings will be inaccurate.

When the test leads are separated, there is no current in the circuit and the pointer drops back to the left end of the scale, where the infinity ohms mark is placed. When there are several inches of air between the test leads, there is a great deal of ohms resistance between them. To measure resistance in a circuit, turn off the power, discharge electrolytic capacitors, disconnect one terminal of the device to be measured, and connect the ohmmeter.

When the test leads are touched to the ends of a resistor of unknown value, the resistance is read directly from the ohms scale. Usually, an ohmmeter has several ranges, with different combinations of series resistance and battery voltage used in each range.

The ohmmeter diagrammed in figure 7-10 is called a series ohmmeter and is usually installed in a case containing multiple-contact switches, voltmeter resistors, and ammeter shunts, forming the multimeter or voltohmmeter that is widely used in testing electronic equipment.

In electrical testing, the ohmmeter is often used as a limited source of electrical power. In commercial multimeters, different makes vary as to the polarity of voltage available at the terminals. If needed, this polarity can be checked with a separate voltmeter.

For measuring very low resistances, a shunt-type ohmmeter is used, figure 7-11. The scale reads from left to right because high resistance permits more current through the meter. Zero resistance in the test-lead circuit permits most of the current to bypass the meter.

Megohmmeter

For insulation testing and similar high-resistance tests, a type of ohmmeter called a *megger* is often used. It contains a high-voltage, hand-cranked generator which produces current through series resistors, the unknown resistance, and a special two-coil mechanism which operates the pointer.

WATTMETER

In dc circuits, watts = volts x amps. To measure watts, a meter must have two coils, one affected by voltage and one affected by amps. The voltage coil is the moving coil, connected across the line so that the magnetic strength of this coil is proportional to the line voltage, figure 7-12. Note that this coil with its series resistor is like the voltmeter previously described. But, instead of a permanent magnet to provide the magnetic field for the moving coil to work with, there are *current coils* to provide magnetic field.

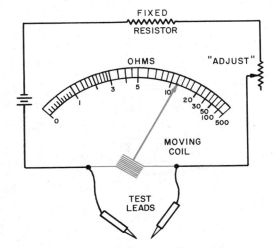

Fig. 7-11 Shunt-Type Ohmmeter

The magnetic strength of these coils is proportional to the current in them, which is the current taken by the device being tested. Both the movable and the fixed coils are air-core coils. This type of meter movement, called the *dynamometer*, is also used in other ac meters.

With the coils connected as shown in figure 7-12, the amount of movement of the moving coil and pointer depends on the strength of both coils. If there is voltage but no current, then the current coils provide no magnetic field to turn the moving coil so the pointer reads "0". Increasing the magnetic strength of either coil increases the turning force, that is, the turning force depends on the product of the magnetic strengths of the two coils. This coil arrangement thus gives us a pointer reading that depends on the product of volts on one coil times amps in the other coil, so the meter is scaled in watts. A wattmeter operates on ac as well as dc because when the current reverses, the magnetic polarity of both coils reverses and the turning force is still in the same direction.

Wattmeters are more necessary in ac measurements than in dc. With dc, watts always equal volts x amps and a wattmeter is not absolutely necessary. In ac circuits, there are occasions in which watts do not equal

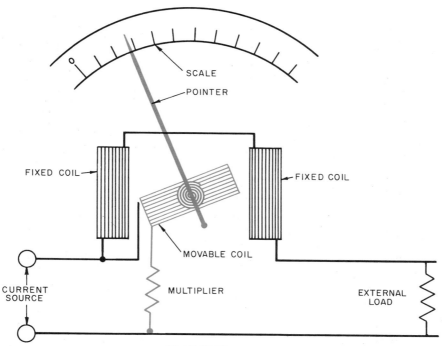

Fig. 7-12 Wattmeter

volts x amps. However, the wattmeter still tells the true power consumption in the circuit.

CONNECTING METERS IN A CIRCUIT: POLARITY

The positive terminal of the voltmeter or ammeter is connected to the positive of the supply source, figure 7-13.

Electrons run through the meter from the negative terminal to the positive terminal of the meter.

Locate the wire marked X in figure 7-13. The electrons in it move toward the meter because the battery provides the driving force. The negative on the meter has meaning only as compared to the positive of the meter. The negative on the battery is to be compared with the positive of the battery. The negative signs on each end of wire X are meaningless if compared with each other.

If a voltmeter is to be connected across resistor AB, where do the positive and negative of the meter go? Electrons run from

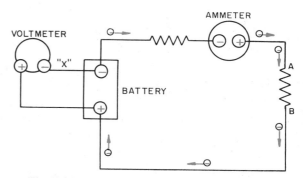

Fig. 7-13 Ammeter and Voltmeter in Circuits

A to B; therefore, A is more negative than B, so connect the negative of the meter to A, positive to B.

A resistor in a circuit is shown in figure 7-14, page 160. The current direction is unknown, but it must be determined. Assume a voltmeter is connected as shown in A, and the meter tries to read backwards, the pointer moving to the left.

Since the meter is connected incorrectly, reverse the meter connections to B so it reads correctly. When the meter reads correctly,

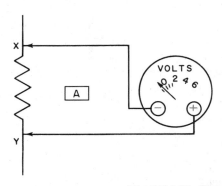

Fig. 7-14 Finding Current Direction with a Voltmeter

electrons run through it from its negative to
its positive terminal. Therefore, they must
have come from Y and flow through the
meter toward X. Y is more negative than X so
electrons run through the resistor from Y to X.

AC METERS

Rectifier-Type Voltmeter

If ac is sent through the moving coil of
a dc meter, the magnetic force on the coil
alternates rapidly so that the pointer tries to
vibrate back and forth about the zero position, giving no useful indication. The dc
meter, figures 7-3 to 7-9, can be used to
measure alternating current or voltage if the
current through the moving coil is rectified —
which means changed from ac to dc. A
rectifier is a device that permits electrons to
flow readily in one direction, but not in the
other. Rectifiers are explained further in
Chapter 17. Semiconductor rectifiers such as
silicon or germanium are used in meters.
Rectifier-type meters are used for low-current
devices such as voltmeters and milliammeters,
but do not operate reliably for high currents.

The Magnetic-Vane Attraction Meter

This meter uses a soft iron vane or
plunger projecting part way into a stationary
field coil, figure 7-15. Current in the field
coil produces magnetic force which pulls the
plunger further into the coil. In this simple

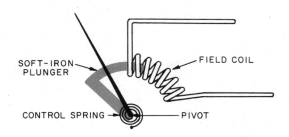

**Fig. 7-15 Simple Magnetic-Vane Attraction-Type
Instrument**

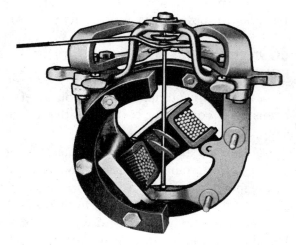

Fig. 7-16 Inclined-Coil Attraction-Type Voltmeter

solenoid-and-plunger meter, the attraction
force is least when the plunger is just entering
the coil and increases rapidly as more of the
soft iron vane enters the coil. This results in
crowding of numbers in the lower end of the
scale and expansion of the upper range. This
movement is mostly used for low-cost ac
ammeters, but can also be used for dc. With a

many-turn coil and a series resistor, the movement can be used for voltmeters.

The Inclined-Coil Meter

The Thompson inclined-coil movement may be seen in high-grade ac ammeters and voltmeters. It provides a long and reasonably linear scale. The basic principle of this movement is that an iron vane which is free to move in a magnetic field tends to take a position parallel to the flux. The cutaway view in figure 7-16 shows a pair of elliptical iron vanes attached to a shaft passing through the center of the stationary field coil. Increasing current in the field coil produces an increasing force tending to align the vanes with the coil flux, turning the shaft and moving the attached pointer across the scale.

The Repulsion-Vane Meter

This instrument, used for both alternating current and alternating-voltage measurements, operates on the magnetic repulsion force between two soft iron vanes which are in the same magnetic field, figure 7-17. One iron vane is attached to the instrument shaft; the other is secured to the stationary field coil. With no current in the coil, the spring holds the movable vane close to the fixed vane. In the field coil, ac magnetizes both vanes with nearby like poles repelling. This repulsion force on the moving vane turns the instrument shaft. This meter has a fairly uniform scale.

The Repulsion-Attraction AC Meter

Figure 7-18 shows a cutaway of the structure of a repulsion-attraction meter, which is used for both ac ammeters and voltmeters.

Figure 7-19 shows instantaneous polarities of the iron vanes. The movable vane is first repelled from the wide end of the

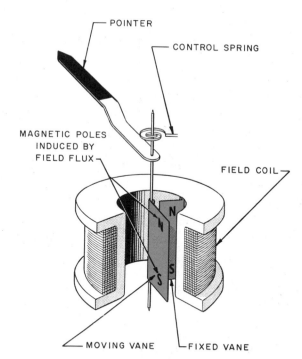

Fig. 7-17 Repulsion-Type, Vane-Type Meter Mechanism

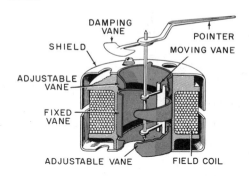

Fig. 7-18 Repulsion-Attraction Type Instrument

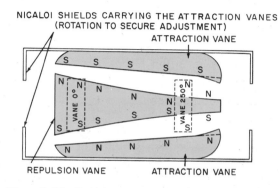

Fig. 7-19 Development of Repulsion-Attraction Magnetic System

middle stationary vane and the repelling force decreases as the movable vane travels along the tapering stationary vane. There is also an attracting force which increases as the ends of the moving vane move closer to the upper and lower stationary attraction vanes. By having these upper and lower attraction vanes of correct size and spacing, it is possible to provide a useful scale length of 250° of angular deflection. The scale distribution is determined by the shape and separation of the vanes; meters can be designed for broadening of the scale at any desired portion.

PRECISION INSTRUMENTS AND STANDARDS

By this time, one might reasonably have inquired about how the standard ampere and volt were determined. One might also wonder who decides the amount of resistance in one ohm.

In the past, scientists have set up several systems of electrical units: electromagnetic units, electrostatic units, and so on. Many people contributed to this work between 1840 and 1890. Our practical units, amperes, volts, and ohms, are based on measurements of magnetic effects.

Convenient sizes were chosen for units of current, potential differences, and resistance, in order to avoid the continual use of very large or very small numbers in calculations. Furthermore, the units were established of such a size that the formula, "E = IR" rather than "E = I x R x some inconvenient number" would express the relation between current, potential difference, and resistance.

The Ampere

In Chapter 2, one ampere was stated to be a flow rate of one coulomb per second, which is true. Historically, the ampere was established before the coulomb. The fundamental definition of an ampere is given by formulas that tell the magnetic force developed by coils of wire with current in them. The most accurate instrument for measuring current is a type of balance that measures the force between two carefully built coils that have the same current in them. Such an instrument, at the National Bureau of Standards, can measure current with an accuracy of about one part in a million. This highly sensitive instrument, time-consuming and requiring special skill in its use, is not used for ordinary standardizing of instruments. It is used only occasionally in checking other primary standards of resistance and voltage.

A formerly used method for standardizing an ammeter involved using a special electro plating device, in which a steady current is allowed to deposit silver from a plating bath, and the silver deposited during a given time is weighed. One ampere will deposit 0.001 118 gram of silver each second. This statement was adopted in 1893 as the definition of the *international ampere* and is now the legal standard.

The Ohm

One highly accurate method for establishing the ohms resistance of a resistor involves equalizing a potential difference across the resistor with the voltage generated by a disk spinning in a magnetic field. The formulas are not too complex, and only the area of the disk, the number of turns of wire in the coil, and the rpm of the disk have to be measured accurately.

This device, like the current balance described above, is used only for establishing the resistance of certain standard coils. These standard resistances, accurate to a few parts in a million, are usable in routine comparison work.

The international ohm was once defined as the resistance of a specifically shaped mercury column. This standard has been abandoned. Now, ten one-ohm coils of wire at the Bureau of Standards, Washington, D.C., are the standards of resistance.

The Volt

The *standard volt* is legally defined by Ohms's Law. One volt produces one ampere in one ohm. The volt is set up as the emf generated when a wire cuts 100 000 000 lines of force per second, but no highly accurate instrument has been devised to measure voltage from this principle.

As a real standard of voltage, there is a type of battery called a *Weston standard cell.* (More details on cells appear in Chapter 11.) A standard cell is used only in a voltage comparison circuit, in which it is permitted to produce not more than a few microamperes for a short time. The Weston standard cell is essentially a mercury-cadmium cell with cadmium sulfate solution as the electrolyte. At 20°C, its emf is a constant 1.018 3 volts.

THE WHEATSTONE BRIDGE

The *Wheatstone bridge* is a frequently used arrangement for accurate measurement of resistance. Its readings are considerably more precise than that of common series ohmmeters. The circuit is often pictured in the shape of a diamond, figure 7-20, with power supplied to two corner points, A and B, and a galvanometer connected to the opposite points, C and D.

In using the instrument, the unknown resistor to be measured (R_X) is connected to B and D. The rheostat (R_3) is adjusted so that points C and D are at the same potential (no current through the galvanometer). S_1 is closed for the final adjustment. In this condition voltage drops on R_1 and R_3 must be equal, as are voltage drops on R_2 and R_X. Since there is no current in the galvanometer, the current in R_1 (I_1) is the same as the current in R_2. The current in R_3 (I_3) is probably different from I_1, but is the same current as in R_X. Given equal voltage drops then:

(1) $I_1 R_1 = I_3 R_3$ and

(2) $I_1 R_2 = I_3 R_X$

(3) if both sides of equation (1) are divided by equal quantities from equation (2) we have

(4) $\dfrac{I_1 R_1}{I_1 R_2} = \dfrac{I_3 R_3}{I_3 R_X}$ the I_1's and I_3's cancel and we are left with

(5) $\dfrac{R_1}{R_2} = \dfrac{R_3}{R_X}$ solving this equation for R_X yields:

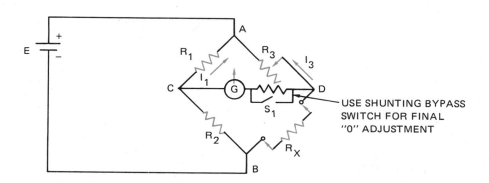

USE SHUNTING BYPASS SWITCH FOR FINAL "0" ADJUSTMENT

Fig. 7-20 Wheatstone Bridge

(6) $R_X = \dfrac{R_3 \times R_2}{R_1}$

Equation (6) shows that the accuracy of the instrument depends on very accurate construction of R_1 and R_2, and very accurate scale indications on R_3.

ELECTRONIC METERS

Measurement of very small voltages or currents, or larger voltages in high resistance circuits requires the use of electronic voltmeters. For many years the vacuum tube voltmeter (called a VTVM) fulfilled this need, and there are many such instruments still in use. Their modern replacements use field effect transistors and/or integrated circuits to isolate the circuit in which a quantity is being measured from the readout device. A conceptual diagram is shown in figure 7-21. Controls are available to select the unit of measure (volts, amperes, or ohms), the polarity of leads, and the scaling of the readout. The readout may be the familiar needle and meter face or lighted digits and a decimal point. An FET multimeter with digital readout is not much more expensive than one with needle readout, but it is much more accurate and more quickly read. Electronic meters are required for sensitive radio and television repair and adjustment.

OSCILLOSCOPES

The *oscilloscope* is a device for drawing a visual graph of voltages or currents with respect to the passage of time. The readout for the oscilloscope is the front of a cathode-ray tube, very similar to a black and white TV picture tube, sometimes called a CRT, figure 7-22. Chapter 8 provides specific details of the operation of the CRT.

Oscilloscopes have many controls, and one scope may differ from another considerably. Only four basic categories of controls, which are found in some form on all oscilloscopes, will be discussed. However, before discussing the controls, a better understanding of a graph of voltage and time is important.

A graph is drawn on a set of grid lines drawn at right angles to each other. Usually electrical quantities such as voltage are marked off on some suitable scale in the vertical direction (called the *vertical axis*). The passage of time is indicated on the *horizontal axis*. Figure 7-23 illustrates some typical graphs of voltage and time. The rectangular grid lines are permanently mounted on a transparent screen in front of the CRT face.

The four main control categories found on all oscilloscopes deal with the (1) brightness, sharpness, and position of the graph; (2) the selection of vertical scaling (vertical size of graph); (3) the selection of horizontal scaling; and (4) time synchronization selection control. The graph on the oscilloscope is often called a *trace* or *pattern*.

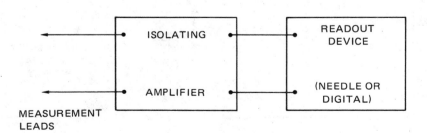

Fig. 7-21 Electronic Meter Diagram

Fig. 7-22 An Oscilloscope Readout

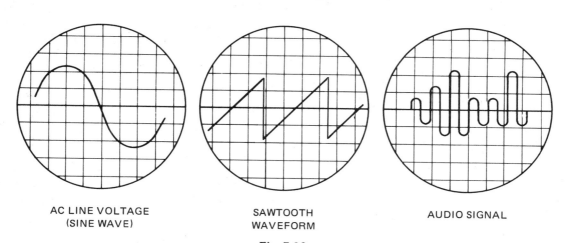

AC LINE VOLTAGE
(SINE WAVE)

SAWTOOTH
WAVEFORM

AUDIO SIGNAL

Fig. 7-23

Trace Convenience Controls

(1) Brightness control makes the pattern brighter and more visible or more faded. A really bright trace shortens the CRT screen life.

(2) Focus control will make the trace broad and fuzzy or sharp and clear. There is usually some inter-effect between the brightness and focus controls. Adjustment of one may make it necessary to adjust the other.

(3) Vertical position control should move the whole pattern up or down on the screen without changing its shape in any dimension. It is useful to center the pattern over a horizontal base line or position the extremity of a trace on a line for accurate vertical measurements.

(4) Horizontal position control moves the entire pattern horizontally without changing its shape. It is useful for centering a pattern, or starting a pattern on a specific line for horizontal time measurements.

Vertical Scaling Controls

(1) Step attenuator control allows the operator to either send the entire signal into the oscilloscope, or divide it, usually by factors of 10, before sending it to the oscilloscope vertical amplifier, figure 7-24. It adjusts the vertical size of the pattern in major steps.

(2) Vertical gain control gives the operator continuous control over the vertical size of the graph, between the step sizes made possible by the step attenuator. Use of these two controls allows the operator to connect common signal to the vertical input terminals and to cause it to appear on the screen

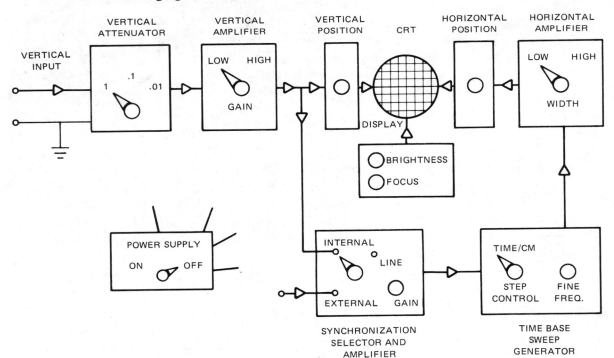

Fig. 7-24 Block Diagram of Oscilloscope Functional Controls

with *any* vertical scaling he desires. The operator measures voltage by connecting a *known* voltage into the oscilloscope vertical input terminals and adjusting the vertical attenuator and vertical gain controls to a convenient scaling. Then the unknown voltage is connected and compared to this scale.

For instance, the operator may determine the ac line voltage to be 120 volts. The operator connects this voltage to the oscilloscope vertical input, adjusts the vertical attenuator and vertical gain controls so that the pattern reaches over 6 major divisions from top to bottom. The number 6 is arbitrary, 2 divisions, 10 divisions, or any other could have been used. The oscilloscope is now calibrated to read 20 volts per vertical division on the scale. Connecting the unknown voltage to the scope, the operator discovers that its pattern is 4.5 divisions from top to bottom. The unknown voltage must be 4.5 x 20 = 90 volts. Although some expensive oscilloscopes have accurately scaled controls for voltage measurements, any oscilloscope will allow the operator to set his own scale with an outside voltage source. All inexpensive oscilloscopes *require* this calibration procedure for accurate work. The voltage readings cannot be more accurate than the calibration source. Many scopes have a special terminal on their faces providing a convenient calibration source.

Horizontal Scaling Controls

(1) Horizontal gain control (or width control) will stretch or shrink the horizontal length of the pattern without changing its vertical dimensions. It is useful in setting the pattern to a convenient overall length for measurement or calibration purposes.

(2) In time-base or sweep controls, the time-base generator produces a signal for moving the electron beam forming the trace back and forth across the front of the CRT. If the generator is set to an extremely slow speed, the operator can actually watch a bright spot move from left to right across the screen repeatedly. As the generator is speeded up, the spot moves faster and faster until the operator only sees a continuous line. The speed of the spot is controlled in major steps by a time-base (sweep) control often labeled in time per C.M. units. Smaller changes in the spot speed are affected by the fine frequency adjustment. The word sweep is often associated with these controls because the electron beam sweeps across the front of the tube and is visible, then is shut off and reaimed at the left-hand side of the tube for the next pass across, much like a broom moves across a floor and is lifted over and back before making the next working pass. The generator of inexpensive oscilloscopes can sweep the front of the tube anywhere from 1/10 of a second to approximately 1/100 000 of a second.

Horizontal time scaling is completely controlled by the operator. If the time for one sweep were set at 1/30 of a second and the horizontal width control were adjusted to make the trace 3 major divisions long, then each major division would represent 1/3 of 1/30 second or 1/90 of a second.

Common inexpensive oscilloscopes can easily make accurate time measurements to a hundred-thousandth of a second. Problems such as the length of time automobile ignition points remain open, the time for a radar signal to reach its target and return, the time necessary for an electronic calculator to add or subtract, or the time for an unknown ac signal to complete a cycle are all solved by applying this instrument.

Time Synchronization Controls

Sync-Selector. In order to draw a stable pattern on the face of the scope, the vertical input signal must be in perfect synchronization with the horizontal sweep. While the vertical input signal moves the trace up and down, the horizontal sweep generator moves it from left to right. In order to keep the two together, the operator has a *sync-selector knob* which allows the horizontal sweeps to be controlled in step with signals outside the sweep generator. There are at least 3 positions of this knob:

(a) In line synchronization the horizontal sweep keeps a regular rhythm with respect to the ac power line.

(b) In internal synchronization part of the vertical input signal is fed to the horizontal sweep oscillator for timing purposes.

(c) In external synchronization the operator can synchronize the sweep oscillator in step with some external signal or event. For example, in automobile troubleshooting the mechanic synchronizes off from the firing of the number one spark plug.

Sync-amplifier gain. This control is usually physically close to the sync-selector knob. It allows the operator to either reduce or amplify the selected synchronization signal before putting it to work in the sweep generator. It assures that the sync signal is of the proper magnitude to accomplish its purpose. If in attempting to use the scope, the pattern keeps traveling across the screen, adjustment of this control will help stabilize it.

When measuring an unknown voltage, or measuring an unknown frequency, the scope is first calibrated by a known outside signal, scaling is determined, and then the unknown signal is connected. The measurement is taken by comparison of the effect of the unknown signal with that of the known one.

Expensive Oscilloscopes

More expensive oscilloscopes generally have three features that are distinct advantages over the type of scope discussed in detail in this chapter.

(1) Dual trace allows the operator to show two patterns at once on the same screen. It makes comparisons much easier, particularly in critical timing applications.

(2) Calibrated amplifiers and attenuators and sweep times are important because when the scope is in the calibrated position, the readings associated with its dials are accurate, allowing direct voltage and time measurements from the screen. This eliminates the necessity of the two step measurement process.

(3) In oscilloscopes having triggered sweep, the electron beam will not trace across the CRT face until an external trigger releases it. The trigger is an electrical signal which may or may not be continuously repeated. Having made one sweep, the trace

will not sweep again until triggered. This feature allows the operator to mount a camera over the screen face and take a picture of electrical events which may occur only once. Other scopes require a continuously repeated signal pattern, drawn over and over again on the screen to be a visible, stable and useful pattern. If one learns how to operate an inexpensive nontriggered oscilloscope there is no particular problem in adjusting to the operation of the more expensive and complicated ones. On the other hand, if one had only had experience and training on a calibrated, dual trace, triggered scope, he would not know how to take accurate measurements with less sophisticated oscilloscopes.

POINTS TO REMEMBER

- A dc galvanometer consists of a small coil of wire, moved by magnetic action in the field of a permanent magnet, with spiral hairsprings to bring the pointer back to zero.

- An ammeter is a low-resistance meter consisting of a galvanometer and shunt (parallel low resistance).

- Ammeters should be connected in series with the device in which current is to be measured.

- A voltmeter is a high-resistance meter consisting of a galvanometer plus a series resistor.

- A voltmeter can be connected directly to a voltage source and must be across (in parallel with) the device on which voltage is to be measured.

- An ohmmeter includes a galvanometer, dry cells, and series resistors. It measures resistance placed between its test leads. It must be used on dead circuits, not on resistors that have current in them from some other source.

- A wattmeter contains a voltage coil across the line, and a current coil in series with the line. A wattmeter reads true watts, ac or dc.

- There are more electrons on the negative terminal of a device (meter, resistor, battery, etc.) than there are on its positive terminal. (Use this idea to determine either polarity or electron current direction.)

- Electrical measurements are based on the accuracy of mathematical formulas and the reliability of high-precision instruments.

- Rectifier-type meters use a dc meter plus a rectifier to measure ac quantities.

- Iron vane-type meters operate on magnetic forces. They are reliable for ac measurements and operate inaccuately on dc.

- When measuring unknown currents or voltages, start with the highest range scale available on the meter to reduce the possibility of damage.

REVIEW QUESTIONS

1. Ammeters, voltmeters, and ohmmeters are all similar in construction, being modifications of a simple basic instrument called a galvanometer. Describe a galvanometer and explain how this instrument functions.

2. Will a dc ammeter record an ac current? Why or why not?

3. State the purpose of hairsprings in a galvanometer, shunt in an ammeter, and series resistor in a voltmeter.

4. Diagram the internal circuit of a voltmeter, an ammeter, an ohmmeter, and a wattmeter.

5. Why should an ammeter be connected in series with a load rather than in parallel with it?

6. The user of an ohmmeter finds that when the test-probes are shorted, the adjustment will not make the pointer move over to the zero mark as it should. The pointer stops at about R = 5. What is wrong?

7. Calculate the resistance of a shunt, to convert a 100-microampere meter with a 40-ohm moving coil, to a 10-milliampere meter.

8. Calculate series resistor to convert the above 100-microampere meter to a voltmeter, scaled 100 millivolts full-scale.

9. Calculate series resistor to convert the same 100-microampere meter to a voltmeter scaled to 100 volts.

10. (a) Using data from Fig. 7-7, calculate the voltage across the moving coil, and the voltage across the shunt, when 2 amps pass through the meter, using $\left(+2\right)$ and $\left(-\right)$.
    (b) When 10 amps pass through $\left(+10\right)$ and $\left(-\right)$ terminals, find voltage across moving coil and voltage across shunt.

11. Two resistors, A and B, are connected in parallel and this combination is placed in series with a third resistor, C. This entire group is connected across a 120-volt dc supply. The current in C is 5 amps and the current in B is 3 amps. The resistance of A is 20 ohms.
    (a) Determine the voltage across A, the resistance of B, and the resistance of C.
    (b) If the resistor A is accidentally open-circuited, find the new voltage across resistor B and resistor C.

12. A 0-150-volt voltmeter has a resistance of 2 000 ohms per volt. It is desired to change this voltmeter to a 0-600-volt instrument by the addition of an external multiplier (series resistor). What would be the resistance in ohms of this external multiplier?

13. A pair of #28 copper wires in a telephone cable is accidentally short-circuited, figure 7-25. On connecting a Wheatstone bridge to the accessible ends of the wires, the bridge gives R_1 = 100, R_2 = 327.8, R_3 = 100. Calculate distance from the accessible end of the cable to the point where the pair of wires are shorted (Temp. 70°F./21°C).

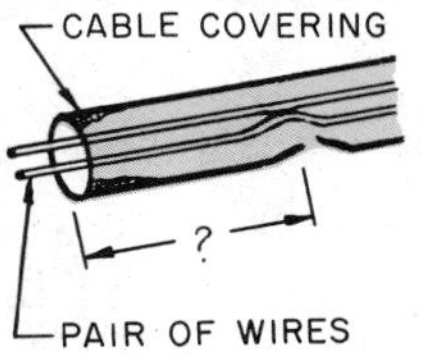

CABLE COVERING

?

PAIR OF WIRES
Fig. 7-25

14. Identify 3 common uses of the oscilloscope.

15. Describe features that are available on more expensive oscilloscopes.

16. An oscilloscope is adjusted to show 2 complete cycles of the 60 hertz line frequency. Then, without changing horizontal adjustments, an unknown signal was applied to its terminals. The unknown signal showed ten complete cycles on the screen. What is the frequency of the unknown signal?

RESEARCH AND DEVELOPMENT

Experiments and Projects on Measuring Instruments

INTRODUCTION

Measurement is a fundamental part of all industry. Its application to electricity and electronics is particularly important in conducting experiments and testing new devices, in measuring and recording the operation of functioning equipment, and in locating trouble when making repairs. In Chapter 7, the construction, operating principles, and uses of the measuring instruments commonly used in the industry are covered. A number of experiments are suggested to give you an opportunity to use and become familiar with the operation of measuring instruments. Suggestions are made for the designing and building of simple instruments embodying some of the principles used in instrument construction.

EXPERIMENTS

1. To learn how to measure current with an ammeter.

2. To learn how to measure emf with a voltmeter.

3. To learn how to measure resistance with an ohmmeter.

4. To learn how to read a watt-hour meter and compute the cost of electricity.

5. To learn how to measure the power in a circuit with a wattmeter.

6. To learn how to display a signal wave form on the oscilloscope.

PROJECTS

1. Design and construct a galvanometer.

2. Design and construct an ammeter.

3. Design and construct a measuring device using an electric meter movement to register the measurement of a quantity or the like, for example, a liquid-level gauge, temperature gauge, or tachometer.

4. Design and construct a voltmeter.

5. Design and construct an electron charge indicator.

6. Design and construct a multirange direct-current voltmeter and milliameter.

7. Design and construct a capacitor substitution box and continuity checker.

EXPERIMENTS

Experiment 1 USING AN AMMETER

OBJECT

To learn how to use an ammeter to measure current in a circuit.

APPARATUS

1 - DC ammeter
1 - AC ammeter
1 - Dry cell, 6-volt or dc power supply, ac power supply
1 - Resistor, 10-ohm, 10-watt (R_1)
1 - Resistor, 15-ohm, 10-watt (R_2)
1 - Resistor, 20-ohm, 10-watt (R_3)
1 - Resistor, 30-ohm, 10-watt (R_4)
1 - Lamp receptacle
1 - Lamp, 100-watt, 120-volt
1 - Lamp, 150-watt, 120-volt
1 - Glocoil, 660-watt, 120-volt
1 - SPST switch
    Mounting stands for resistors

PROCEDURE

Part 1 Direct-Current Ammeter

1. Connect the switch and resistors as shown in figure 7-26.

2. Disconnect the circuit between R_1 and R_2 and connect the dc ammeter in the circuit, figure 7-27.

   Note: Ammeters are always connected in series with other components, never in parallel.

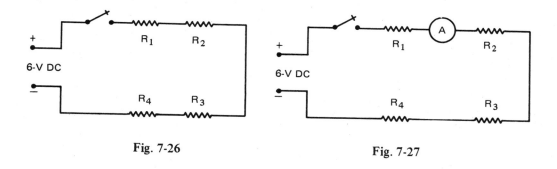

Fig. 7-26   Fig. 7-27

3. Turn on the current and record the current indicated on the meter.

4. Turn off the current, remove the ammeter and connect R_1 and R_2 again.

5. Using the same procedure, find the current between R_2 and R_3, R_3 and R_4, and finally between R_4 and the negative terminal to the dc supply.

OBSERVATIONS

Use a table similar to the one shown here to record your observations.

OBS. NO.	AMPERES
1	
2	
3	
4	

Part 2 Alternating-Current Ammeter

6. Connect the switch, lamp receptacle, and ac ammeter as shown in figure 7-28, page 174.

7. Insert a 100-watt lamp in the receptacle and turn on the current.

8. Read the meter and record the current.

9. Repeat the same operations using the 150-watt lamp and then, the 650-watt glocoil.

OBSERVATIONS

Use a table similar to the one shown here to record your observations.

OBS. NO.	AMPERES
1	
2	
3	

QUESTIONS

1. What precautions are necessary when using an ammeter in a circuit?

2. How do the connections for an ac ammeter differ from those for a dc ammeter?

3. How did the reading for the total resistance in the dc circuit compare with the sum of the separate reading? How do you account for your answer?

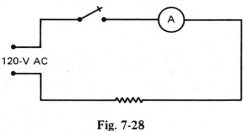

120-V AC

Fig. 7-28

Experiment 2 USING A VOLTMETER

OBJECT

To learn how to use a voltmeter to measure emf in a circuit.

APPARATUS

1 - DC voltmeter
1 - AC voltmeter
1 - Dry cell, 6-volt or dc power supply, ac power supply
3 - Resistors, 10-ohm, 10-watt
3 - Lamp receptacles, 120-volt
3 - Lamps, 60-watt, 120-volt
1 - SPST switch

PROCEDURE

Part 1 Direct Current Voltmeter

1. Connect the switch and resistors as shown in figure 7-29.

2. Turn on the current and touch the test probes of the meter to the connections of R_1, as indicated in figure 7-30, and record the voltage.

> Always use a voltmeter with a higher voltage range than the voltage being measured.
>
> When using dc voltmeters, connect positive terminal to positive side of the source, and negative terminal to the other side.

3. Connect test probes across the terminals of R_2, as in figure 7-31, and take a reading. Next, connect the probes across both R_2 and R_3 and take a reading.

OBSERVATIONS

Use a table similar to the one shown here to record your observations.

OBS. NO.	VOLTS
1	
2	
3	

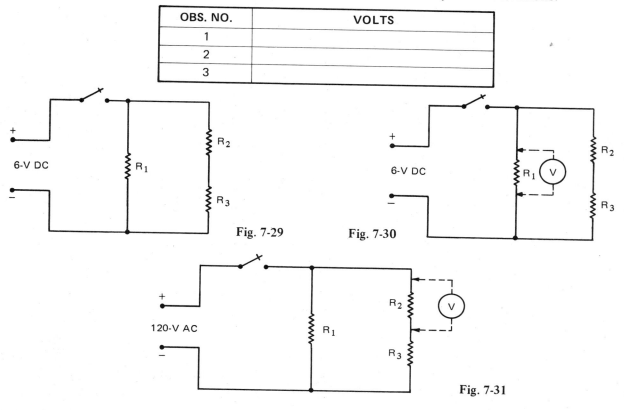

Fig. 7-29   Fig. 7-30

Fig. 7-31

Part 2 Alternating-Current Voltmeter

4. Connect the switch and lamp receptacles as in figure 7-32.

> Be particularly careful to hold the test probes on the insulated part when testing high-potential alternating current. Make certain that the voltmeter has a larger capacity than the voltage you are about to measure.

5. Connect the test probes across L_1, as shown in figure 7-33, and turn on the current. Read the meter and record the voltage.

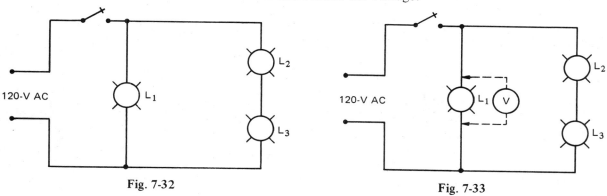

Fig. 7-32   Fig. 7-33

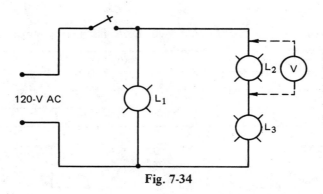

Fig. 7-34

6. Next, make connection with the probes across L_2 as shown in figure 7-34. Read voltage and record. Also, find voltage across L_3.

OBSERVATIONS

Use a table similar to the one shown here to record your observations.

OBS. NO.	VOLTS
1	
2	
3	

QUESTIONS

1. How does the method of connecting a dc voltmeter differ from that of connecting an ac meter?

2. Why is it good practice to hold test probes on the insulated part when measuring high potential alternating current?

3. Compare the voltage across R_1 and the reading taken across R_2 and R_3. How do you account for the result?

4. How did the voltage across L_1 compare with the voltage across L_2?

5. Was the voltage across L_1 and the ac source the same? If they were different, explain why.

6. Why should you always observe polarity when measuring dc voltages?

Experiment 3 USING AN OHMMETER

OBJECT

To learn how to use an ohmmeter to measure resistance.

APPARATUS

1 - Ohmmeter

1 - Resistor, 100-ohm, 1-watt (R_1)

1 - Resistor, 39-ohm, 1-watt (R_2)

1 - Resistor, 22-ohm, 1-watt (R_3)

1 - Resistor, 40-ohm, 10-watt (R_4)

1 - Resistor, 20-ohm, 10-watt (R_5)

1 - Resistor, 10-ohm, 10-watt (R_6)

Mounting stands for resistors

PROCEDURE

1. Connect resistors as shown in figure 7-35.

2. Zero the meter. To do this, hold the contact ends of the test leads together and observe the meter reading. It should read zero on the right-hand side of the ohms scale. If the pointer does not read zero, adjust the zero-ohms knob until it does. The ohmmeter is now ready to measure resistance.

3. Measure the resistance of R_1 and record the reading, figure 7-36. Note: Be sure to read the instrument on the correct scale and the proper multiplier.

4. Repeat the same procedure for each of the six resistors.

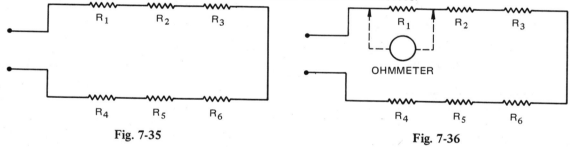

Fig. 7-35   Fig. 7-36

5. Measure the resistance of the total circuit, figure 7-37, and record the result.

When using an ohmmeter to measure resistance in a circuit that has been energized, be certain to break the continuity of the circuit or disconnect one end of the element being measured from the circuit.

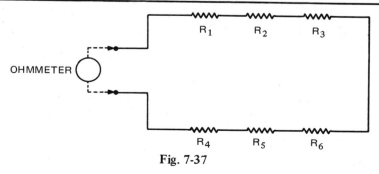

Fig. 7-37

OBSERVATIONS

Use a table similar to the one shown here to record your observations.

OBS. NO.	READING IN OHMS
1	
2	
3	
4	
5	
6	
7	

QUESTIONS

1. Why must the circuit be disconnected before an ohmmeter is used to measure an element of the circuit or the entire circuit?

2. Why is it necessary to zero the ohmmeter before taking a reading?

3. What is the source of power in the ohmmeter circuit?

4. Why is a variable rheostat necessary in an ohmmeter?

5. Which of the following meters is used as a basic instrument in the ohmmeter: millivoltmeter, ammeter, wattmeter, milliammeter, or galvanometer?

6. Now that you have measured resistors, explain what they do in an electrical circuit.

Experiment 4 THE WATT-HOUR METER

OBJECT

To learn how to read a watt-hour meter and to understand how the cost of electric energy is computed.

APPARATUS

1 - Watt-hour meter, commonly known as a kilowatt-hour meter (kwh)

PROCEDURE

1. Examine the meter in the shop and learn how to read it.

2. Read the meter in your home. Record the reading and compare it with the last statement from the power company.

3. Subtract the last meter reading from the one you recorded.

4. Multiply the difference by the rate per kwh charged by the power company.

QUESTIONS

1. What do you actually buy from the power company?

2. Why is this commodity sold in kwh?

3. What is a watt-hour? A kilowatt-hour?

Experiment 5 USING A WATTMETER

OBJECT

To learn how to measure power in a circuit with a wattmeter.

APPARATUS

1 - Wattmeter
1 - AC ammeter (0-50 A)
1 - AC voltmeter (0-150 V)
1 - Lamp receptacle
1 - SPST switch
1 - Load devices—lamp, heater element, motor, choke coil, transformer

PROCEDURE

1. Locate the current terminals of the wattmeter, connect to them as you would connect an ammeter, that is, in series with the load. You can plug into the current source, but the wattmeter will not read anything until the voltage coil of the wattmeter is connected.

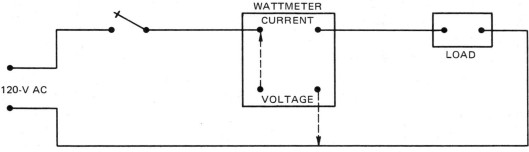

Fig. 7-38

2. The voltage terminals of the meter are to be connected across the line, just as you connect a voltmeter. (See dotted lines, figure 7-38. If you find that the pointer of the wattmeter tries to move to the left (reading backwards), then reverse the connections to the voltmeter terminals. Observe maximum current rating of wattmeter.

3. Use a lamp, or heater coil, as load. Compare wattmeter reading with product of volts times amps. Do you expect them to be equal? Explain your answer.

4. Use a motor, or choke coil, or 120-volt primary of a transformer, as load. Does watts equal volts times amps? Later, in Chapter 14, you may find additional reasons why watts does not always equal the product of volts times amps.

OBSERVATIONS

Use a table similar to the one shown here to record your observations.

OBS. NO.	WATTMETER READING
1	
2	
3	

QUESTIONS

1. Diagram the internal circuit of a wattmeter.

2. Upon what does the amount of movement of the moving coil and pointer depend?

3. Explain why the scale of the meter reads in watts.

4. Why are wattmeters more necessary in ac measurements than in dc measurements?

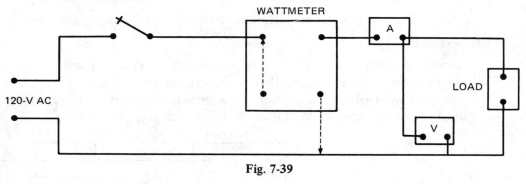

Fig. 7-39

Experiment 6 VIEWING AN AC SIGNAL

OBJECT

To learn how to use the oscilloscope to display an ac signal and measure its frequency.

APPARATUS

1 - Nontriggered oscilloscope
1 - Signal generator
1 - Crystal microphone

PROCEDURE: PART 1

Often technicians know what the pattern of a particular signal should look like. They use the oscilloscope to see if the actual pattern obtained in a given circuit conforms to the one expected. The commercial line voltage supplied to schools and homes should have a sine waveform as in figure 7-22, page 165. This is easily shown.

1. Be sure the scope is isolated from earth or commercial-power ground by using an isolation transformer between its line cord and the wall convenience outlet. Turn on the scope and adjust brightness and focus controls.

2. Set the vertical step attenuator to divide the signal at least by a factor of 10.

3. Connect the line voltage to the vertical input terminals and look at the effect on the screen. (The pattern may be an entanglement of crossed lines.)

4. Set sync-selector control to "line" and adjust sync amplifier gain at approximately 25 percent of its travel in a clockwise direction.

5. Adjust the fine-frequency sweep control and watch the pattern on the screen. It should be possible to adjust the pattern to show one, two, or three complete alternations. The entanglement of crossed lines, moving patterns between the stable positions of the pattern as the control is adjusted, should be ignored. These represent the fact that the sweep frequency cannot keep up with the vertical signal in an easily interpreted way for some settings of the sweep frequency control.

6. Adjust *all* the other controls on the scope and note their effect on the pattern.

PROCEDURE: PART 2

Measuring an unknown frequency.

1. Set up the scope to view the line waveform as directed above using internal synchronization instead of line sync. Adjust the sweep frequency controls to show one complete alternation.

2. The commercial power line alternates at 60 hertz, so it takes 1/60th of a second for the trace to show this alternation from left to right.

3. Adjust the horizontal position and horizontal gain control (width) to spread this 1/60 second trace across 10 major divisions of the screen. The scope is now calibrated to measure time with the scaling of 1/600 second per major division of trace movement on the screen.

4. Disconnect the line voltage and connect a microphone to the vertical input terminals. Hum a constant low voice note into the microphone and adjust the vertical attenuator and vertical gain controls for ease of viewing. (Do not touch the horizontal controls or calibration will be lost.)

5. While humming a constant note, count the number of alternations on the screen. Remember that this number of alternations is occurring in 1/60th of a second of time. If a 60-hertz waveform alternated once

during this time period, and there are now five alternations on the screen, then your voice frequency must be five times as fast or have a frequency of 300 hertz. (*Hertz* (Hz) is the official title for cycles per second.)

6. Raise or lower the pitch and loudness of your hum and observe its effect on the pattern.

QUESTIONS

1. List each major control on the oscilloscope and briefly describe its purpose.

2. What two variables are usually displayed in an oscilloscope pattern?

3. How can the oscilloscope be used to measure volts, time, and frequency?

4. What happens to the oscilloscope pattern if the test leads are reversed at its vertical input terminal?

PROJECTS

A SIMPLE GALVANOMETER

The galvanometer is a sensitive measuring instrument that will indicate the presence of current, the strength of the current, and the direction in which it is moving. In this device, a small quantity of electrons flowing through the coil will cause the compass needle to move, indicating the presence of current electricity. A change in polarity will cause the compass needle to move to the opposite pole. This change of polarity is directly related to the "left-hand rule."

Figure 7-40 suggests a size for a wooden form. A plastic box or form of a similar size will function very well. Thirty-three rows of #28 magnet wire should be wound on each side. A continuous coil should cross the form in the center on the bottom side.

MATERIALS

Compass
Magnet wire, #26, #28, or #30 AWG
Fahnestock clips, small size
Plastic, wood, or cardboard tube

PROCEDURE (For Rectangular Type)

Make a working drawing of the form you choose to support the wire. Show details of construction, dimensions, and any innovations you wish to include.

Note: A small, clear plastic box of appropriate size may be used for a form if enough material is removed from one side to receive the compass and secure the wire, and from the opposite side to secure the other end of the coil.

Prepare the form. If it is made of wood, do not use steel nails.

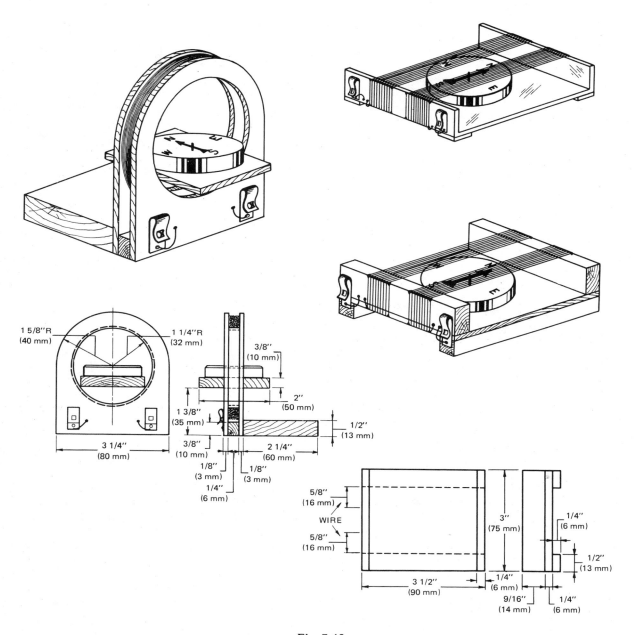

Fig. 7-40

Wind the wire over the form. Notice the shallow notches at the beginning and ending of the windings, and how each end of the wire is secured through two small holes.

Attach the binding posts with brass screws and secure the ends of the wire to them with solder.

Place a compass in the center with the compass needle North and South and the coil, East and West.

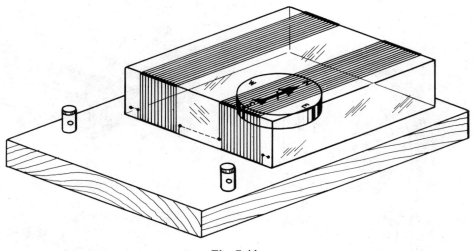

Fig. 7-41

PROCEDURE (For Circular Type)

Secure a good quality cardboard tube, about 2 1/2 inch (64 mm) outside diameter.

Cut off a piece with parallel sides, 1/4 inch (6 mm) wide. Lay out and make other pieces. Glue the ends to the tube.

Fill the 1/4 inch (6 mm) square cavity nearly full of #30 magnet wire. Assemble, test, and evaluate.

MAGNETIC-VANE ATTRACTION-TYPE AMMETER

This ammeter will indicate the presence of direct or alternating current in a circuit. It is more sensitive when used to measure dc. The meter illustrated in figure 7-42 will measure current from one-half ampere to eight. The high point will vary with the adjustment of the counterbalance.

You will find that the distance between the graduations on the scale becomes progressively greater as the amperes increase, thus making the measurement above two amperes easier to read and, consequently, more accurate.

MATERIALS

1 - Pc. enameled magnet wire, #12 AWG, 60" (1.5 m) long

1 - Pc. sheet iron, #22 gauge, 2 1/2" x 3 1/2" (65 mm x 90 mm)

2 - Binding posts or Fahnestock clips

2 - Brass wood screws, #6 x 1/2" RH

2 - Brass escutcheon pins, 1" (25 mm) long

1 - Pc. copper wire, #16 x 2" (50 mm)

1 - Pc. Spring brass wire, #22 x 4" (100 mm)

1 - Pc. wood, 1/2" x 4" x 6 1/2" (13 mm x 100 mm x 165 mm)

1 - Pc. wood, 3/4" x 4" x 4" (20 mm x 100 mm x 100 mm)

1 - Brass or fiber washer

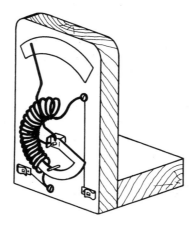

Fig. 7-42

PROCEDURE

Plan your procedure for making each part shown in figure 7-43, page 186.

Select material for the upright and base. Cut to size, assemble, and finish.

Allowing about two inches of wire on each end, wind the coil on a 7/16" (11 mm) smooth steel rod with ends free from burrs. Keep the turns of wire close together and be careful to protect the enamel on the wire. An abrasion of the enamel might cause a short circuit.

Lay out the contour of the coil and carefully bend it until it conforms to the pattern.

Bend the eye on each end and adjust the coil for position on the upright.

Lay out the armature and fabricate it. To insure accuracy when drilling the holes, use a spacing block between the front and back part as in figure 7-44, page 186.

The hole drilled in the base to receive the escutcheon pin pivot must be perpendicular to the upright and a press fit.

Cut a piece of good quality white paper large enough for the scale. Draw two concentric arcs on it one-half inch apart. Calculate the radius using the fulcrum pivot as the center.

Assemble the unit and test it with a dry cell. The armature should be sucked into the coil all or nearly all the way. It should not touch the coil at any point.

Calibrate your meter with another ammeter of known reliability. To do this, connect them in series with a load. An adjustable load can be provided with a bank of lamps of different wattage connected in parallel. Another method is to use a variable resistor in place of the lamp bank.

Evaluate.

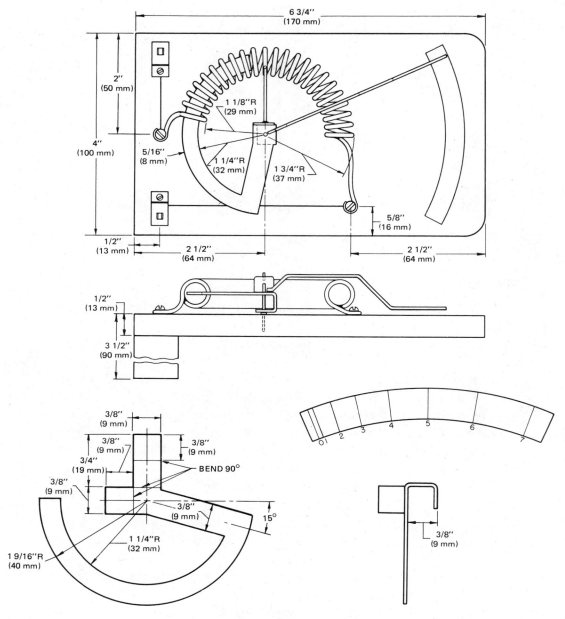

Fig. 7-43

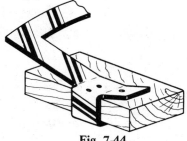

Fig. 7-44

SOLENOID PLUNGER ATTRACTION-TYPE AMMETER

This solenoid plunger-type ammeter, figure 7-45, is comparatively easy to build. It allows considerable variation in design to acquire greater sensitivity. Various size coils, plungers, and linkage could be designed and tested. The meter suggested has a range of one-fourth ampere to six. This range will vary with adjustment of spring tension.

MATERIALS

1 - Panel, tempered hardboard 1/8" x 6" x 10" (3 mm x 150 mm x 250 mm)

1 - Base, wood, 3/4" x 3" x 6" (20 mm x 75 mm x 150 mm)

1 - Needle and fulcrum arm, #12 gauge steel wire or sheet aluminum, 16 or 18 gauge

1 - Fiber tube, 5/16" (8 mm) ID x 3" (75 mm) long

1 - Plunger, soft iron, 5/16" x 2 3/4" (8 mm x 70 mm)

1 - Spring, music wire, 4- to 6-gauge

2 - Disks, fiber or thin plywood, 1 1/16" (27 mm) diameter, hole to be press fit on tube

Magnet wire, #26 AWG sufficient for 8 layers

2 - Machine screws, #6 x 32 x 1" (M3.5 x 25 mm L) RH steel

4 - Machine screws, #4 x 40 x 3/8" (M3 x 9 mm L) RH steel

4 - Nuts, #6 x 32 (M3.5)

4 - Nuts, #4 x 40 (M3)

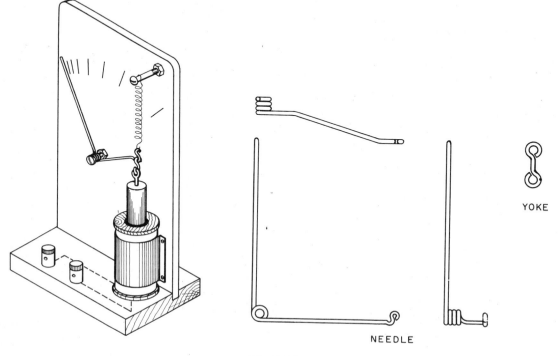

Fig. 7-45

1 - Washer, #6 hole x 3/8" diameter (M3.5 x 9-mm diameter)
1 - Coil support, tinplate or aluminum, 2 5/8" x 4" (67 mm x 100 mm)
2 - Wood screws, #6 x 1/2"
2 - Fahnestock clips
1 - Screw eye, 1/8 (3 mm) ID
1 - Yoke, #18 gauge wire

PROCEDURE

Plan your procedure for making each part. Consider innovations for improvement of design and construction.

Make and finish each part. For the spring refer to figure 5-39, page 111.

Note: Assemble and test the action of the plunger and needle linkage. The plunger should move freely in the tube and the connections of the linkage should not bind.

Calibrate your meter with another ammeter of known reliability. To do this, connect them in series with a load. An adjustable load can be provided with a bank of lamps of different wattage connected in parallel. Another method is to use a variable resistor in place of the lamp bank.

To establish a satisfactory range on the dial, it will be necessary to vary the tension of the spring. You may have to make several size springs. This can be accomplished by changing the diameter of the spring, the distance between the coils, or the size of the wire.

Note: The meter range can be changed by using another spring of a different tension.

Evaluate.

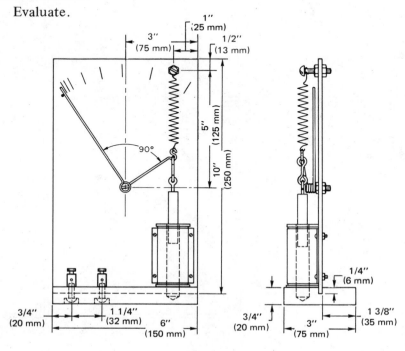

Fig. 7-46

EXTINCTION-TYPE VOLTMETER

This extinction-type voltmeter, figure 7-47, is a fairly reliable instrument and one that is rather inexpensive to build. It operates on the principle that a neon bulb of a specific type will cease to glow when the voltage drops below a certain point. The potentiometer in the circuit permits adjustment of the voltage until the lamp is extinguished. At this point, the voltage can be determined according to a calibrated scale.

MATERIALS

1 - Neon bulb, type NE-2 (N_1)
1 - Potentiometer, 250 000-ohm (R_1)
2 - Carbon resistors, 51 000-ohm, 1/2-watt (R_2, R_3)
1 - Clear plastic box
2 - Test probes

PROCEDURE

Do some research on neon bulbs. Find out what types are available from supply-house catalogs.

Check the manufacturer's data for each bulb and calculate the external resistance required to produce a voltmeter of the range you desire.

Procure the neon lamp, resistors, and potentiometer, and then find a clear plastic box suitable to house the parts, or make a container of your own design.

Fit the potentiometer into the box or container in a manner which will provide room for a dial-type voltage scale.

Assemble the parts according to figure 7-48.

Calibrate the meter with standard ac and dc voltmeters. Provide a separate scale for each kind of current. Make one scale red and the other black.

Evaluate.

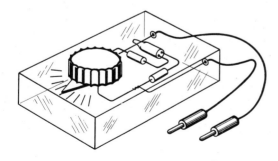

Fig. 7-47

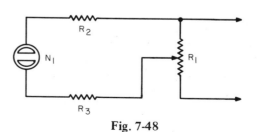

Fig. 7-48

DESIGN AND CONSTRUCT A MULTIRANGE DIRECT-CURRENT VOLTMETER AND MILLIAMMETER

This multipurpose meter has a range of 0-50 microamperes, 0-1, 0-10, 0-100, and 0-1 000 milliamperes. It also has a range of 0-1, 0-10, 0-100, and 0-200 volts, figure 7-49.

With a little imagination and a certain amount of skill, an acceptable meter can be designed and built. This particular meter is a high sensitivity instrument suitable for work in radio and electronic circuits.

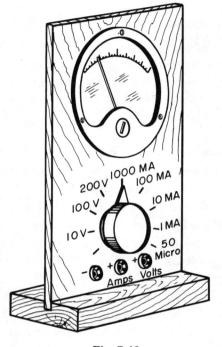

Fig. 7-49

MATERIALS

1 - Microammeter, 0-50 microamps, resistance about 2 000 ohms
    Note: Change meter scale to read 0-10
1 - Switch, 9-position rotary, shorting type — *Note:* 12-position switch with pinset for 9 is satisfactory (Mallory 1211 L)
1 - Shunt, Sh_1 - 100-ohm resistor, for 0-1 mA range
1 - Shunt, Sh_2 - 10-ohm resistor, for 0-10 mA range
1 - Shunt, Sh_3 - 104″ #30 copper wire, for 0-100 mA range
1 - Shunt, Sh_4 - 10 1/2″ #30 copper wire, for 0-1 000 mA range
1 - Resistor, R_1 - 18 000-ohm, 1/2- watt or larger for 0-1 volt range
1 - Resistor, R_2 - 200 000-ohm, 1/2- watt or larger for 0-10 volt range
1 - Resistor, R_3 - 2-megohm, 1/2-watt or larger for 0-100 volt range
1 - Resistor, R_4 - 4-megohm, 1/2-watt or larger for 0-200 volt range
1 - Panel
1 - Case
3 - Tip jacks

PROCEDURE

Study the wiring diagrams, figure 7-50(A-C), and material specifications. Obtain the parts and make a plan for assembling them on a panel.

Note: Allow sufficient space on the panel to mount it in a case.

Mount the meter, the switch, and tip jacks, figure 7-49.

Make the shunts and attach to meter and proper switch terminals, figure 7-50(A).

Note: Make all leads to the switch terminals as short as possible. If switch contact is used as a tie point, bend contact so switch does not close on this position. If switch is not left open on this position, meter will not operate according to scale as switch will act as a low-resistance shunt.

Connect series voltage resistors R_1, R_2, R_3, and R_4 as shown in figure 7-50(A).

After all connections are securely made and checked, ask your instructor to examine your work and test the meter.

Should you have to change the value of any parts, follow this procedure: To calibrate the milliammeter side of your meter, connect it in a circuit with a meter of known reliability as shown in figure 7-50(B).

Methods of Making Adjustments

Shunts — Shorten shunt wire or decrease shunt resistance to lower the needle on the scale. Connecting a resistance of higher value in parallel with shunt resistor will lower its value.

Note: The shunt must be in the circuit while the current is on or the meter will be destroyed.

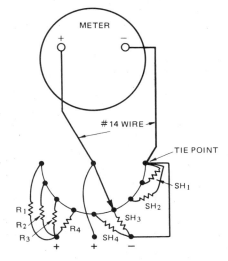

A BACK OF METER

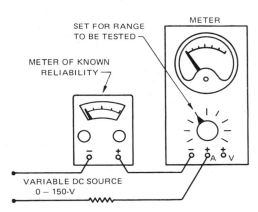

B CIRCUIT TO CALIBRATE MILLIAMETER SCALE

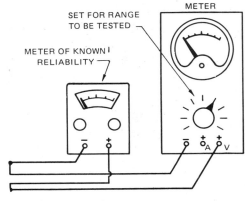

C CIRCUIT TO CALIBRATE VOLTMETER SCALE

Fig. 7-50 (A-C)

To calibrate the voltmeter side of your meter, connect it in a circuit with a meter of known reliability as shown in figure 7-50(C).

To make the needle read higher on the scale, connect one or more very high resistors in parallel with a series resistor. This will decrease the resistor's effect.

Evaluate.

A suggestion: A meter of satisfactory quality for dc power circuits can be made using any milliammeter or higher microammeter. A possible source is through the surplus property agencies.

Note: In attempting to determine the suitability of a meter, it should be noted that the meter dial calibration is not always a true indication of the meter movement. In many cases where external shunts are used, the full-scale value of the meter movement is indicated on the lower right-hand corner of the meter face. For example, FS = 50 μa dc might well be the meter movement used in the project.

Always check this rating before installing a meter in a circuit.

CAPACITOR SUBSTITUTION BOX AND CONTINUITY CHECKER

This device, figure 7-51, is a convenient source of capacitors to select from when testing for malfunctioning parts in a circuit. Included is a continuity checker, figure 7-52.

MATERIALS

1 - Mounting chassis box, size to be planned by student
1 - Rotary switch, Mallory non-shorting-type, 1 000 (S_1)
1 - Pointer knob
1 - Dry cell holder
1 - Miniature pilot light assembly for miniature screw base lamps
1 - Lamp, 1.2-volt, Mfg. type 112 (L_1)
1 - Dry cell, size D, 1 1/2-volt, (B_1)
2 - Insulated tip jacks
2 - Standard test leads
1 - Tip jack
1 - Phone tip plug
1 - Capacitor, 0.001 at 400-600 volts (C_1)
1 - Capacitor, 0.005 at 400-600 volts (C_2)
1 - Capacitor, 0.01 at 400-600 volts (C_3)
1 - Capacitor, 0.02 at 400-600 volts (C_4)
1 - Capacitor, 0.05 at 400-600 volts (C_5)
  - Capacitor, 0.1 at 400-600 volts (C_6)
1 - Capacitor, 0.5 at 400-600 volts (C_7)

1 - Capacitor, 20. at 50 volts (C_8)
1 - Capacitor, 20. at 300 volts (C_9)
1 - Capacitor, 40. at 450 volts (C_{10})

PROCEDURE

Obtain the capacitors and arrange them on a piece of blank paper. Plan the circuit connections for the capacitors and switch, and the test light and dry cell.

Determine the dimensions required for the box chassis.

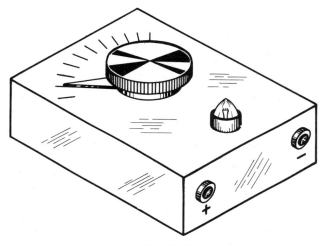

Fig. 7-51

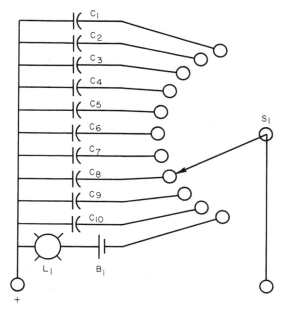

Fig. 7-52

Note: A similar device can be planned and assembled with a series of resistors, the range in value of resistors to be determined according to the average need.

If you prefer a single device and if you can select a set of five capacitors and a set of five resistors which will satisfy your needs, consider planning and assembling a combination unit as suggested in figure 7-53.

If a prefabricated chassis is not available, make a pattern layout for a chassis complete with location of holes for the switch, pilot light assembly, and tip jacks.

Make the chassis, assemble the parts, test, and evaluate.

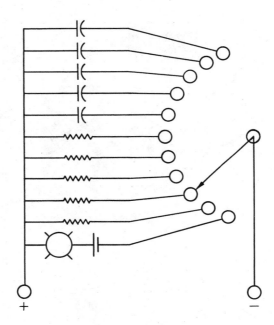

Fig. 7-53

Chapter 8
Electrical effects in liquids and gases

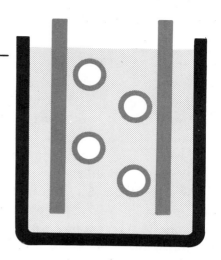

ELECTRONIC NATURE OF METALS AND METAL COMPOUNDS

Atoms of all materials contain electrons and protons. In pure elements, these negative and positive particles are equal in number in the atom, making the atom electrically neutral. Elements which have atoms with one, two, or three electrons in the outermost shell are called *metals.*

The material dug out of the earth in an iron or copper mine does not look like iron or copper. It may have useless rock mixed with it. A more important reason for this is that most metals are found combined with nonmetallic elements. The word combined means something more drastic than a simple mixing or stirring together. The combination of metal with nonmetal, called a *chemical compound,* contains atoms that are not electrically neutral. The metal atoms are positively charged; the nonmetal atoms are negatively charged, and the electrical attraction of these opposite charges holds the compound together. To separate the desired metal from the nonmetal which may not be wanted, the positive-charged metallic atoms must be supplied with enough electrons to make them neutral. This frees them from the electrical attraction that held them in the compound.

Electrically charged atoms are so much different from neutral atoms that the charged atoms are given another name: *ions.*

There are many compounds of elements which do not contain ions because there are ways of holding atoms together other than simple electrical attraction. But these other compounds need not concern us now. In a compound that consists of ions, every atom has an electrical imbalance, which causes the compound to look and act differently from the pure elements. If a small metal ball is charged by friction, a few billion electrons might easily be taken from the chunk of metal. But the metal has so many atoms that only one atom out of a billion billion is missing an electron. Therefore, the appearance and general properties of the metal ball are not changed.

Some common ionic compounds containing positive metal ions and negative nonmetal ions include ordinary salt, called *sodium chloride,* which consists of ions of sodium (a metal) and chlorine (nonmetal). Calcium chloride and magnesium chloride are saltlike materials whose names indicate what they contain. When iron rusts, iron atoms lose electrons to oxygen atoms from the air, forming iron oxide.

When ordinary salt or other saltlike material is dissolved in water, the ions attain as much freedom of movement as the water molecules themselves. (Liquids can be poured or stirred because their molecules slide by one another readily.) With equipment as shown in figure 8-1, it can be shown that the presence of charged ions causes water to become a good conductor. With ordinary water in the glass, there will be a little current, but not enough to light the bulb. When salt is added, the lamp brightens as the salt dissolves.

CONDUCTION IN LIQUIDS

Conduction in liquids differs from conduction in solids in these ways:

1. In a liquid there are no individually free electrons, as there are in a metal wire.

2. Conduction in a liquid is a two-way movement of charged ions, which are much bigger than electrons. Positive ions drift toward the negative wire; negative ions at the same time are attracted toward the positive wire. This movement is the electric current in the liquid, figure 8-2.

The opposite movement of the ions separates the two parts of the originally dissolved compound; this electrical separation is called *electrolysis.* Any conducting liquid is called an *electrolyte.*

Ions of a solid can be given freedom of movement by heating the solid until it melts to a liquid. Lightweight metal magnesium is made commercially by the electrolysis of melted magnesium chloride.

Similarly, all aluminum is made from aluminum oxide by electrolysis. Aluminum oxide does not dissolve in water, nor does it melt readily. However, it will dissolve at a high temperature in a melted mineral-like mixture of fluorides, called cryolite, figure 8-3. The term *cathode* is given to the electrode terminal where electrons enter a device. The *anode* is the electrode where they leave. Aluminum ions (positively charged) pick up electrons at the cathode and become ordinary metallic aluminum; oxygen ions (negatively charged) attracted to the carbon anode combine with the carbon, burning it away.

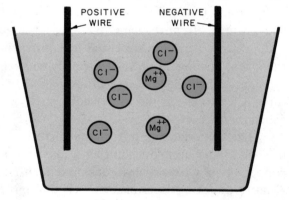

Fig. 8-2 Movement of Ions in a Liquid

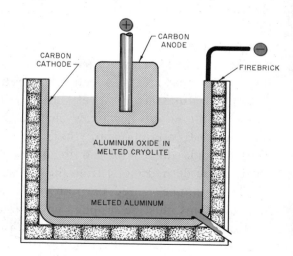

Fig. 8-3 Production of Aluminum by Electrolysis

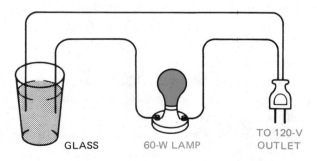

Fig. 8-1 Liquid Conduction

Electroplating

In any *electroplating* process, copper, silver, nickel, etc., the solution must contain ions of the metal that is to form the coating. Metal ions are all positive, so the object to be plated is connected to the negative wire. The positive terminal is usually made of the same metal that is to form the coating. Plating processes are devised and controlled by chemists rather than by electricians. Plating solutions contain other ingredients than the dissolved metal compound. They are added to prevent corrosion of the work to be plated, prevent poisonous fumes, and aid in forming a smooth coating.

Iron objects are often first plated with copper, then with a finished surface of nickel, chromium, or silver. In the commercial electroplating of copper onto iron, a poisonous compound called cuprous cyanide is used to make the plating solution (the electrolyte). The object to be plated, which must be an electrical conductor, is connected to the negative terminal of a dc source and immersed in the copper cyanide solution. The positive terminal of the dc supply is connected to a copper bar in the solution. When the circuit is complete, positively charged copper ions in the liquid move toward the object to be plated. When these copper ions touch the negatively charged object, they pick up electrons and become neutral atoms of ordinary copper, forming a copper coating over the negative object. The diagram, figure 8-4, shows copper ions marked Cu, which is the chemical symbol for copper. The cyanide ion is made of an atom of carbon, C, and an atom of nitrogen, N, hence the symbol CN for cyanide.

When the cyanide ions bump into the positive copper bar at the right, copper atoms on the surface lose electrons and become positive ions. The negative cyanide ion (CN-) helps push an electron off a copper atom, making it a copper positive ion, which is then attracted into the solution by the negative ions. The electrons lost by the copper atoms on the surface of the bar drift toward the dc generator. Just as many copper ions dissolve from the copper bar as are plated onto the object to be plated, so the solution stays at constant strength.

Sometimes another copper compound, copper sulfate, is used in copper plating.

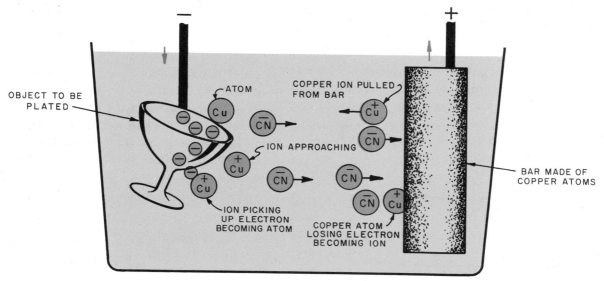

Fig. 8-4 Electroplating Unit

Practically all the copper used commercially has been purified by an electroplating process using copper sulfate solution. Bars of impure copper from the smelting furnaces are used as anodes (positive) which dissolve during the plating; pure copper is formed on the negative plates. Impurities either stay in the solution or do not dissolve. The copper wire industry has made electrolytically purified copper necessary, because impure copper is high-resistance copper. Zinc and some other metals are purified in similar fashion.

CONDUCTION IN GASES

Under proper conditions, vapors or gaseous elements can become conductors. Fluorescent lights, neon signs, electric arcs, and some types of electronic tubes make use of gaseous conduction.

Gases as Insulators

At atmospheric pressure, air and other gases are as close to being ideal insulators as any known material. With air or other gases as the insulator between two charged plates, figure 8-5, the current is practically zero while a moderate voltage is applied. At sufficiently high voltage, the gas suddenly becomes a conductor, with the amount of current limited only by the rest of the circuit.

At pressures lower than atmospheric pressure, this breakdown in insulating ability occurs at a lower voltage. If the pressure is increased, as in compressed air, the applied voltage necessary to start conduction has to be increased proportionally. For example, a higher voltage is required to fire a spark plug

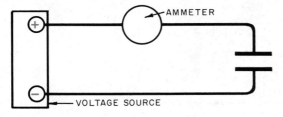

Fig. 8-5 A Test for Insulating Qualities of a Gas

under compression in an engine than is required to create the spark in the open air.

How Conduction in a Gas Starts

If the gas is inside a glass tube at low pressure, with wires sealed into the tube, the application of just moderately high voltage (generally 50 volts or more) is enough to start conduction. In the air, a useful arc is started by touching, then separating, a pair of conducting contacts. An example is a welding arc or carbon arc.

Gas becomes a conductor through *ionization*. Electrons are torn away from gas atoms, causing the gas atoms to become positively charged ions. The loosened electrons are highly movable, and conduction in the gas is due mainly to electron flow (as in metals). Movable positive ions do exist (as in liquids). Even so, the nature of conduction in a gas differs greatly from conduction in a solid or liquid.

How Ionization Occurs

(1) The most important process involved in ionization is *electron impact*. The process may be started by an electron torn free from the negative wire, or more likely by an electron set free from an atom by processes described under steps (2) and (3) which follow. Once an electron is set free, it picks up speed as it is repelled by the negative wire and attracted toward the positive wire. If it bumps into an atom while it is still moving slowly, it merely bounces off and again starts toward the positive terminal. When it is moving fast enough, it will collide with an atom violently enough to knock one or more electrons loose from the atom, figure 8-6. Then there are two loose electrons that start accelerating toward the positive wire. They collide with atoms. Each jars another loose, resulting in four loose electrons. Their collisions loosen more electrons, and in a thousandth of a second, a million electrons may have been released by these collisions.

As a comparison, picture what would happen in an apple tree heavily loaded with apples, in the following situation. An apple will be jarred loose if hit by another apple that has fallen 6 inches. Accidentally, one apple is jarred loose near the top of the tree. The chain reaction causes a bushel of apples to hit the ground.

Gaseous conduction produces light as a result of collisions that are not violent enough to tear electrons loose. A moderately fast-moving electron, by collision, may give some of its energy to the electrons of the atom, producing an *excited state* in the atom. This energy-loaded excited atom usually gets rid of its energy by emitting light. The color of the light is characteristic of the energy state of the atom.

(2) A second process that ionizes gas is *radiation*. Cosmic rays ionize a few atoms, and this is often the source of the first electron to initiate the collision process. After conduction has started, electron disturbances in atoms produce not only visible light, but also much higher-frequency radiation (ultraviolet). This radiation is absorbed by other atoms, giving them enough energy to cause electrons to separate.

(3) Another ionizing process is *heat*. A hot gas conducts better and starts conducting more easily than a cold gas. Heat is movement of atoms and molecules. At a high temperature, collisions of atoms may become violent enough to dislodge electrons. Many of the individual atoms are at higher temperature than the average, and these temporarily speeding atoms cause ionizing collisions. An ordinary flame contains many ions.

Low-Pressure Gas Conducts Better Than High-Pressure Compressed Gas

In gas at atmospheric pressure, a free electron collides with gas atoms so frequently that it has no chance for enough unhindered travel to pick up enough speed to ionize an atom by collision, figure 8-7.

At lower pressures, gas molecules are farther apart, giving more room and opportunity for an electron to gain enough kinetic energy for an ionizing collision, figure 8-8. Gas tube designers are, therefore, concerned with calculating the *mean free path* of electrons, meaning the average distance traveled by a particle between collisions. This path must be long enough to permit the electrons

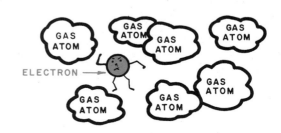

Fig. 8-7 Gas at High Pressure Hinders Conduction

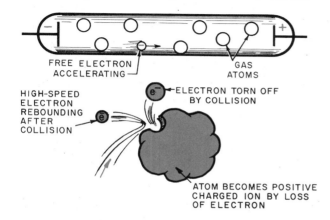

Fig. 8-6 Ionization of Gas

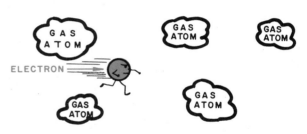

Fig. 8-8 Gas at Low Pressure is a Good Conductor

to pick up enough speed between collisions so that the collisions will be violent enough to ionize the gas. Speed attained also depends on voltage. Hence, the application of higher voltage will make ionization more likely to occur.

CONDUCTION AND IONS IN NATURE

The frequently visible Northern Lights (aurora borealis) are believed to be due to conduction in the very thin upper part of the atmosphere. This display seems to be caused by electrons given off by disturbances on the sun, traveling in toward the earth.

High-speed protons coming toward the earth from outer space strike air molecules and produce intense radiation that ionizes various layers of the atmosphere. These conducting layers are able to reflect radio signals, reflections of which are often useful.

Lightning is a high-voltage spark. Sometimes, conditions not violent enough to produce lightning cause a continuous discharge from steeple tops or ships' masts. A strong electric field ionizes the air at sharp-pointed electrodes; the discharge is visible if other illumination does not interfere. This type of discharge, called corona, occurs in all sorts of high-voltage equipment and can be a serious source of power loss in high-voltage transmission lines.

Conduction in Vacuum

If gas is pumped from a glass tube until the pressure in the tube is reduced to approximately that of 0.001 millimeters (760 millimeters of mercury = atmospheric pressure), the brilliant glowing of the conducting gas is no longer seen. Gas molecules are so far apart that few ions are formed, and conduction consists almost entirely of electron flow.

Years ago, those experimenting with conduction in gases and vacuum had no idea of the existence of electrons. They gave the name *cathode rays* to what was coming from the cathode in their tubes. In 1869, Hittorf, a German, described the glowing of glass struck by these rays and showed by experiments that the rays traveled in straight lines. Crookes, an Englishman, found that ray paths could be bent by a magnet, they could be focused, the rays heated the objects they struck, and speed of the rays depended on the applied voltage. Perrin, a Frenchman, in 1895 showed that the rays carried a negative charge. By 1897, J.J. Thompson, an Englishman, completed a series of experiments in which he was able to measure the ratio of weight to electric charge of the negative particles. From this work, Thompson is credited with finalizing the discovery of electrons; tiny, lightweight, negative-charged particles, present in all materials, figure 8-9.

In 1895 Wilhelm Roentgen, experimenting with cathode-ray tubes, found that when the cathode rays struck metal or glass, a new kind of radiation was produced. These new invisible rays passed through air, paper, and wood very well. They passed through thin metal better than through thick, through flesh better than through bone, and through aluminum better than through lead. The rays caused fluorescence in minerals, affected photographic plates, and were not bent by magnets. Roentgen called them *X rays*. Within a few months, they were used in the field of medicine.

Present day cathode-ray tubes used in oscilloscopes and TV picture tubes use the same principles of electron-ray deflection that Thompson used in his experiments.

In oscilloscopes, the electron beam accelerated from the cathode is deflected horizontally at a known rate. The voltage whose trace is to be observed is used to deflect the beam vertically at the same time, thus producing a time graph of the observed voltage, figure 8-10 and page 164–168.

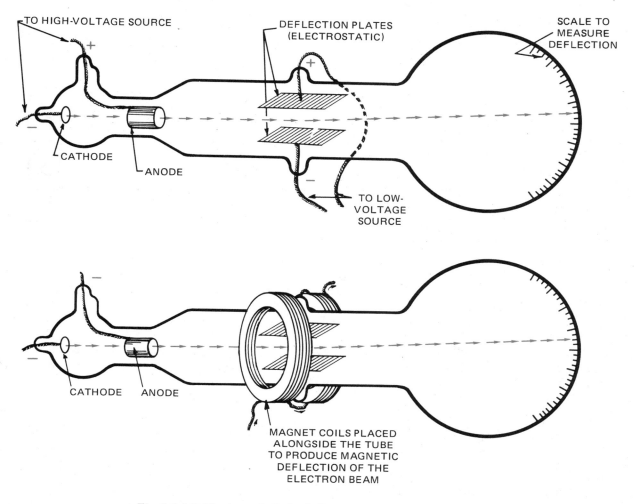

TO HIGH-VOLTAGE SOURCE

+

DEFLECTION PLATES (ELECTROSTATIC)

+

SCALE TO MEASURE DEFLECTION

CATHODE

ANODE

−

−

TO LOW-VOLTAGE SOURCE

−

CATHODE   ANODE

MAGNET COILS PLACED ALONGSIDE THE TUBE TO PRODUCE MAGNETIC DEFLECTION OF THE ELECTRON BEAM

Fig. 8-9 J.J. Thompson's Cathode-Ray Tube, 1896 (Simplified)

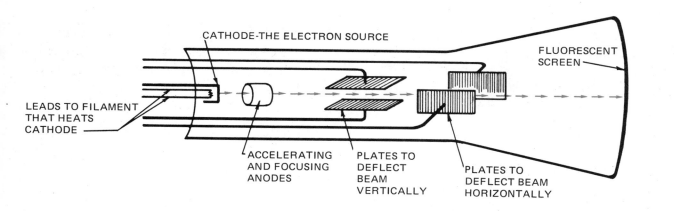

CATHODE-THE ELECTRON SOURCE

FLUORESCENT SCREEN

LEADS TO FILAMENT THAT HEATS CATHODE

ACCELERATING AND FOCUSING ANODES

PLATES TO DEFLECT BEAM VERTICALLY

PLATES TO DEFLECT BEAM HORIZONTALLY

Fig. 8-10 Cathode-Ray Tube for Oscilloscope

In TV picture tubes, magnet coils are used to sweep the beam horizontally and vertically. The intensity of the beam is charged to produce lights and darks in the picture.

There has been some concern about the possibility of harmful X rays being produced by the 20 000-volt electrons that strike the face of the TV picture tube. This voltage is not high enough to cause highly-energetic and penetrating X rays. Recent experiments have shown that if any X rays are produced, they are all absorbed in the glass and in the first inch of air in front of the tube.

Thomas Edison had a chance to discover electrons, but failed. One of his assistants accidentally connected a meter to a dead-end wire sealed into one of Edison's experimental electric lamps and found a small current, figure 8-11.

The heated lamp filament was emitting electrons, but nobody knew that at the time (1883). The loose electrons, attracted to the positive wire through the meter, caused the current. This event was reported. However, the discovery was not used until 20 years later when Fleming constructed a two-element valve to use as a rectifier. (The two elements are the filament, which emits electrons, and the plate, which catches them when the plate is positive. *Valve* is the British term for vacuum tube.)

POINTS TO REMEMBER

- Electrical conduction in liquids is the movement of both positive and negative ions. There are no individually loose electrons as in metal.

- Positive and negative ions are formed by a type of chemical combination in which one metallic element (or group) transfers electrons to a nonmetallic element (or group).

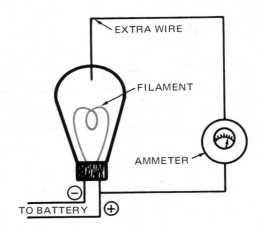

Fig. 8-11

- If the above compound can dissolve in water, these charged ions become freely movable in the solution.

- In electroplating processes, the solution contains ions of the plating metal. The article to be plated is connected to the negative terminal of the current source. A bar of the plating metal is connected to the positive terminal.

- Metal taken from the plating solution to form the plate is replaced in the solution by metal dissolved from the positive bar.

- Liquid conduction moves positive ions one way, negative the other, resulting in a permanent separation of the parts of the compound. This decomposition is called electrolysis.

- Gases are good insulators, within limits. When sufficient voltage is applied, sudden ionization changes them to conductors.

- Ionization of a gas is the loosening of electrons from gas molecules. The positive-charged gas atoms (or molecules) are the positive ions. The electrons are the movable negative particles.

- Conduction in a gas consists mainly of electron movement. As electrons collide with gas atoms, a continual new supply of electrons is shaken free from the atoms.

- Gases at low pressure conduct more rapidly than gases at high pressure.

- At low, near-vacuum pressures, electrons from the cathode travel out in straight paths, originally called cathode rays. The discovery of electrons resulted from the measurement of their properties in cathode-ray tubes.

REVIEW QUESTIONS

1. How do metals differ from nonmetallic elements in the structure of their atoms?

2. Most metals are found combined with nonmetallic elements. What holds such a combination of elements, called a compound, together?

3. What are ions?

4. What would happen if the wires leading from the anode and cathode to the source of current in a copper electroplating vat were reversed?

5. Name two ways in which conduction in a liquid differs from conduction in a solid.

6. Define electrolysis.

7. What is meant by the terms cathode and anode?

8. Why is an object to be electroplated connected to the negative terminal of the current supply?

9. How is copper used in the copper-wire industry made pure?

10. Why is a higher voltage required to fire a spark plug under compression in an engine than is required to fire a similar spark in the open air?

11. Explain the changes necessary for gas to become a conductor of electricity.

12. How does ionization in gases occur?

RESEARCH AND DEVELOPMENT

Experiments and Projects on Electrical Effects in Liquids and Gases

INTRODUCTION

Electrons can move in a liquid, in gases which are normally insulators, and in a vacuum. After studying these phenomenon in Chapter 8, it would be well to review Chapter 5 before doing the experiments which follow. The projects are of the experimental and construction type. Some of the projects will require considerable research in the science of chemistry as well as electricity.

EXPERIMENTS

1. To observe the conduction of a liquid solution.
2. To learn how to plate a metallic object with copper.
3. To observe the effects of electrons passing through carbon vapor.

PROJECTS

1. Set up an electroplating unit and plate an object.
2. Plan and build an anodizing unit. Anodize a piece of aluminum.
3. Plan and build a low-voltage electric-arc device using carbon rods for electrodes.

EXPERIMENTS

Experiment 1 CONDUCTION IN LIQUIDS

OBJECT

To observe the conduction of a liquid solution.

APPARATUS

1 - Glass or plastic jar
1 - Lamp receptacle
1 - Lamp, 40- to 60-watt
2 - Conductors - sheet copper
2 - Brass screws and nuts
1 - SPST switch
   Distilled water, tap water, salt, kerosene oil.

PROCEDURE

1. Assemble equipment, figure 8-12.
2. Fill jar with kerosene oil.
3. Connect to 120-volt line. *Caution:* Pull the plug before handling any metal parts.

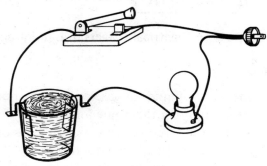

Fig. 8-12

4. Close switch and observe results.

5. Put some table salt in the kerosene. Observe results.

6. Turn off current.

7. Place conductors in a jar of distilled water.

8. Close switch and observe results.

9. Add quantity of table salt.

10. Stir and observe results.

11. Turn off current.

12. Place conductors in a jar of tap water.

13. Close switch and observe results.

OBSERVATIONS

Use a table similar to the one shown here to record your observations.

CONDUCTION IN	LIGHT ON	LIGHT OFF	INTENSITY OF LIGHT
KEROSENE			
SALT AND KEROSENE			
DISTILLED WATER			
SALT AND DISTILLED WATER			
TAP WATER			

QUESTIONS

1. Why is water containing salt a better conductor than distilled water?

2. What was the action of the ions in the solution?

3. Explain the results of the experiment when kerosene oil was used.

4. What happened when the conductors were placed in tap water?

5. What causes electron flow (current) in a liquid?

Experiment 2 ELECTROPLATING

OBJECT

To plate a metallic object with copper.

APPARATUS

1 - Ceramic or glass container
1 - Dry cell or storage battery
1 - Rheostat
1 - DC ammeter (0-30 A)
1 - Copper bar

1 - Metal object
1 - SPST switch
Plating solution: cleansing solution
1 - Metal object to be plated

PROCEDURE

1. Assemble equipment, figure 8-13.

2. Suspend anode (copper bar) in plating solution.

3. Thoroughly clean object to be plated.

4. Check all connections in circuit.

5. Close the switch.

6. Suspend cathode (object to be plated) in plating solution.

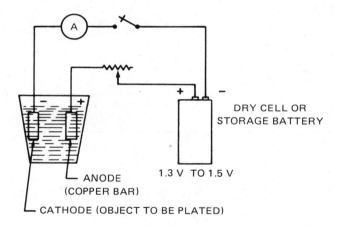

ANODE
(COPPER BAR)

CATHODE (OBJECT TO BE PLATED)

DRY CELL OR
STORAGE BATTERY

1.3 V TO 1.5 V

Fig. 8-13

7. Regulate current supply.

8. Time plating operation.

9. Turn off current.

10. Remove object, wash, and examine.

Note: The current should be turned on before the object is immersed in the solution to avoid chemical replacement problem. Copper sulfate solution will coat some metals when immersed without current.

QUESTIONS

1. What would happen if the wires leading to current supply were reversed?
2. What are ions?
3. Can a piece of wood be electroplated?
4. Explain why the current should be turned on before an object to be plated is placed in a copper sulfate solution.

Experiment 3 OPERATING A CARBON ARC

OBJECT

To observe the effects of electrons passing through carbon vapor.

APPARATUS

1 - Carbon-arc device (page 212–213)
1 - Storage battery, 6-volt or equivalent dc supply
1 - SPST switch
    Arc welding shield or equivalent
    Safety glasses

PROCEDURE

1. Assemble the components according to figure 8-14.
2. Adjust the carbons leaving about 1/8″ (3 mm) between the tips.

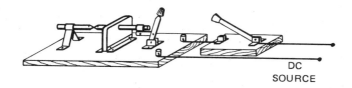

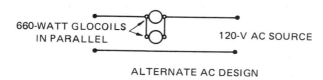

ALTERNATE AC DESIGN

Fig. 8-14

3. Protect your eyes with the welding shield or safety glasses and close the switch.
4. Move the adjustable carbon rod until it touches the other and produces an arc.
5. Adjust the distance between the rods until a brilliant arc is created.

OBSERVATIONS

Use a table similar to the one shown here to record your observations.

POSITION OF CARBONS	DEGREE OF LIGHT
TOUCHING	
ABOUT $\frac{1}{16}''$ (1.5 mm) APART	
ABOUT $\frac{1}{8}''$ (3 mm) APART	
ABOUT $\frac{3}{16}''$ (4.5 mm) APART	
ABOUT $\frac{1}{4}''$ (6 mm) APART	

QUESTIONS

1. In what ways does light from a carbon arc differ from that of an incandescent lamp?

2. Why is direct current commonly used in carbon-arc equipment?

3. Is the adjustment of the distance between the carbon rods critical? Explain.

4. What causes eye irritation when the naked eye is exposed to an arc?

PROJECTS

DESIGN AND BUILD AN ELECTROPLATING UNIT

For many years the electroplating process has been employed to deposit a thin coat of nonferrous metal on another metal. A thin coat of metal such as nickel, chromium, or silver serves to protect and to beautify the plated metal. This same electrochemical process is used to purify copper and other metals.

MATERIALS

1 - Tank - ceramic, glass, or an old storage battery case
1 - DC ammeter (0-15 A)
1 - DC voltmeter (0-5 V)
1 - Rheostat
1 - SPST switch
1 - Base, wood
1 - Panel, hardboard
4 - Binding posts or Fahnestock clips
2 - Support bars, hardwood or plastic

When mixing acid and water, always pour the acid into the water. Wear safety glasses and avoid getting acid on your hands and clothing.

PROCEDURE

Study the drawing, figure 8-15, and materials list.
Procure the parts, such as a meter, rheostat, and tank, etc.
Plan the size of the unit according to your equipment.
Make dimensioned drawings of the part you need to make.
Make the parts, assemble the unit, and evaluate.

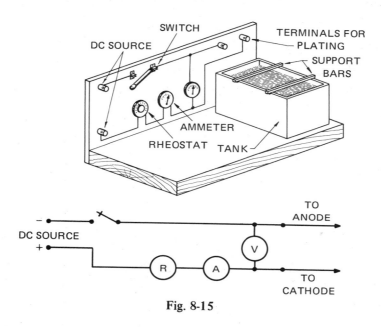

Fig. 8-15

PLAN AND BUILD AN ALUMINUM ANODIZER

The aluminum anodizer is a device used to form, chemically and electrically, a thin coating of oxide on the surface of the metal.

Anodizing beautifies the metal and preserves the finish. An anodizer is a comparatively simple equipment item to design and build.

MATERIALS

1 - Container - glass or plastic, 1-gal.
1 - DC ammeter (0-25 A)
1 - DC voltmeter (0-25 V)
1 - SPST switch
1 - Base, wood
1 - Panel, hardboard
1 - Container cover, wood
2 - Alligator clips, heavy duty
4 - Binding posts, heavy duty
1 - Thermometer
  Sheet lead
  Aluminum wire

PROCEDURE

Study the drawings, figure 8-16, and list of materials.

Obtain the parts, such as meters, container, etc.

Design the base, the back, and the cover for the jar.

Make dimensioned drawings for these parts.

Make the parts.

Assemble the unit.

To prepare an aluminum article for anodizing, all deep surface scratches should be removed with abrasive cloth. The aluminum should then be cleaned with a solution prepared by dissolving 4 oz. (93 g) of caustic soda (sodium hydroxide) in 1 gal. (3.75 L) of water. An alternative (and less dangerous) cleaning agent can be made by mixing 5 oz. (120 g) of Oakite and 3 oz. (70 g) of baking soda (NA_2CO_3) in a gal. (3.75 L) of water. These solutions should be heated to $160°F. - 180°F. (71°C - 82°C)$ for use.

Following the cleaning, the parts should be rinsed in a strong soap solution, carbon tetrachloride, or benzine. The parts should be firmly attached to the racks and dipped in a caustic soda solution that is maintained

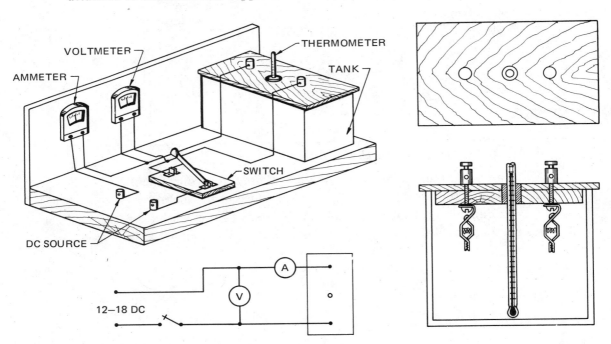

Fig. 8-16

at a temperature of $170°F. (77°C)$. After both the parts and rack have been rinsed, they may be placed in the jar containing the sulfuric acid electrolyte.

The electrolyte is prepared by adding one part of concentrated sulfuric acid (H_2SO_4) to eight parts of water by volume. The specific gravity of this solution should be 1.000-1.125 as checked out with a hydrometer. An oxalic acid solution of 40 g/liter may also be used for the electrolyte.

> Always add acid to water slowly and carefully. Avoid splashing the acid and allow solution sufficient time to cool. Do all mixing in Pyrex glassware because the heat produced may crack other types of glassware. Use a suitable rubber apron and gloves, and wear acid-resistant goggles while mixing and cleaning.

The anodizing process is initiated by applying an electric current of 12-18-volts dc.

The current will range from 1/2 to 20 amperes, depending on the surface area of the parts to be anodized. The current required is usually about 15 to 20 amperes per square foot (1 000 square centimeters). Since some large articles require many amperes, the contact between the aluminum and the holder must be large enough to supply sufficient current. Another consideration is the size of the sheet lead cathode. It must have at least an area as large as the object being anodized. If the lead plate is adjacent to the wall of the tank, it must be twice as large as the article. The time involved will be between 10 and 30 minutes, depending on the thickness of the coat desired.

The anodized parts should be cleaned with running water to remove every trace of acid.

The final coating and/or dye coloring may be applied as desired. The dye solution may be made from commercially available water-soluble dyes. A package of these dyes mixed in one pint (.5 liters) of water provides a good concentration to start experimentation with the depth of tones.

A more lasting finish may be assured by sealing and drying. The sealing solution may be prepared by adding one teaspoon of nickel acetate to a quart (liter) of water. The covering provided by this solution will insure colorfastness and will deter fading. An alternate method would be to use vinegar or acetic acid.

DESIGN AND BUILD A CARBON-ARC LIGHT

This simple carbon-arc light can be used to observe the concentrated brilliance of light created by a carbon arc.

It can also show the intensity of the heat produced by the current.

MATERIALS

Transite board, 1/4" x 2 1/2" x 2 1/2" (6 mm x 65 mm x 65 mm)
Sheet steel, #18 or #20 gauge
1 - SPST knife switch
8 - Wood screws, #6 x 1/2" RH
2 - Wood screws, #6 x 3/4" RH
3 - Machine screws, #6-32 x 1/2" (M3.5 x 12 mm L) RH steel
4 - Machine screws, #6-32 x 3/8" (M3.5 x 9 mm L) RH steel
7 - Machine screw nuts, #6-32 (M3.5) RH steel
2 - Binding posts or Fahnestock clips
2 - Carbon rods, 1/4" or 5/16" x 2" to 3" long (6 mm or 8 mm x 50 mm to 75 mm long)
1 - Base, wood (size to be determined according to individual design)
1 - Wire nut

PROCEDURE

Study the sketches, figures 8-17 and 8-18. Make a full-scale drawing of each part in the size desired.

Plan the major operations necessary to produce each part.

Make the parts. Assemble the device, have it approved by your instructor, and test.

> When experimenting with a carbon-arc device, always wear dark colored glasses or use an arc welder's shield. The concentrated brilliance of the light produced by a carbon arc is harmful to the eyes.

The carbon-arc welder, used for hard soldering and brazing, utilizes the intense heat produced by the carbon arc.

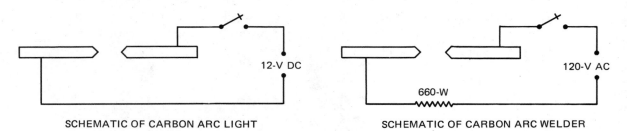

SCHEMATIC OF CARBON ARC LIGHT   SCHEMATIC OF CARBON ARC WELDER

Fig. 8-17

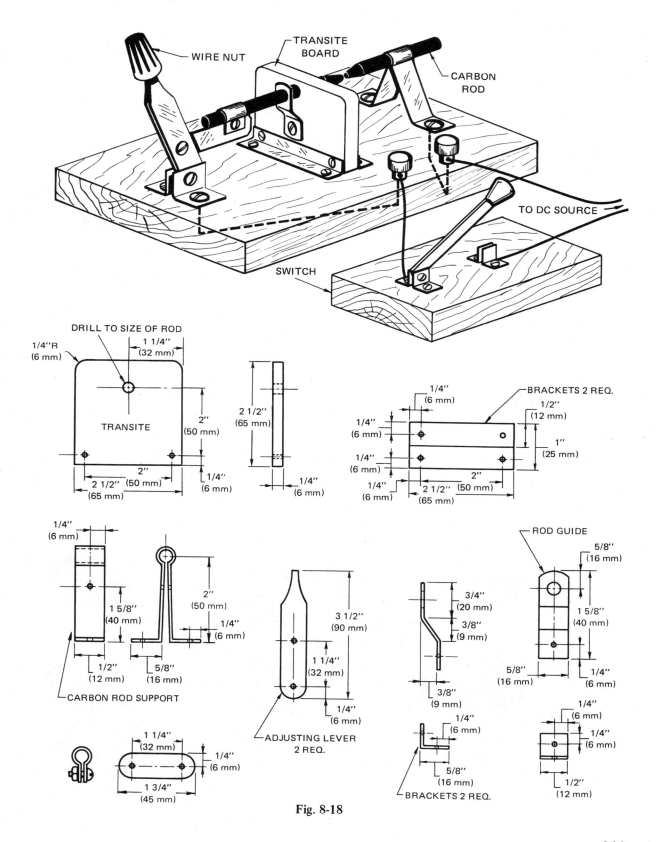

Fig. 8-18

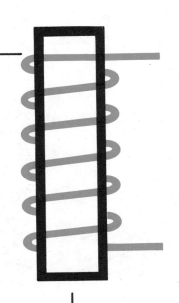

Chapter 9
Induced voltages

ENERGY: MECHANICAL TO ELECTRICAL

The principal source of electrical energy for industry is the *generator,* a device for changing mechanical energy of motion into electrical energy. This energy conversion takes place by the action of magnetic forces, hence the name *electromagnetic.*

An *induced voltage* or *induced current* is one produced by the action of magnetic forces, as contrasted with voltages produced by chemical action or other methods. Chapter 6 described the production of magnetic force by electric current. The effect was first noticed by Hans Oersted in 1819.

During the next few years, many experimenters tried to find out more about the relation of magnetism and electricity. Knowing that electric current can produce magnetism, they tried to see if magnetism could produce electric current. Many experimenters carefully wound a coil of wire around a magnet and waited for the magnet to produce a current in the wire, but nothing happened.

Fig. 9-1 Cutting Slots in Rotor for Alternator (Courtesy General Electric Co.)

In the early 1830's, Michael Faraday and Joseph Henry found out how to produce electric current by using magnetism. At the time, neither man knew that the other was working successfully on the same problem. They discovered this:

> In order to produce electric current in a wire, or preferably, a coil of wire, there must be *motion* of the coil or the magnetic field.

A simple way to demonstrate this induction process is shown in figure 9-2. A piece of copper wire is connected to the terminals of a sensitive meter and moved downward through a magnetic field. The wire cuts across the lines of force. A strong horseshoe magnet (preferably alnico) should be used. The meter may be either a milli-voltmeter, milliammeter, microammeter, or galvanometer, preferably a zero-center type.

While the wire is moving, a voltage or emf is produced. The emf tends to drive electrons from A toward B, figure 9-2. This emf, induced by the movement of the wire across the field, causes a current if a complete circuit exists. If the magnet and wire are kept stationary, no emf is produced. Movement is necessary.

When the wire is moved upward through the magnetic field, figure 9-3, the meter needle is deflected in the direction opposite to its previous motion. This shows that the induced emf and induced current have been reversed in direction.

When the wire is moved lengthwise through the field, as from A to B and back again, no emf is produced. If the wire is moved in a direction parallel to the lines of force, as from south toward north or from north toward south, no induced emf is generated. The wire must move so that it cuts across the lines of magnetic force. This cutting is a quick way of describing the motion that must occur if any voltage is to be produced. The field is just as strong after the wire has passed through as it was before. Remember that the term "cutting the lines of magnetic force" is only a symbolic expression.

Where does the electrical energy come from? In the demonstration shown in

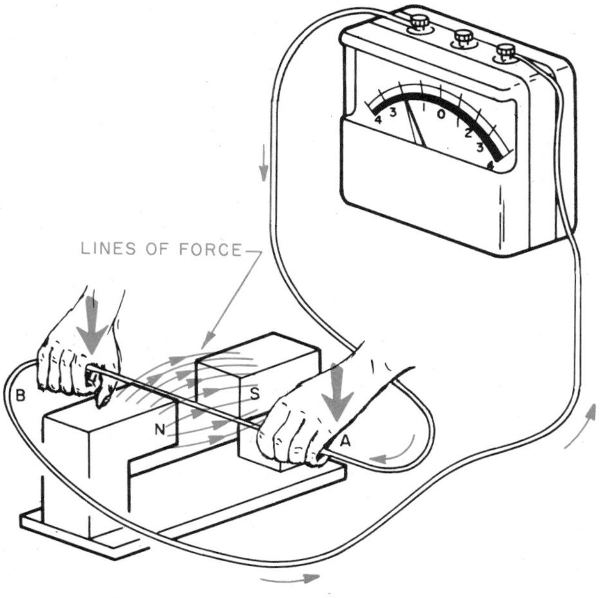

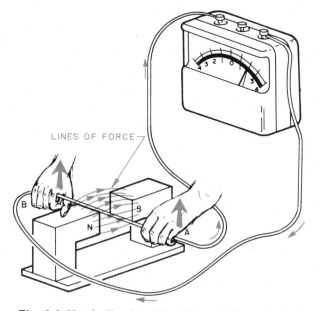

Fig. 9-2 Hands Forcing Wire Downward Through Magnetic Field

Fig. 9-3 Hands Forcing Wire Upward Through Magnetic Field

figures 9-2 and 9-3, the electrons are already in the wire. The energy to move them comes from the person performing the demonstration. The mechanical energy used to push the wire through the field is converted to electrical energy. Mechanical energy must be put into the generating device at the same rate (watts) as electrical energy is produced.

This energy-conversion process used in generators is a reversible process. The same equipment can change electrical energy to mechanical motion. If we connect the wire in figure 9-2 to a 12-volt battery instead of to a galvanometer, the current in the wire will cause the wire to be thrown upward or downward through the magnetic field. The direction of motion depends on current direction.

A rotating machine that uses mechanical energy and produces electrical energy is called a generator. The same machine can act as a motor, putting out mechanical energy when it is driven by electrical energy. The old term, *dynamo,* means an energy converter, usable as either a generator or a motor.

LEFT-HAND GENERATOR RULE

The relation of direction of motion of wire in a field to the direction of emf induced may be briefly described this way: with the thumb, index finger and middle finger of the left hand placed at right angles to each other, figure 9-4, the index finger (or first finger) gives the direction of the field, the thumb gives the direction of motion of the wire, and the center finger gives the direction of the induced current.

This rule is not an explanation. It is merely one of the ways of determining one of the above directions when the other two are known.

Figures 9-2 and 9-3 show the generation of an emf by moving a wire so that it cuts across a magnetic field. It is just as practical to produce emf by moving the magnetic field so that it cuts across stationary wires.

An emf is produced in a stationary coil, figure 9-5, when a bar magnet is withdrawn from or inserted into the coil. The lines of force, moving along with the magnet, cut across the wires of the coil.

With several magnets of various strengths available, one may show that a greater voltage is produced by a stronger magnet.

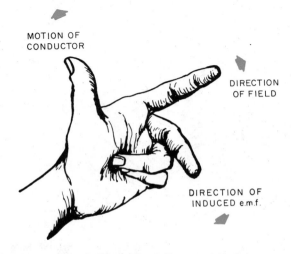

MOTION OF CONDUCTOR

DIRECTION OF FIELD

DIRECTION OF INDUCED e.m.f.

Fig. 9-4 Left-Hand Generator Rule

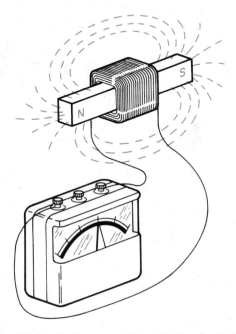

Fig. 9-5 Producing an emf in a Stationary Coil

By trying the effect with several coils differing in number of turns, it is seen that more turns produce more emf. This last effect may be explained in either of two ways:

1. A coil of ten turns has ten loops that are in series with each other. If one volt is produced in each turn, a total of ten volts is produced in ten turns.

2. The amount of voltage depends on the amount of cutting of lines of force. One line of force cutting across ten wires does as good a job as ten lines cutting across one wire. The total amount of cutting across is the same.

While the above equipment is available, it is easy to demonstrate that moving the magnet slowly produces a small emf and faster motion causes more emf.

The amount of induced emf depends on the rate the lines of force are cut with wires. The induced emf in a coil is proportional to the product of these three factors which determine the rate of cutting: (1) *number of lines of force;* (2) *number of turns of wire;* and (3) *speed of movement* (relative motion of wires and field).

The relationship of voltage measurements to magnetic field measurements is this:

One volt is produced when 100 000 000 lines of force per second are cut by wire.

If 50 000 lines of force are cut by a coil of 2 000 turns in one second, the total cutting is 50 000 x 2 000 = 100 000 000 lines cut per second. One volt is induced.

If 300 000 lines of force are cut by a coil of 5 000 turns in two seconds, the total cutting is 300 000 x 5 000 = 1 500 000 000 lines in two seconds. That is a rate of 750 000 000 lines cut each second, which, being 7 1/2 times as much as 100 000 000, produces 7 1/2 volts.

Figure 9-6 shows the withdrawal of the magnet from the coil. The magnetic field of the north pole is moving to the right, cutting across the wires of the coil. To determine the direction of induced emf by using the left-hand rule (figure 9-4), remember that the hand rule is based on relative motion of the wire. Pulling the magnet to the right is equivalent to moving the coil of wire to the left. To use the hand rule, the thumb must therefore point to the left, representing the relative motion of the wire through the field. To find the direction of emf over the top of the coil, the thumb points to the left, the first finger (for field) upward, and the center finger then gives the current direction toward the observer at the top of the coil. At the bottom of the coil, the thumb still points to the left. The field is downward and the center finger points away from the observer, giving the current direction around the coil as shown by the arrows on the wire in the diagram.

LENZ'S LAW

There is a fundamentally more important way of determining the current direction that may be described with the help of figure 9-7, page 218. Electrical energy is produced at

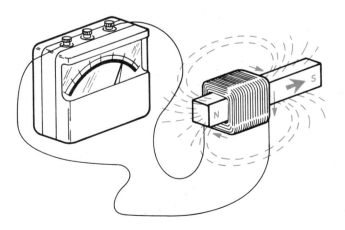

Fig. 9-6 Direction of emf in Relation to Movement of Magnet

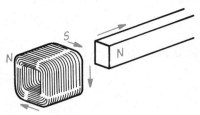

Fig. 9-7

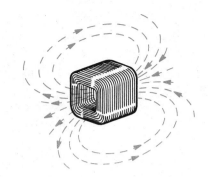

Fig. 9-8 Magnetic Field About a Coil

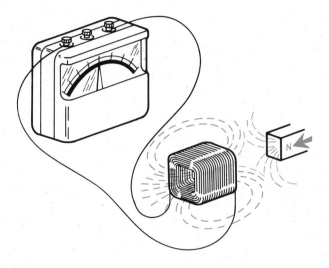

Fig. 9-9

the expense of mechanical energy; the hand that removes the magnet from the coil must do some work. The magnet does not push out of the coil by itself, someone must pull it out.

The coil makes it difficult to remove the magnet by creating magnetic poles of its own. These poles tend to draw the magnet into the coil. Figure 9-7 shows the magnet a little more removed from the coil. The induced current in the coil is in a direction which develops poles on the coil as shown (recall left-hand rule for a coil). The poles try to pull the magnet toward the coil.

If the motion of the magnet is reversed, that is, pushed into the coil, the induced current reverses also. This develops poles on the coil that try to repel the approaching magnet.

This general idea, recognized years ago by Lenz, is summarized in *Lenz's Law,* which states:

> An induced current opposes the motion that causes it.

Figures 9-5 to 9-7 can also be used to illustrate another view of Lenz's Law, which will be useful in determining induced current direction in more complex machinery. Note that in figure 9-5 the coil is surrounded by the magnetic field of the bar magnet. The field is shown in figure 9-8.

The removal of the magnet is removal of this magnetic field. The induced current in the coil tries to maintain a field of the same strength and direction as the field that is being removed. (Apply the hand rule to the coil in figure 9-7. Determine the direction and shape of field of the coil due to the current in it. See that poles, current, and field direction of figures 9-7 and 9-8 agree.)

Now consider a coil affected by the approach of a magnet, figure 9-9. Originally, the coil had no magnetic field. As the lines of force of the approaching magnet enter the coil, the coil develops a field of its own that tends to restore conditions in the coil to the original zero-field condition, that is, tends to cancel out the oncoming field.

Another way of stating Lenz's Law is:

> Induced voltages and induced currents oppose change of magnetic field.

In the preceding discussion, the terms *induced voltage* and *induced current* may seem to have been used interchangeably. One should understand that the relative motion of wire and magnetic field always induces a voltage, or emf. This induced emf causes a current if there is a closed circuit.

THE TRANSFORMER

In the last few pages, the effects of moving a magnet near a coil of wire were described. Actually, the important motion is the motion of the magnetic field of the magnet. By using an electromagnet, we can make a magnetic field appear by turning on the current in the magnet coil and remove the field by shutting it off. This can be a more efficient way of generating an emf than by mechanically moving the magnet iron.

Suppose a coil of wire (coil no. 1 in figure 9-10) is wound around a soft iron bar and is connected to a battery through a pushbutton switch. Coil no. 2 is wound around the same iron bar, but connected to a voltmeter. Both coils are of insulated wire, and there is no direct connection of coil no. 1 to coil no. 2.

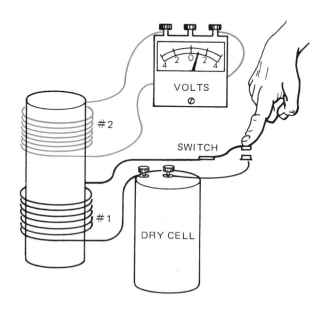

Fig. 9-10 Producing Induced Current

As soon as the switch is closed, a current starts in coil no. 1, magnetizing the iron, that is, bringing a magnetic field into coil no. 2. During the split second that the magnetic field is building up, the emf induced in coil no. 2 will affect the voltmeter. As soon as the iron is magnetized, its field is no longer increasing or moving, and no continued emf is induced in the second coil.

At the moment the switch is opened, the soft iron becomes demagnetized. Its magnetic field, in effect, shrinks back into the iron, which has the same effect on the second coil as does the removal of a magnet from the coil.

The amount of induced emf in the second coil depends on the rate the lines of force are cut, as already mentioned. To produce a high emf, build a primary (no. 1) coil that will magnetize the iron strongly and quickly, and use a large number of turns of wire in the secondary coil. (Remember the three factors of determining emf mentioned earlier in this chapter.)

Energy Transfer by Magnetic Field

A *rotating generator* transforms mechanical energy to electrical energy through the rotating magnetic field. In the coils-on-iron device called a *transformer,* electrical energy from one circuit is transferred to the electrons in another circuit by means of the moving magnetic field. This transfer of energy can be done without waste of energy because there is no friction of mechanically moving parts. An efficient transformer is built like the one in figure 9-11, page 220, with one coil wound over the other on the same iron bar. The transformer is enclosed in an iron shell so that a strong magnetic field can be formed readily.

When working with transformers, remember that some produce high voltages. Check with the instructor before connecting transformers into a circuit or attempting voltage

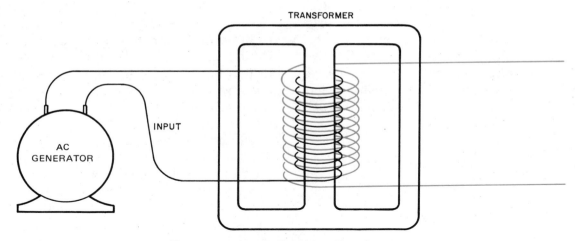

Fig. 9-11 Current Induced in a Transformer

measurements. Always use extreme caution when working with or around high voltages. They are very dangerous.

The coil that delivers energy into the transformer is called the *primary coil;* the energy is taken out of the transformer through the *secondary coil.* The real purpose of the transformer is to produce energy at a different voltage than the primary (input) voltage. The increase or decrease of voltage is determined by the number of turns of wire on each coil. The wire diameter must be big enough to carry the maximum current expected in each coil.

Figure 9-11 shows the primary coil connected to an ac generator to show the greatest use of transformers. An alternating current is continually changing. Alternating currents increase in one direction, then decrease, then increase and decrease in the opposite direction. The magnetic field produced in a coil is continually changing in amount and direction. When this changing magnetic field sweeps through the secondary coil, it induces a new alternating current of the same frequency (hertz or cycles per second) as the alternating current in the primary.

The number of volts, however, that is induced in the secondary depends on the relative number of turns on each coil and the primary voltage, as follows:

$$\frac{\text{Volts Secondary}}{\text{Volts Primary}} = \frac{\text{Turns Secondary}}{\text{Turns Primary}}$$

If the secondary has ten times as many turns of wire as the primary, the secondary voltage will be ten times the primary voltage.

Problem: A transformer has 250 turns on its primary coil and 600 turns on its secondary; it is intended for 120-volt input. Find voltage output.

$$\frac{\text{Volts Secondary}}{\text{Volts Primary}} = \frac{\text{Turns Secondary}}{\text{Turns Primary}} \quad \frac{V}{120} = \frac{600}{250}$$

V = 120 x 2.4 = 288 volts on secondary. Such a transformer, which increases the voltage, is called a *step-up transformer.* By its use, we are not getting something free. Energy that did not previously exist is not being created. In order to get energy out of a transformer, just as much energy must be put into it. Using the numbers from the above problem as an example, if the 288-volt secondary is connected to a 144-ohm resistor, the current in the secondary circuit 288 volts/144 ohms = 2 amps. The secondary circuit is using energy at a rate, in watts, of 288 volts x 2 amps = 576 watts.

To get 576 watts out of the transformer, 576 watts must be put into the transformer. (Or a little more than 576 in case the transformer is not 100 percent efficient.) To put in 576 watts from a 120-volt source, 120 volts x X amps = 576; X = 576/120 = 4.8 amps in the primary circuit. These ideas can be summarized this way:

> primary watts = secondary watts.
> primary volts x amps = secondary volts x amps.

Transformers are reversible in this sense: we could apply a 288-volt alternating current to the 600-turn secondary coil, and induce an output voltage of 120 volts in the 250-turn coil that was called the primary. However, transformer coils are designed to operate at particular maximum voltages, so 200 volts should not be applied to the coil that was designed to operate at 120 volts. This would cause the transformer to overheat.

If a steady dc is applied to the transformer, it merely heats the primary coil and produces no electrical output. A pulsing dc, which frequently increases and decreases in amount, produces a magnetic field which likewise increases and decreases, so a pulsing dc on the primary does induce an alternating current in the secondary winding. The so-called output transformer on a radio is supplied with a pulsing dc from tubes or transistors and produces the ac that causes sound in the speaker.

Other Transformer Uses

Pulling the trigger on the quick-heating, transformer-type solder gun energizes the primary coil of a step-down transformer. The secondary coil is one or a few turns of copper strip or tubing. Figure 9-12 shows one form of construction, the core being wound of one long thin strip of iron. If the primary coil has 240 turns, with 120 volts applied to it, then 0.5 volt is induced in the one-turn secondary. The resistance of the secondary may be about 0.00125 ohm, of which 0.001 ohm is the wire soldering tip, and 0.00025 ohm is the heavy copper bar. The current in the secondary = $\frac{0.5 \text{ volts}}{0.00125 \text{ ohm}}$ = 400 amps. Use watts = I^2R to find the rate of heating of the tip: $(400)^2$ x 0.001 = 160 000 x 0.001 = 160 watts.

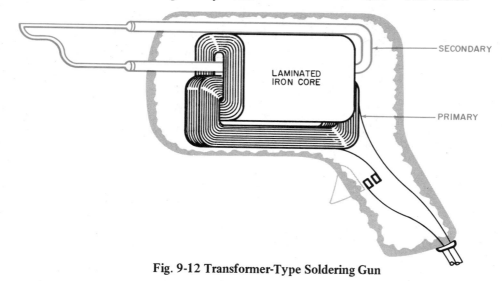

Fig. 9-12 Transformer-Type Soldering Gun

Frequently, a measurement of ac amperes in a circuit is desired under conditions where it is inconvenient to open the circuit and insert an ordinary ammeter. The clamp-on type of meter, figure 9-13, uses the ability of the alternating current to induce a voltage. When the laminated iron hook is opened and placed around a conductor, that conductor acts as the one-turn primary of a step-up transformer. The voltage (and current) induced in the secondary coil by the alternating magnetic field around the conductor is proportional to the current in the conductor. The output of the secondary coil is applied to a moving-coil type of meter

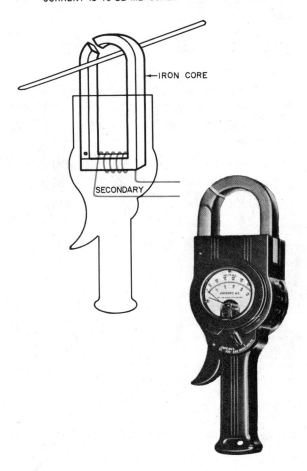

SINGLE WIRE IN WHICH
CURRENT IS TO BE MEASURED

←IRON CORE

SECONDARY

Fig. 9-13 Transformer in a Clamp-On Type Meter

through resistors and a rectifier. The meter is scaled to read the current in the primary conductor.

It is advisable for the reader to work some review problems on transformers at this point, in preparation for the following discussion.

AUTOTRANSFORMER

The *autotransformer* is a simple modification of the ordinary two-coil transformer, figure 9-14. (Autotransformers have nothing to do with automobiles.) The primary and secondary coils in this transformer have some or all of their turns in common.

Figure 9-15(A) shows the common type of construction of transformers, with both coils wound on the same iron core, one on top of the other. Notice that at any instant the secondary current (provided the secondary is connected to a load so there will be current) is in opposite direction to the primary current around the core. The reason for this is that the induced current (secondary) is always trying to oppose the magnetic effects of the primary according to Lenz's Law.

In figure 9-15(B), the two coils have been removed from the iron core so they can be viewed separately. They could illustrate the following problem. If the 12-volt secondary of a transformer is connected to a 1.2-ohm resistor, find the current in the secondary winding and in the 120-volt primary winding. The diagram gives the answers; see if you agree. Notice the colored wire that has been added, connecting the two coils together. There will be no current in it, for it does not complete any circuit. At present it simply ties the two coils together so there is no potential difference between the two coil ends.

Looking at figure 9-15(C), assume one terminal of an ac voltmeter is connected to the bottom wire on the primary coil and a

Banks of Autotransformers Connecting 230 000-Volt Distribution to 115 000-Volt Distribution. The Total Power is 40 000 Kilowatts. (Pacific Gas and Electric Co., San Mateo, Calif.)

Fig. 9-14

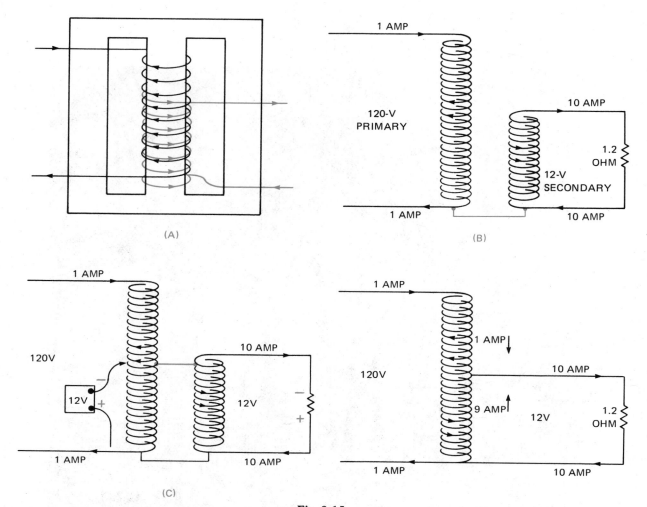

(A)

(B)

(C)

Fig. 9-15

little insulation has been scraped off the primary winding so the other voltmeter wire can slide along the coil. Measure the voltage between the bottom wire and various points along the coil. Since the total 120 volts applied is distributed all along the coil, a 12-volt reading can be found about one-tenth of the way up the coil. The instantaneous polarity of this voltage shown by the positive and negative marks at the meter, is determined from the choice of instantaneous current direction. Electrons are being forced from the top of the coil toward the bottom.

Still looking at figure 9-15(C), decide what happens if another colored wire is connected from the 12-volt point on the primary to the top of the secondary coil. If the top of the secondary is 12 volts more positive than the bottom, a 24-volt potential difference would be applied to the added colored wire. The chosen direction of primary current determines the direction of secondary current. The induced voltage at this instant drives electrons on to the top wire of the secondary circuit, making that top wire negative in comparison with the bottom wire. The potential difference between minus 12 and minus 12 is zero, so there is no reason for current in this added wire. It suggests the possibility that if the primary coil to the left moves sideways and melts in with the equal number of turns of the lower 12-volt section marked on the secondary, the result would look like figure 9-15(D).

In figure 9-15(D), the 1.2-ohm resistor is still connected to points having a 12-volt potential difference (PD), so its current is 12/1.2 = 10 amps. It is taking energy at a 120-watt rate (12 volts x 10 amps = 120 watts). The primary still has 1 amp in it, so it can deliver the 120 watts. Why are there 9 amps on the lower part of the coil? In sketch (C) the primary carried 1 amp downward, and the secondary carried 10 amps upward instantaneously. This equals 9 amps upward. Or 9 amps, induced by the 1 amp in the other nine-tenths of the coil, join with the 1 amp to make the 10 amps. This is called a *step-down autotransformer* (self-transformer).

The schematic drawing at the top of figure 9-16 shows a variable or adjustable autotransformer. It has a sliding contact so any desired portion of the coil can be used, letting the user control the voltage

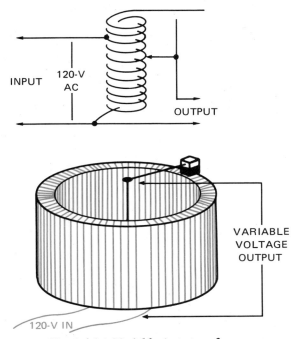

Fig. 9-16 A Variable Autotransformer

output. The laminated iron core of the small 7-amp transformer looks like a 3-inch (75 millimeter) long piece of 4-inch (100 millimeter) diameter pipe, with a coil on it as in figure 9-16. Turning the adjusting knob slides a graphite brush along the coil. Autotransformers can also step up voltage. To make it reasonable, instead of connecting the 12-volt secondary parallel with part of the primary, as was done in figure 9-15(C), slide the secondary down and to the left, connecting it in series with the 120-volt primary so it looks like figure 9-17, with series voltages adding.

For many uses of transformers, the two-coil type may be necessary to maintain isolation of the secondary circuit from the primary. In this case the autotransformer is inappropriate. Autotransformers are very efficient and less expensive to build than the two-coil type of the same power rating. 3:1 autotransformers connect the 345 000-volt Niagara-New York line to previously-built 115 000-volt systems.

AUTOMOTIVE IGNITION SYSTEM

Although figures 9-10 and 9-11 were intended to introduce the power transformer, they could also illustrate the basic principle of the automobile ignition system. Reviewing figure 9-10, as current increases and decreases in the primary coil, a voltage is induced in the secondary coil which depends on the number of turns in the secondary and the number of magnetic lines of force that sweep across that secondary in a short time.

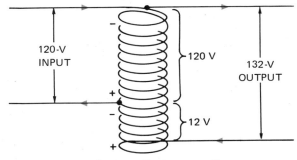

Fig. 9-17 Step-Up Autotransformer

The ignition system creates a high-voltage spark at the proper time to ignite the compressed gasoline vapor mixture in the engine cylinders. Twelve volts in the primary circuit is stepped up to about 20 000 volts in the secondary circuit, which includes the spark plug. Figure 9-18 shows a pair of coils. They are seldom called a transformer. They are usually called an induction coil or simply, the coil. The primary coil has a few hundred turns of wire; the secondary coil has several thousand. Look at the primary circuit, shown in black in figure 9-19.

Assume the key-switch is turned on and the engine is operating. The rotating cam is geared with the engine to rotate at half the rpm of the engine. In the diagram, the ground symbol, $\perp$, represents a connection of wire to the conducting steel frame of the car. With the contact points closed as shown, battery energy makes electrons flow in the direction

of the arrows marked on the primary circuit. In some cars, the positive terminal of the battery is grounded. In that case, the current direction would be the opposite of that shown, but coils and contacts will behave the same way.

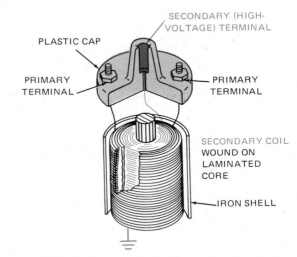

Fig. 9-18   Induction Coil for Automobile Engine Ignition

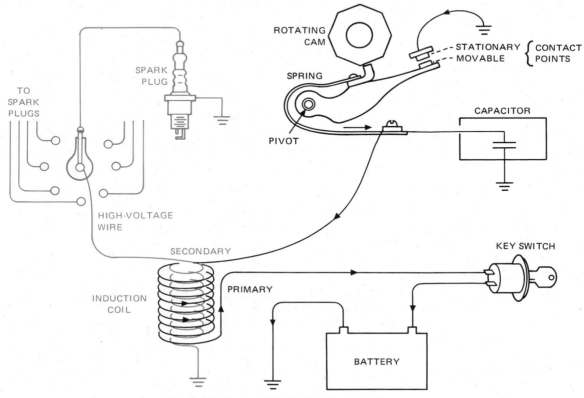

Fig. 9-19 Automobile Engine Ignition System

At the first closing of the contacts, current in the primary starts and builds up slowly because the expanding field induces an opposing voltage in the primary coil itself. (Lenz's Law always working.) The slowly expanding magnetic field induces only a small and useless voltage in the secondary coil. When the contact points are pushed open by the rotating cam, useful things happen. Opening of the contact points breaks the primary circuit and the primary current stops. The collapsing magnetic field induces a usefully large voltage in the secondary coil, provided the current in the primary stops suddenly enough.

The capacitor helps the primary current stop suddenly. This capacitor (sometimes called the condenser) consists of two long strips of aluminum foil rolled up with waxed paper between them. The waxed paper acts as insulation. Inside the capacitor rectangle the symbol for a capacitor is shown. It represents the two conducting sheets. Electrons cannot flow through a capacitor because of the insulation, but the surface area of the metal foil can serve as a storage space for electric charges. Electrons can flow onto one of the sheets if, at the same time, electrons are pulled off the other. The number of electrons involved in this action is limited by the size of the capacitor and the voltage available. As a mechanical comparison, it is like a spring. The more force used, the farther it can be moved. However, there is always a limit.

Refer again to the primary circuit in figure 9-19. Whether or not there is a capacitor there, when the contact points start to open, the current in the primary coil starts to lessen. The collapsing magnetic field induces a voltage in the primary as well as the secondary. The induced voltage in the primary tends to keep electrons flowing through the primary coil and the contact points. (Lenz wrote: induced voltage opposes change.) This induced voltage, helping the battery voltage, starts a

little arc as the points first open if there is no capacitor in the circuit. If an arc forms, the vaporized metal in the arc permits enough conduction to gradually stop the primary current. A gradual reduction in current means that the magnetic field contracts slowly, inducing only a small emf in the secondary coil. A small emf is not enough to make sparks at the spark plug. Therefore, the engine will not start.

With the capacitor in parallel with the contact points, the arc does not form at the contacts. When the points start to open, electrons, traveling from the car frame through the points to the coil, meet increasing resistance at the opening points. The empty aluminum sheets in the capacitor appear for an instant to have a great deal of space for electrons. For just an instant it offers a low-resistance route. Electrons from the car frame rush on to the grounded capacitor plate while electrons from the other capacitor plate are kicked off to complete a circuit through the primary coil. In a short time, the negative foil of the capacitor has loaded up with electrons and the upper sheet is positively charged, having been stripped of as many electrons as the applied emf can remove. During this time, the points open farther without starting a conducting arc. Cold air between the points is a good insulator. Suddenly, electrons cannot jump across the gap at the points, or move into or across the capacitor. They stop moving completely. The magnetic field collapses abruptly. The high rate of collapse induces a high emf in the secondary coil. This is enough to cause a spark at the gap in the spark plug sufficient to ignite the compressed gasoline vapor in the cylinder head. This causes the engine to run.

The secondary circuit (in color) is at any one instant a series circuit of secondary coil, rotor arm, and one spark plug. The rotor is placed on the cam and turns with it. The cam in the diagram has 8 lobes (high points) for an

8-cylinder engine. Whenever a lobe opens the points, the rotor contacts the appropriate spark plug contact. The rotor, cam, and contacts are in one assembly called the distributor. Figure 9-19 might give a wrong idea as to relative size of these parts. The cam and contact points have been drawn about 20 times too large in proportion to other items so they can be seen.

POINTS TO REMEMBER

- The voltage produced by generators and transformers is called induced electromagnetic force (emf). This name is given to the emf that is produced by the motion of wires across a magnetic field, or by the motion of a magnetic field across wires.
- This induction process in a generator converts mechanical energy into electrical energy.
- The amount of induced emf depends on the strength of the magnetic field, the number of turns of wire in the device, and the speed of movement.
- To produce one volt, lines of force must be cut by wire at the rate of 100 000 000 per second.
- Lenz's Law states: an induced current opposes the motion that causes the induced current. Induced voltages and currents oppose change of magnetic field.
- For a transformer:

$$\frac{\text{Volts Secondary}}{\text{Volts Primary}} = \frac{\text{Turns Secondary}}{\text{Turns Primary}}$$

and

$$\frac{\text{Primary}}{\text{(volts x amps)}} = \frac{\text{Secondary}}{\text{(volts x amps)}}$$

- In the induction coil of a gasoline engine ignition system, a high-voltage pulse is induced in the multiturn secondary coil when the magnetic field of the dc primary quickly collapses.

REVIEW QUESTIONS

1. How is an induced voltage, or an induced current, produced?
2. What is the left-hand generator rule? How is it used? What are its limitations?
3. The amount of induced emf depends on the rate of cutting of magnetic lines of force with wires. Name three factors that will cause an increase in the amount of voltage induced.
4. A wire is moved through a magnetic field as sketched below in figure 9-20. What is the direction of the induced emf?
5. State Lenz's Law.
6. Is it advisable to use a step-up transformer as a step-down transformer? Justify your answer.

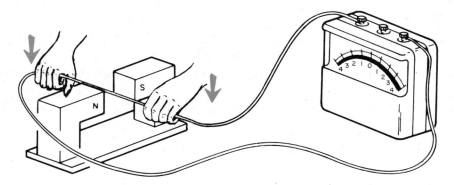

Fig. 9-20

7. What would happen if a steady dc were applied to a radio transformer?

8. In a generator, where does the electricity come from?

9. In 1912, some people took short trips in an electric automobile, powered only by storage batteries and an electric motor. The necessity of frequent battery charging helped make these cars obsolete. Would a generator, belt-driven from the wheels of a trailer, charge the batteries and make a longer trip possible?

10. A transformer has a 300-turn primary and a 60-turn secondary. 120 volts (ac) is applied to primary. Find the voltage output of the secondary. Is this output ac or dc?

11. A certain transformer puts out 24 volts when the input is 120 volts. The 24-volt secondary is connected to a 4-ohm lamp. Find the current in the secondary circuit. Find the current in the primary circuit. Which coil (primary or secondary) needs larger diameter wire?

12. A distribution transformer on a street pole has an input of 4 800 volts and an output of 120 volts, figure 9-21. Which coil (primary or secondary) has more turns of wire? What is the ratio of turns on the two coils? When the household customers take 80 amperes from the 120-volt secondary, how much current is required in the primary?

13. When a welding autotransformer in the shop puts out 100 amps at 48 volts, how much current is there in the 120-volt primary circuit?

14. A transformer for a neon sign puts out 20 milliamperes at 7 500 volts. The 120-volt input winding has 400 turns. Calculate the:
    (a) number of turns on secondary.
    (b) current in primary.
    (c) watts, both from input and output volts and amps.

15. Assume that in figure 9-19 the positive pole of the battery is grounded and the negative wire goes to the ignition switch. Trace out the current path in the primary and explain what happens in the primary circuit, including the function of the capacitor.

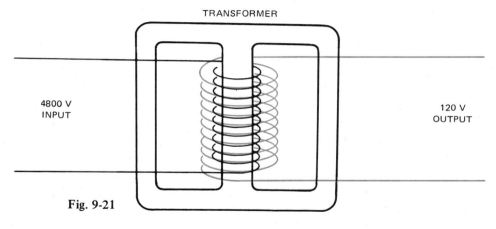

TRANSFORMER

4800 V
INPUT

120 V
OUTPUT

Fig. 9-21

RESEARCH AND DEVELOPMENT

Experiments and Projects on Induced Voltage

INTRODUCTION

You have learned about the production of magnetic force by electric current. In Chapter 9, you studied about the discovery of induced current and the importance of inductance in producing electromotive force as well as the principles, rules, and Lenz's Law concerning induced voltages and induced current. Several experiments have been arranged. They will help you understand how induced voltage and current are produced. The projects are concerned with the production and use of induced emf.

EXPERIMENTS

1. To observe induced current direction and magnetic polarity in a coil.

2. To find out if the amount of induced voltage depends on the number of turns of wire in the coil where the voltage is induced.

3. To measure the primary and secondary power and efficiency of a transformer.

4. To compare the current produced by an ac transformer operating in a radio.

PROJECTS

1. Operate a magneto.

2. Assemble, test, and disassemble an experimental transformer.

3. Design and build a low-voltage transformer.

EXPERIMENTS

Experiment 1 INDUCED CURRENT BY MOTION OF MAGNETIC FIELD

OBJECT

To observe induced current direction and magnetic polarity in a coil.

APPARATUS

1 - Galvanometer
1 - Small hollow coil
1 - Bar magnet

PROCEDURE

1. Connect the coil and galvanometer so that they form a complete circuit, figure 9-22.

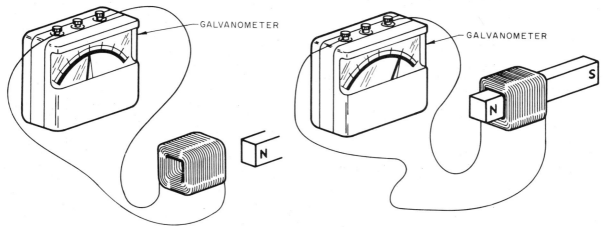

Fig. 9-22

2. Insert one end of the magnet into the coil and then withdraw it. While you are withdrawing the magnet, observe the effect the distance the magnet is moved has on the movement of the galvanometer needle.

3. Find out if the approach of the north pole has the same effect as the approach of the south pole.

4. Observe what happens as the north pole of the magnet comes near the coil. Also determine from meter deflection and by looking at the coil, the direction of the current produced in the coil.

5. Make a diagram and mark the direction of the current produced in the coil and the magnetic polarity on the end of the coil that faces the approaching north pole.

6. Make a similar diagram showing direction of current found, and magnetic coil polarity produced, when the north pole of the magnet is moving away from the coil.

7. Make another diagram with the south pole of the magnet approaching the coil and one for the removal of the south pole.

QUESTIONS

1. What happened to the needle on the meter when the north pole of the magnet was inserted into the coil?

2. Did the meter read the same when the south pole was inserted?

3. What magnetic polarity does the coil current produce on the end that faces the approaching north pole?

Experiment 2 USING A COIL TO PRODUCE THE CHANGING MAGNETIC FIELD

OBJECT

To find out if the amount of induced voltage depends on the number of turns of wire in the coil where the voltage is induced.

APPARATUS

1 - Galvanometer or zero-centered milliammeter or microammeter
1 - Coil, 50-turn
1 - Coil, 100-turn
1 - Coil, 200-turn
1 - Coil, 300-turn
1 - Coil, 400-turn
1 - DC voltmeter (0-150 V)
1 - AC voltmeter (0-150 V)
1 - Experimental transformer
1 - Iron bar
1 - SPST switch
1 - Transformer core
   DC power supply

PROCEDURE

1. Connect the 200-turn coil to the dc source so that a switch will control the current, figure 9-23.

2. Place the 400-turn coil on the bench so that the coils are end-to-end, with the 400-turn coil connected to the galvanometer, figure 9-24.

3. Close the switch and observe the galvanometer. Open the switch and observe the galvanometer.

4. Remove 400-turn coil and connect 50-turn coil to galvanometer; place 50-turn coil close to 200-turn coil as before. Turn dc on and off in 200-turn coil.

5. Observe the similarity in behavior and the difference between this arrangement of coils and the first one.

6. Place a small round iron bar inside the coils (200-turn and 50-turn), figure 9-25. Turn dc on and off in the 200-turn coil. Observe the behavior.

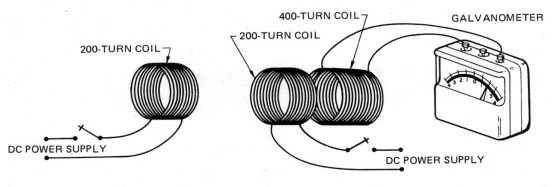

Fig. 9-23   Fig. 9-24

7. Connect a variety of combinations of primary and secondary coils on an experimental transformer core. See figure 9-26.

   The primary coil should be attached to a low-voltage ac source. Measure that voltage. Measure the secondary voltage in each combination of coils starting with the high setting on the voltmeter, then moving down the range selector until secondary voltage can be read accurately.

8. Use a table similar to the one shown here to record the results of your experiments with various coil combinations.

TRIAL	VOLTS	RATIO $\frac{\text{PRIMARY VOLTS}}{\text{SECONDARY VOLTS}} =$	NO. TURNS	RATIO $\frac{\text{PRIMARY TURNS}}{\text{SECONDARY TURNS}} =$
	Pri._____ Sec._____		Pri._____ Sec._____	
	Pri._____ Sec._____		Pri._____ Sec._____	
	Pri._____ Sec._____		Pri._____ Sec._____	
	Pri._____ Sec._____		Pri._____ Sec._____	

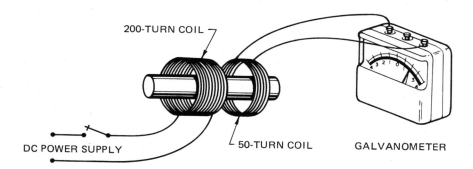

Fig. 9-25

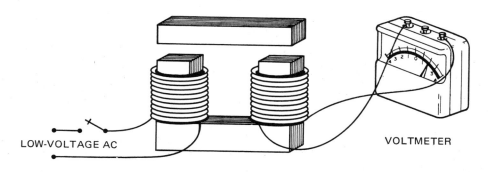

Fig. 9-26

QUESTIONS

1. When you closed and opened the switch in Step 3, was the galvanometer action the same each time? Explain.

2. How does this action compare to the insertion and removal of a magnet from a coil, as you have done before?

3. How did the galvanometer reading in Step 4 differ from Step 3? What is similar about galvanometer behavior to that observed with 400-turn coil? What is different?

4. What evidence have you that the amount of induced voltage depends on the number of turns in the coil where the voltage is induced?

5. What is the relationship of volts ratio to turns ratio?

6. Explain how the action of a transformer with its alternating currents relates to the previous action of the moving magnet on the galvanometer.

7. What happened when you placed the round iron bar inside the 200-turn and 50-turn coils?

Experiment 3 PRIMARY AND SECONDARY POWER AND EFFICIENCY OF A TRANSFORMER

OBJECT

To measure the primary and secondary power and efficiency of a transformer.

APPARATUS

2 - Transformers, 120-volt to 6-volt or 120-volt to 12-volt
2 - AC ammeters (0-1 A)
1 - AC ammeter (0-5 A)
1 - AC voltmeter (0-150 V)
1 - Lamp (10-watt to 60-watt)
1 - Lamp base
1 - SPST switch

PROCEDURE

1. Set up two transformers in a circuit as shown in figure 9-27.

2. Add ammeters and a small lamp to the circuit, figure 9-28.

3. With the lamp operating, record the input and output of volts, amps, and watts going into the first transformer.

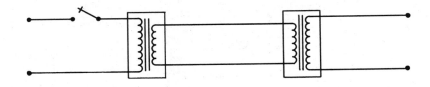

Fig. 9-27

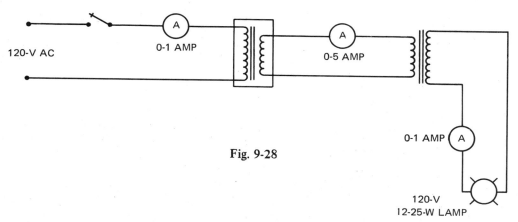

Fig. 9-28

4. Calculate the efficiency of the first transformer.

5. For the second transformer, record the input and output of volts, amps, and watts. Calculate the efficiency of the second transformer.

6. How efficient is the entire circuit?

PROJECTS

OPERATE A MAGNETO

More than 100 years ago, Michael Faraday discovered that magnetism could produce electricity. He found that if a magnet is moved past a wire, electrical current will start through the wire. Magnetos operate on this principle: magnets are caused to produce electricity through mechanical movement. Early telephones used a magneto to ring bells. Magnetos are used in the ignition systems of small gasoline engines.

MATERIALS

1 - Magneto
1 - Bell
1 - Lamp
1 - Lamp receptacle
1 - Oscilloscope

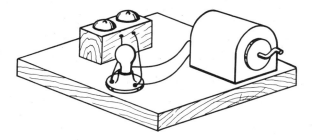

Fig. 9-29

PROCEDURE

Mount the parts on a wooden base, figure 9-29. Connect the circuit according to figure 9-30.

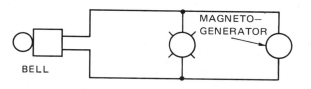

Fig. 9-30

Operate the magneto with bell and lamp in circuit, and with the lamp out.

Connect an oscilloscope, operate the magneto, and observe the waveform with voltage peaks.

ASSEMBLE, TEST, AND DISASSEMBLE AN EXPERIMENTAL TRANSFORMER

In the coils-on-iron device called a transformer, electrical energy from one circuit is transformed to the electrons in another circuit by the moving magnetic field, figure 9-31. This transfer of energy can be done very efficiently because there is no friction of mechanically moving parts. It is an application of mutual inductance.

Through the medium of electromagnetic induction, the transformer changes the value of electromotive force in the circuit.

MATERIALS

1 - Primary coil
1 - Secondary coil
1 - Laminated iron core
1 - AC voltmeter (0-50 V)
1 - SPST switch
   AC power source

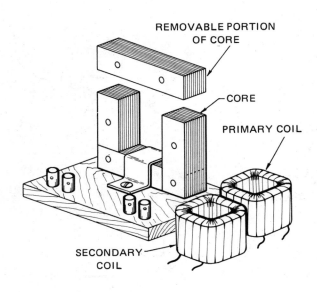

REMOVABLE PORTION OF CORE

CORE

PRIMARY COIL

SECONDARY COIL

Fig. 9-31

PROCEDURE

Assemble the component parts according to the sketch in figure 9-32.
Connect the primary winding to the ac power supply.
Turn on the power and read the voltmeter.
Disassemble the parts and return them to the proper storage cabinet.

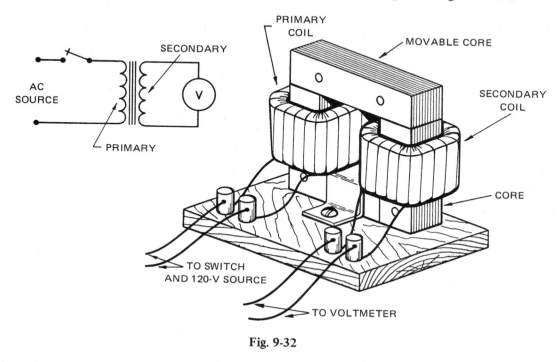

Fig. 9-32

DESIGN AND BUILD A LOW-VOLTAGE TRANSFORMER

A serviceable transformer for use in experimental work and other
purposes, such as operating electric trains, low-voltage motors, and signal
systems, can be designed and constructed with the required watts output
and one or more secondary voltages, figure 9-33.

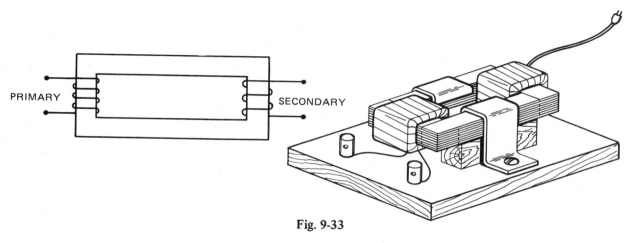

Fig. 9-33

MATERIALS

Magnet wire for coils
Sheet iron for core laminations, #26 gauge
Band iron for supports, 1/8" (3 mm) x 3/4" (20 mm) x required length
Cotton tape
Cotton or plastic tubing
Male attachment plug
Appliance cord
Wood for base
Binding posts or Fahnestock clips
Assorted screws
Insulating varnish

PROCEDURE

Decide what capacity transformer you require. Analyze the problem of constructing one.

Calculate the size of the core. Calculate the size, number of turns, and amount of wire required for the primary and the secondary coil.

Make dimensioned drawings of the parts to be fabricated, then make the parts.

Wind the coils, finish with cotton tape, and varnish. For information in winding coils, refer back to Chapter 6.

Assemble the parts and test the transformer with a voltmeter.

Note: Specifications for transformers are given in figure 9-34, one is to be fabricated and assembled from purchased parts.

When working with transformers, remember that some produce high voltages. Check with the instructor before connecting transformers into a circuit or attempting voltage measurements. Always use extreme caution when working with or around high voltages. They are very dangerous.

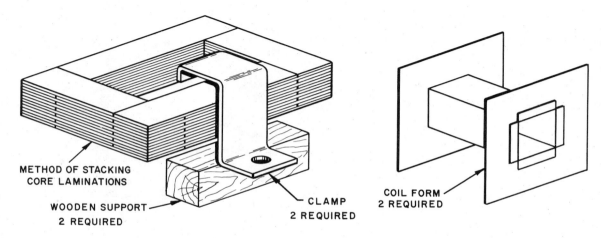

Fig. 9-34

SPECIFICATIONS

For a 40-watt, 115-volt step-down transformer, parts to be made:

Primary coil — 600 turns, #22 AWG magnet wire, Formvar

Secondary coil — 77 turns, #14 AWG magnet wire, Formvar

Core laminations, 26-gauge sheet iron

Note: Obtain metal from an old transformer, if possible.

The core should be built of laminations one inch (25 millimeters) wide. It should be one inch (25 millimeters) thick when complete. The outside dimensions are 3 1/2" (89 mm) long x 2 7/8" (73 mm) wide. For a transformer which will deliver 6, 8, 12, and 14 volts, taps should be taken off the secondary coil as follows:

6 volts, tap at 33 turns

8 volts, tap at 44 turns

12 volts, tap at 66 turns

14 volts is 77 turns at end of coil

The primary coil could have 8 layers, with 75 turns to the layer, minus one on the last layer.

The secondary should have 3 layers with 26 turns to the layer, minus one on the last layer.

For a 65-watt, 120-volt step-down transformer, parts (with exception of coils and base) to be purchased: (Range 6 and 12 V)

Primary coil - 600 turns, #24 AWG magnet wire, Formvar

Secondary coil — 66 turns, #14 AWG magnet wire, Formvar

Core - square stack, E-type laminations. See dimensions, figure 9-35.

For a transformer which will deliver 6 and 12 volts, a tap should be taken off the secondary at the thirty-third turn.

Coils can be wound on a shop-made or commercial-type bobbin. See dimensions, figure 9-36, page 240.

The coils may be "scramble wound" or carefully wound in layers.

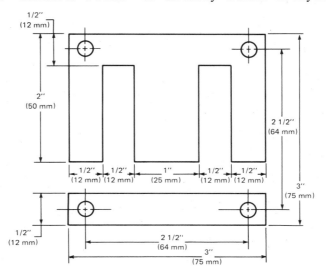

Fig. 9-35

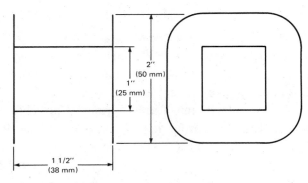

Fig. 9-36

MATERIALS

52 - E-type laminations, 26-gauge, 4-hole, 1" x 1 1/2" (25 mm x 38 mm) core, 2" x 3" (50 mm x 75 mm) silicon steel

52 - I-type laminations, #26-gauge, 2 hole, 1/2" x 3" (12 mm x 75 mm) silicon steel

1 - Front cover (steel), 1" (25 mm) standard, with flange

1 - Rear cover (steel), 1" (25 mm) standard, with flange

1 - Bobbin, 1 1/2" x 2" (38 mm x 50 mm) with 1" (25 mm) steel hole

1 - Base, wood, 3/4" x 3 1/2" x 4 1/2" (20 mm x 89 mm x 114 mm)

1 - Plug, phenolic or rubber

1 - Cord, 16-gauge, lamp-type, length optional

1 - Rubber grommet

1 - Primary coil, 600 turns, #24 AWG magnet wire, Formvar

1 - Secondary coil, 66 turns, #18 AWG magnet wire, Formvar
Insulating paper
Cotton tape
Plastic tape

3 - Fiber washers, #8 (M4) hole, flat

3 - Fiber washers, #8 (M4) hole, extruded

3 - Solder lugs, #8 (M4) hole

7 - Machine screws, #8-32 x 1 1/4" (M4 x 30 mm L), RH steel

7 - Machine screw nuts, #8-32 (M4)

3 - Wing nuts, #8-32 (M4)

Chapter 10
Electromagnetic generators

Chapter 9 described the generation of voltages by moving wires through a magnetic field or by moving a magnetic field through a coil of wire. More about the use of this effect in dc and ac generators will now be discussed.

As mentioned before, a generator changes mechanical energy into electrical energy. So, a generator must be driven by some other energy source, such as steam, a gasoline or diesel engine, or a windmill.

DC GENERATORS

A few automobiles use a small dc generator to keep the battery charged, most

use an alternator. These components can operate only when the engine is running. Industrially, large dc generators produce the dc that is necessary in metal refining. Some welding machines consist of a dc generator driven by an ac motor. The dc is used to perform the arc welding, figure 10-1. The largest dc generators provide dc for adjustable speed motors where stepless and precise speed adjustment is necessary. Nearly all railway locomotives use a diesel engine to run a dc generator which supplies energy to dc motors that drive the locomotive wheels.

Direct current generators use stationary field magnets, the induced current being generated in rotating coils. The basic principle of producing dc may be shown by following in detail what happens when one loop of wire is rotated in a magnetic field, figure 10-2.

Fig. 10-1 Motor Generator Type of DC Welding Machine

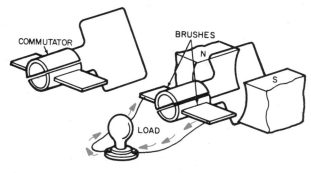

Fig. 10-2 A Simple DC Generator Attached to a Load

241

In order to produce a direct (one-way) current in the outside circuit that is served by the generator, the ends of this loop are fastened to semicircular metal strips which are insulated from each other. These metal strips rotate along with the loop. These two half-circle segments form the part of this generator called the *commutator*.

Two stationary conductors (often blocks of graphite) called *brushes* are held in contact with the commutator. These brushes carry the generated dc to the outside circuit.

The following series of diagrams shows successive positions of the steadily rotating loop. The letters N and S represent poles of a horseshoe magnet.

At the instant of rotation represented by position (1), figure 10-4, no emf is produced. This is because the wires are moving parallel to the field and there is no cutting of lines of force. (Half of the loop is shown in color for later identification.)

As the loop moves from position (1) to position (2), lines of force are cut at an increasing rate even though the rotation rate is steady. Notice how A, B, C, and D in the diagram below show the relative number of lines cut during small equal time intervals between (1) and (2), figure 10-3.

At position (2) the sides of the loop are cutting lines of force at the maximum rate. The induced current in the loop (found by either left-hand rule or by Lenz's Law) is a flow of electrons directed toward the brush

on the right and away from the brush on the left. This forcing of electrons from the rotating coil toward the right-hand brush makes this brush negatively charged. The removal of electrons from the left-hand brush makes it positively charged. In the stationary wiring of the external circuit, electrons flow from the negative brush to the positive brush.

Between positions (2) and (3), current continues to be induced in the same direction, but decreases in amount as the wire approaches position (3). At this position the two segments of the commutator are short-circuited. Since at this instant no emf is being generated, there is no ill effect. At position (3) no lines of force are being cut.

Between positions (3) and (4), increasing voltage and current are produced. This is just as happened between positions (1) and (2). Notice that the color side of the loop, previously moving downward through the field, is now moving upward. The current direction in the color side has been reversed. Compare diagrams for positions (2) and (4).

However, the color side of the loop is now taking electrons from the left brush, forcing them around through the white side of the loop toward the right-hand brush. The charges on each brush are the same as before.

Reviewing the diagrams which follow, note that in the rotating loop itself the generated current alternates in direction. Watch the color half of the commutator as it rotates. In position (2), the color segment is negative and supplies electrons to the right-hand brush. In position (4) the color segment has become positive, but just as it becomes positive it slides away from the right-hand brush and contacts the left brush. Likewise, the white segment of the commutator changes position as the

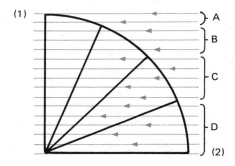

Fig. 10-3

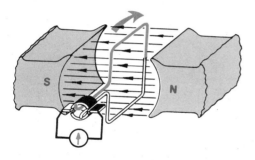

Fig. 10-4 Rotating Loop, Position 1

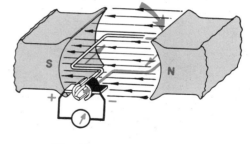

Fig. 10-5 Rotating Loop, Position 2

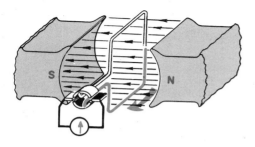

Fig. 10-6 Rotating Loop, Position 3

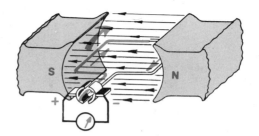

Fig. 10-7 Rotating Loop, Position 4

current direction in the loop changes, so that the right-hand brush is always supplied with electrons and the left-hand brush is always losing electrons into the rotating loop. Thus the electron flow through the outside load is kept one-directional. This is the purpose of the commutator.

The graph, figure 10-8, page 244, shows how the voltage (or current) increases and decreases in amount during one rotation of the loop. This type of current is called *pulsing dc* as contrasted to *steady dc* obtained from a battery.

A single loop of wire was shown in figures 10-4 to 10-8 to show the principle of dc generation. Because one loop of wire produces such a small amount of voltage, a coil of several turns is used on a real generator. A coil of 100 turns makes 100 times as much voltage as a one-turn coil, at the same rpm and magnetic-field strength.

If the generator coil is wound on an iron core, the added iron creates a stronger magnetic field for the coil to operate in. With the same field magnet, adding the iron core between its poles increases the number of lines of force by several times. Increased magnetic field strength results in increased generation of emf.

One disadvantage of the simple generator still needs correcting. This coil of wire, rotating in a strong field, still produces an emf that rises and falls to zero twice during each rotation, figure 10-8. Mechanically, this is like a one-cylinder gas engine. Electrically, current that varies in this fashion causes difficulties in the generator and in the circuit it feeds.

To produce a somewhat smoother emf and current, use two coils, each with its own pair of commutator segments, with the two brushes placed so that they connect

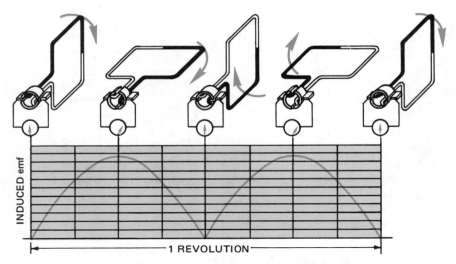

Fig. 10-8 Graph of Current Produced in One Revolution of Rotating Loop

to the coil that has the greater emf induced in it. The output voltage is shown in figure 10-12. Compare figure 10-12 to figure 10-8.

Figure 10-13 shows a still better arrangement. This arrangement is better because it uses all the emf generated in the coils, instead of disconnecting one coil as is done in the arrangement in figure 10-11. To show connections of wires, most of the length of the commutator bars is cut away in this diagram. The coils are wound on an iron core, as in figure 10-10, but the iron core is not shown here. Assume that these coils are rotated between magnet poles and the emf generated in the coils takes electrons from a brush that touches commutator segment no. 1 and electrons are pushed on to a brush that rests on segment no. 3.

Trace the path of electrons through these coils and find the two parallel paths. Starting from segment no. 1 follow the color arrows around the coil, arrive at segment 4 and go from segment 4 around another coil, arriving at segment 3 and its outgoing brush. Following the dotted arrows, trace downward from segment no. 1 around a coil that brings us to segment 2, then around the horizontal coil to finish on segment 3. The circuit is shown in figure 10-14.

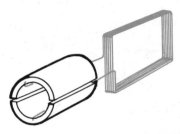

Fig. 10-9 Rotating Loop with Several Turns of Wire to Form a Rotating Coil

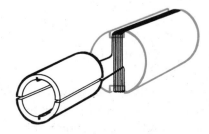

Fig. 10-10 Rotating Coil with Iron Core

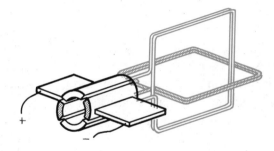

Fig. 10-11 Rotating Coils Attached to a Segmented Commutator

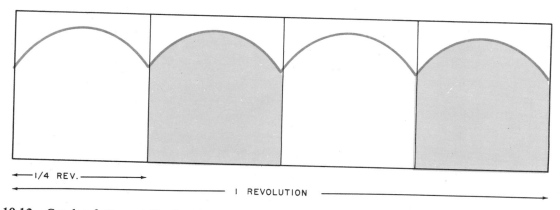

Fig. 10-12 Graph of Current Produced in One Revolution of Rotating Coils with Segmented Commutator

In figure 10-13, the wiring that first appeared to be two coils is actually four coils, a series pair in parallel with another series pair. The voltages generated in a series pair of coils combine as shown in figure 10-15. As more coils and commutator segments are used, the total emf is much steadier.

DC ARMATURES

Figure 10-13 shows an example of a *drum winding* for a dc armature. The varieties of drum windings used in dc generators and motors depend on voltage or current requirements, number of field magnet poles, number of commutator segments used, and other factors. A complete discussion of them would make another complete textbook. Ordinarily, generators and motors use more coils than were shown in figure 10-13, and more commutator bars. Various series-and-parallel arrangements for the coils are used. Direct current armatures look complicated to one not familiar with them, for the same reason a roadmap looks complicated to a six-year old. If you can read a map for a 20-mile (32-kilometer) trip, you can read a map for a 2 000-mile (3 219-kilometer) trip. The 2 000-mile (3 219-kilometer) trip is not 100 times more difficult to understand; the travel time is simply longer.

Small dc armatures are easy to obtain for study and disassembly; there are millions

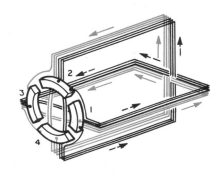

Fig. 10-13 Illustration of Path Electrons Travel in a DC Generator

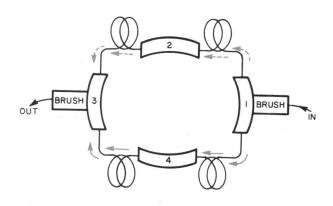

Fig. 10-14 Illustration of Relationship of Coils to Commutator Segments and Current Path

of them in use in automobiles and aircraft. Armatures from dc motors and from universal (ac-dc) motors used in vacuum cleaners and hand drills are similar in construction.

Field Structure for Generators

In a few low-power applications, permanent magnets provide the magnetic field for the generator. The voltage of such generators is used for controls or signaling, such as an electrical rpm indicator (tachometer).

Ordinarily, the field structure is of the general shape shown in figure 10-17. The circular frame, or yoke, is often of cast steel, fitted with pole pieces of laminated iron.

This stationary iron is magnetized by current in the field coils that surround the pole pieces. In most generators, this current is supplied by the generator itself. Only a small fraction of the generated power (about 3 percent or less) is used to supply this magnetizing, or *exciting,* current. When a generator supplies its own excitation current, it is called a *self-excited* generator.

Field Connections for Self-Excited Generators

1. Shunt (or parallel): In small generators the field coils are usually connected to the brushes of the generator. This puts the field coils in parallel with the external load. These shunt coils consist of many turns of small wire and carry only a small current, figure 10-19(A).

2. Compound: In many large dc generators the field magnet iron is magnetized by the combined effect of two sets of coils. The shunt coil is a set of high-resistance (many turns, small wire) coils in parallel with the load circuit. The coil in series with the load consists of a few turns of heavy wire, usually arranged so that the load current aids the shunt field coil in magnetizing effect, figure 10-19(B).

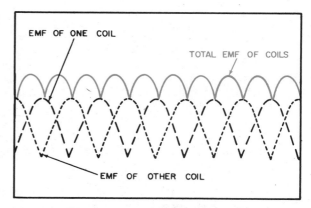

Fig. 10-15 Graph of EMF in Coils

Fig. 10-16 DC Armature

Fig. 10-17 DC Generator Two-Pole Field

The successful starting up of a generator depends on the existence of residual magnetism in the field iron. *Residual magnetism* is a small amount of magnetism remaining from the effect of previous current in the field coil. When the armature of a generator starts rotating, a very low voltage is generated in it due to the weak field in which the armature rotates. This low voltage causes a small current in the

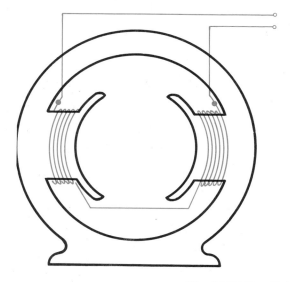
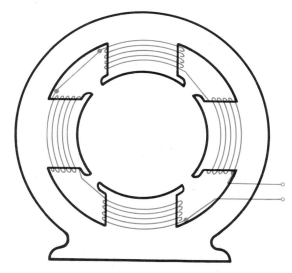

Fig. 10-18 Two-Pole, Four-Pole Field Windings

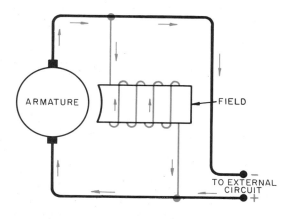

(A) Shunt Generator

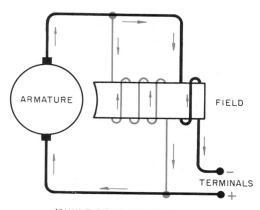

(SHUNT FIELD ACROSS THE ARMATURE)

(B) Compound Generator

Fig. 10-19

shunt field coils, increasing the strength of the field slightly. As the field becomes stronger, the generated voltage increases slightly. This increased voltage causes more current in the field, increasing the field strength still more. As a result, still more voltage is generated in the armature. Maximum field strength is limited by magnetic saturation of the iron.

AC GENERATORS

As noted already, the electromotive force generated in a rotating coil in a magnetic field is an alternating electromagnetic force.

If the rotating coil is connected to a pair of complete circular sliding contacts (called slip rings) an alternating current can be taken from the rotating coil through brushes to the external circuit. Some small generators are built in this fashion, figure 10-20, page 248.

In all large ac generators, including those that generate alternating voltage for commercial power, a cylindrical magnet is rotated inside the stationary coils of wire. Figure 10-21, page 248, shows the simplest form of such an ac generator. The cylindrical rotor is magnetized along a diameter. The useful output current is generated in the stationary coil as the rotating magnetic field sweeps across the wires.

Sketches 1-7, figure 10-22, show the direction of induced voltage in the coil at succesive instants during one rotation of the magnet. In position (1) there is no induced voltage because lines of force are not cutting the coil. During the first quarter-turn, the voltage increases, reaching a maximum as the poles sweep close by the wires, then decreases during the next quarter-turn. The direction of the induced current can be found by using Lenz's Law: "An induced current opposes charge of magnetic field." In this example, from sketch 1 to 3, the original magnetic field is upward $\uparrow \begin{smallmatrix}N\\S\end{smallmatrix}$. The induced current tries to maintain this original magnetic field direction. Applying the left-hand rule for magnetic field of a coil we find the direction of induced current in the coil. At position (4) the induced voltage becomes instantaneously zero because the lines of force of the magnet are not cutting across the wires of the coil. At (4), the rotor magnet establishes a downward $\begin{smallmatrix}S\\N\end{smallmatrix}\downarrow$ field, which the induced current in steps (5) and (6) tries to maintain. As the magnet continues rotating, the induced current continually surges back and forth. One complete rotation of this simple 2-pole magnet produces one cycle of induced alternating current. Rotating the magnet at a speed of 60 revolutions per second (3 600 rpm) produces ordinary 60-hertz ac for distribution to homes and industry.

As an aid in describing useful details of the behavior of alternating currents, these continuously varying voltages and currents

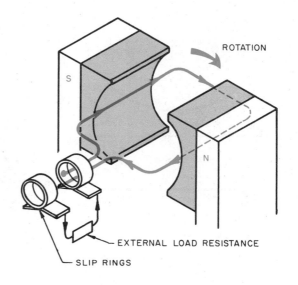

Fig. 10-20 Simple AC Slip-Ring Generator

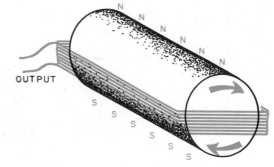

Fig. 10-21   Cylindrical Rotor for AC Generator

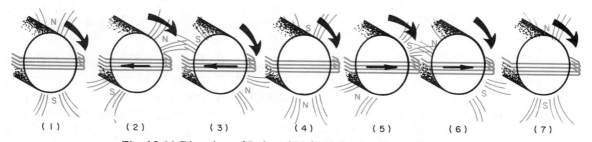

Fig. 10-22 Direction of Induced Voltage During One Turn of Rotor

are shown in graphical form, as in figure 10-26. On the vertical scale of this graph is shown the amount of voltage assumed to be generated. The numbers (1) to (7) in figure 10-26 are to compare with numbers in figure 10-22. At position 1, generated emf is zero, and at a time between 2 and 3, at the moment the magnet poles passed close to the generator wires, the maximum volts (assumed 150) are generated. At 4, emf sinks to zero. Between 4 and 7, the voltage is plotted below the zero axis to show its change in direction. Between time-instants 5 and 6, the generated voltage is at 150 for an instant, which is another voltage maximum. The plus and minus signs on the vertical scale at the left refer only to directions of voltage (minus 150 is just as strong as plus 150).

The horizontal time-scale could be marked off in fractions of a second. If the graph were to show the changes in voltage for so-called 60-Hertz ac (which means 60 cycles per second), then the time for one cycle is 1/60 second. Usually, instead of a time-scale in seconds, a scale of degrees is used. This is to be compared with degrees of angular rotation in the simple 2-pole

Fig. 10-23 Stator Winding, 450 000 kW Alternator

Fig. 10-24 4-Pole Rotor for Stator

Fig. 10-25 Steam-Turbine Rotors for 125 000 kW Alternator

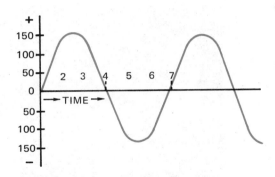

Fig. 10-26 Graph of Alternating Volts

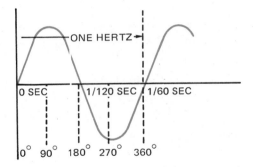

Fig. 10-27 AC Sine Graph

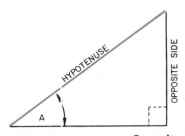

Fig. 10-28 Sine A = $\dfrac{\text{Opposite}}{\text{Hypotenuse}}$

generator. One complete rotation (360°) of the magnet produces one complete cycle (360°) of electron vibration.

Due to its shape, the graph, figure 10-27, is called a *sine graph*. The voltages produced in the ac generator increase and decrease in proportion to the sine of the angle of rotation. You may have already used a ratio called sine of an angle in right-triangle calculations. In this right triangle, the sine of angle A is the number found by dividing $\dfrac{\text{side opposite the angle}}{\text{hypotenuse}}$. For example, if in this triangle, the opposite side is 3 inches long and the hypotenuse is 5 inches, then the sine of Angle A is 3/5 = 0.6.

In order to apply the sine number for angles larger than 90°, a better definition of the sine is needed.

Think of a line, L, that rotates around a pivot point, like the spoke of a wheel. Give attention to the angle (angle A) that opens as the line sweeps away from the right end of the horizontal reference axis, figure 10-29.

The sine of Angle A is the ratio $\dfrac{a}{L}$. The vertical distance from the free end of line L to the horizontal axis is "a". This amounts to the same definition as the right triangle's $\dfrac{\text{side opp}}{\text{hyp}}$.

Let L rotate past 90°. The sine of angle A is the ratio $\dfrac{a}{L}$, in which "a" is still the vertical distance from the free end of L to the horizontal axis.

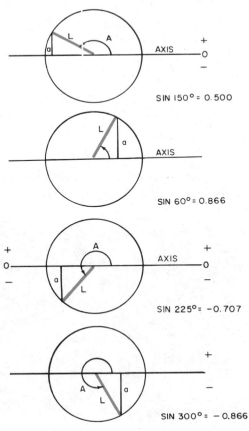

Fig. 10-29 Sine of Angle A = $\dfrac{a}{L}$

As L continues to sweep around, past 180°, the sine of the angle is still $\frac{a}{L}$ with "a" still the vertical distance from the end of L to the axis. But "a" is now below the horizontal axis, in the graph's negative direction. The sine $\frac{a}{L}$ is now a negative quantity, figure 10-29.

As L passes the three-quarter mark, sine A is still $\frac{a}{L}$ with "a" defined as before. Now, "a" is in the downward direction, indicated by writing "a" as a negative number, so the sine $\frac{a}{L}$ is negative.

Returning to the ac generator, the rotating magnet could be a permanent magnet, but all large generators use an electromagnet.

A small dc generator, called the *exciter,* supplies dc for the rotor magnet. Use of an electromagnet permits control of magnetic field strength, thereby permitting regulation of output voltage while keeping the frequency constant. This rotating-field type of ac generator is often called an *alternator.* These are used extensively in automobiles.

Frequency

Many steam-turbine driven alternators use a 2-pole magnet, rotating at 3 600 rpm to produce 60-hertz ac. Some alternators use a 4-pole field magnet running at 1 800 rpm. Since the frequency (cycles per second) of the ac is determined by the number of poles that sweep past the coil wires per second, 4 poles going by 1 800 times per minute produce the same number of alternations as 2 poles going by 3 600 times. The name *hertz* (Hz) means cycles per second. The relationship of poles, frequency, and revolutions may be written:

Hertz (cycles per second) equals pairs of poles times revolutions per second.

For the simple 2-pole magnet, which has one pair of poles, 60 hertz = 1 pair times 60 revolutions per second (60 revolutions per second is equal to 3 600 rpm).

In order to produce 60 hertz ac with a 6-pole magnet (3 pairs of poles), 60 hertz equals 3 pair times how many revolutions per second?

20 revolutions per second fits this requirement, or, in other words, the 6-pole field must be run at 1 200 rpm.

Steam turbines operate most efficiently at high speeds, so the high (3 600) rpm is advisable. Water-wheel driven alternators need to have a large slower-moving water wheel, so a rotating field with many poles is used. For example, to find the number of poles on a 150-rpm, 60-hertz alternator: 60 hertz = X pairs x 2.5 revolutions per second; X = 60/2.5 = 24 pairs, which is 48 poles.

Regardless of the rpm of the field magnet, the sine graph of figure 10-27 represents the generated ac, with each cycle divided into 360°. From now on, interpret these so-called electrical degrees as convenient subdivisions of a cycle.

The Alternator in the Automobile

The recent development of cheap, reliable high-current silicon rectifiers made it possible to use alternators plus rectifiers, instead of dc generators, to keep automobile batteries charged. A *rectifier,* also called a *diode,* is a device that permits electron flow through it only in one direction, even though an alternating voltage may be applied to it. Briefly, it changes ac to dc. There is more information about diodes in Chapter 17.

The automotive alternator is similar in construction to the rotating-field alternators described previously. To make it work well at low speeds, it needs several pairs of poles on the rotor. Large alternators have many coils arranged in machined slots in the rotor, figure 10-31, to develop several

Fig. 10-30 Motor-Generator Set

pairs of poles on it. A cheaper method of construction was developed for the small alternator. Instead of arranging coils as in figure 10-31, one coil is wound around the laminated iron rotor as in figure 10-32. The magnetic field shown in figure 10-32 is useless for generating currents in the stator; the field must be reshaped by surrounding this rotating coil with an iron shell. Figure 10-33 shows, in a cross section of this shell, how the magnetic lines of the coil are collected by the iron and

brought around to form poles at the surface of the rotor. Figure 10-34 shows the essential feature of the surface of the rotating iron shell. The magnet poles indicated in figure 10-33 actually form on the stubby triangular fingers projecting from each half of the shell. The resulting magnetic field between the north fingers and south fingers is enough like the field between the poles shown in figure 10-31

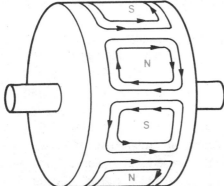

Fig. 10-31 Alternator Rotor with Several Pairs of Rotating Poles

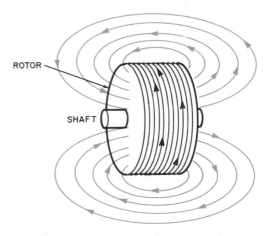

Fig. 10-32 Single-Coil Laminated Iron Core Alternator Rotor

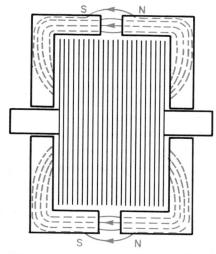

Fig. 10-33 Cross Section of Iron Shell of Rotor

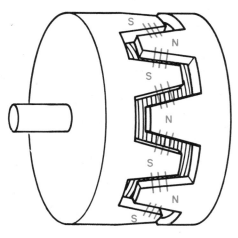

Fig. 10-34 Surface of Rotating Iron Shell of Alternator

so that this simple-to-build structure, figure 10-34, serves very well as a multipole rotating field. Automotive alternators have 4 to 7 pairs of poles on the rotor. Figure 10-35 is a more accurate sketch of the actual appearance of a rotor. The ends of the coil in the rotor are brought out to a pair of insulated metallic slip rings. The brushes that bring direct current to the rotor rest on these rings.

The rotor is surrounded by stationary coils held by the end housing, figure 10-36, page 254. The slip-ring end housing carries the brushes and the rectifiers. To make good use of the multiple poles on the rotor, the

stator assembly requires many coils. The whole assembly consists of three sets of coils; one set of coils for a 14-pole rotor is shown in figure 10-37. All three are shown in figure 10-38, page 254. The seven little coils in one set, figure 10-37, are in series, so the voltage output of each set is seven times the voltage generated in a single coil. The three sets are overlapped in such a way that each set of coils produces its maximum voltage at a time when the other two sets are producing small voltages. In this alternator, the purpose of the three-winding arrangement is to make the output rectified (dc) current steady, rather than a rough off-and-on pulsation.

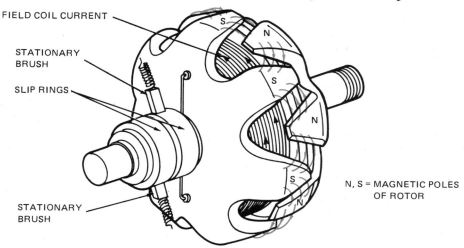

Fig. 10-35 Rotating-Field Magnet for Automobile Engine Alternator

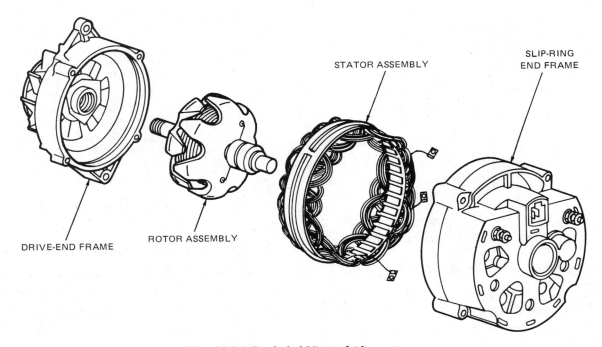

STATOR ASSEMBLY

SLIP-RING END FRAME

DRIVE-END FRAME

ROTOR ASSEMBLY

Fig. 10-36 Exploded View of Alternator

The large commercial generators described previously also have three sets of coils, which produce three alternating currents which are out-of-step with each other. The commercial reason for the three currents is that they make more economical distribution of energy possible. It is also required for the operation of three-phase ac motors. Details of ac motor operation are in Chapter 16.

Alternator stator windings are illustrated in figure 10-38. It shows three coil ends brought out and marked A, B, C, with the other ends fastened to each other. Figure 10-39 is a *schematic diagram* instead of a picture. It is not intended to show what the coils look like. It is intended to show the connections clearly and simply. And the 120° separation of the coils is intended as a

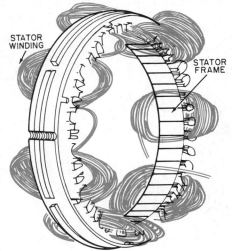

STATOR WINDING

STATOR FRAME

Fig. 10-37 One Set of Stator Coils

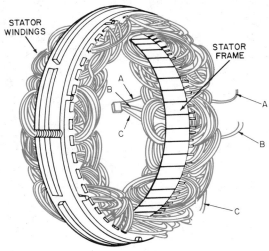

STATOR WINDINGS

STATOR FRAME

A

B

C

A

B

C

Fig. 10-38 Complete Stator Assembly

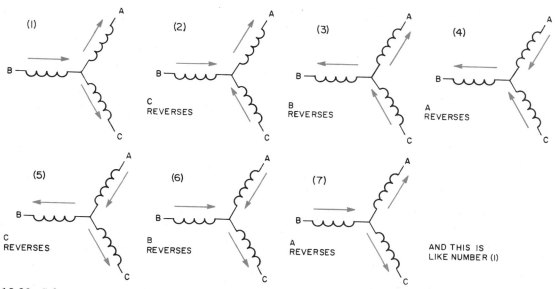

Fig. 10-39 Schematic Diagram Showing Connections and Relationship of Coil Currents as Rotor Turns (Three-Phase Generator)

reminder that the timing of the alternations in each coil is equally out-of-step with the alternations in the other two coils. The coil currents take turns at reversing at equal time intervals. Figure 10-39 shows successive instantaneous views of current direction, taken at equal time intervals.

The alternating currents in alternator wires A, B, and C are rectified to dc by six diodes, two for each wire, Figure 10-40 pictures a diode about three times its actual size. Figure 10-41, page 256, shows all six of the rectifying diodes (the circled R's). Understanding the function of one pair covers the rest. As arranged in figure 10-41, each diode will let electrons flow through it to the right, but it stops electrons trying to get through to the left. The stator coil marked B feeds alternating current to the junction between two rectifiers at the top. When coil B is pushing electrons out of the generator toward the rectifiers, they cannot flow through the rectifier at the left. They do flow, however, through the one at the right, pushing electrons toward the negative terminal of the dc output and maintaining a negative charge on that

terminal. When coil B's current reverses, electrons flow from the rectifier assembly into coil B. These electrons are pulled off the positive terminal (keeping it positively charged) through the rectifier at the left and then into coil B. The two rectifying diodes at the top conduct alternately. The diode at the left lets coil B suck electrons off the positive terminal; the diode at the right lets coil B push electrons on to the negative terminal.

The same thing as above happens to the other two coils and their pairs of rectifiers. Within each pair, the rectifiers conduct alternately. The three stator coils take turns at

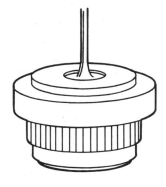

Fig. 10-40 Silicon Diode

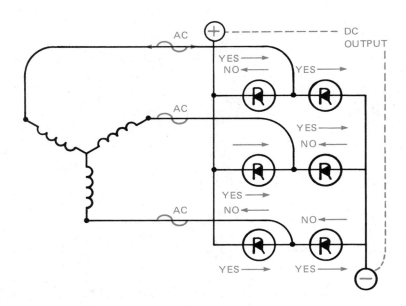

Fig. 10-41 Schematic Diagram of Arrangement of Diodes and Their Relationship to Coils

supplying electrons to the negative terminal and removing them from the positive terminal. The negative terminal is never left without an active electron supply, for at least one and sometimes two of the coils will be pushing electrons toward it. Likewise, at all times at least one coil is removing electrons from the positive terminal. Electrons can flow from the negative terminal toward the positive terminal only through the external load because the rectifiers themselves block any attempt of electrons to flow back through the rectifiers.

The external load connected to the generator output consists of the battery to be kept charged, the ignition system, any lights and accessories in use, and the field coil of the generator. Because the voltage output of a generator, either ac or dc, depends on the speed at which it is run by the engine, a voltage-regulating system is required with all automotive generators. Various models of regulators differ in so many small details that they cannot all be described here. But they are all alike in this respect: they control the voltage output by controlling the current in

the field coil of the generator — either the stationary field coils of a dc generator or the rotating field coil of an alternator. The control is accomplished by the electromagnet of a relay. The current in the relay coil is proportional to the voltage applied to it. Full battery voltage is applied to the generator field by the relay contacts when the voltage of the car battery is low due to discharged battery, low engine speed, or too many lights, lighters, radios, etc., in use. High field current will make the generator produce the maximum voltage possible at the speed it is operating. A small rise in voltage will increase the pull of the regulator electromagnet coil, which operates contacts that put a resistor in series with the field coil, cutting down on the field current. The third step in voltage control occurs when voltage rises a little higher. At that point field coil current is cut off entirely. During normal operation, contacts in the voltage regulator are opening and closing many times per second, increasing and then decreasing the field current so that a predetermined average voltage output is maintained.

POINTS TO REMEMBER

- Generators produce electrical energy; they are driven by other energy sources.

- Rotation of coils in a stationary magnetic field induces alternating voltage in the coils. A rotating commutator changes this alternating voltage to one-way (dc) voltage pulses in the external-load circuit of a dc generator.

- Useful dc armatures carry several coils in a series-parallel arrangement. There are two or more parallel paths through the armature, each path consisting of several coils in series.

- The magnetic field of the dc generator is usually supplied by current from the generator itself. This process is called self-excitation.

- Rotation of a magnet inside stationary coils induces alternating voltage in the stationary coils. This structure forms the usual ac generator (alternator).

- The rotating field magnet of large alternators is magnetized by dc from a small dc generator.

- The automobile alternator generates ac which is rectified by diodes, giving a dc output.

- Alternator:

$$\frac{\text{revolutions}}{\text{second}} = \frac{\text{hertz}}{\text{pairs of poles on rotor}}$$

REVIEW QUESTIONS

1. What is the purpose of a generator?

2. What is the difference between a dc generator and a dc motor?

3. What is the purpose of the commutator in a dc generator? How does a dc generator differ from an ac generator?

4. From figure 10-42, determine which brush is positive and which brush is negative.

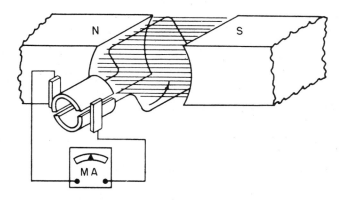

Fig. 10-42

5. How does the current from a dc generator differ from the current supplied by a battery?

6. See figure 10-22. When the magnet is rotating, is there any torque tending to move the coil?

7. How does the current from an ac generator differ from the current of a dc generator?

8. The speed of a 6-pole alternator is 1 200 rpm. Determine the frequency of the output of the generator.

9. A 25-hertz alternator has 2 poles. Determine its speed in rpm.

10. Why is 60-hertz alternating current service used in preference to a frequency of 25 hertz in most areas of the United States and Canada?

11. What does sine of an angle mean?

12. When a 12-pole (6 pairs) automotive alternator is running at 5 000 rpm, what is the frequency of the ac generated?

RESEARCH AND DEVELOPMENT

Experiments and Projects on Electromagnetic Generators

INTRODUCTION

A generator changes mechanical energy into electrical energy by moving wires through a magnetic field, or by moving a magnetic field through a coil of wire. In Chapter 10, the principles and characteristics of direct- and alternating-current generators were explained.

The experiments which follow are arranged so you can set up equipment to produce direct and alternating current, and so you can examine commercial-type equipment used to produce current electricity. In the project section, suggestions are made for the examination of commercial-type generators as well as the designing and constructing of models that produce current.

EXPERIMENTS

1. To assemble an experimental dc generator, study principles of operation, and record results.

2. To study the construction of an automotive-type generator, to operate the generator, and measure current output.

3. To study the polarity characteristics of ac and dc generators.

4. To study the waveform of the output of a dc generator and of an ac generator.

PROJECTS

1. Examine a commercial-type generator.

2. Dismantle an automobile generator and a small ac-dc motor, such as used in a vacuum cleaner or electric hand tool. Examine the armatures and coils. Note the construction of each.

3. Design and build a simple dc generator to illustrate the production of current electricity.

4. Design and build a simple ac generator.

EXPERIMENTS

Experiment 1 THE DIRECT-CURRENT GENERATOR

OBJECT

To assemble an experimental dc generator, study principles of operation, and record results.

APPARATUS

1 - DC generator kit
1 - Galvanometer
1 - Miniature lamp receptacle
1 - Miniature lamp, 1 1/2-volt

PROCEDURE

1. Examine each part of the generator, read the instructions, and assemble it.

2. Connect a galvanometer to the terminals of the generator as shown in figure 10-43.

3. Rotate the generator shaft slowly in a clockwise direction. Observe the needle on the galvanometer.

4. Rotate the generator shaft rapidly and observe the galvanometer reading.

5. Rotate the generator shaft in a counterclockwise direction and observe the needle on the galvanometer.

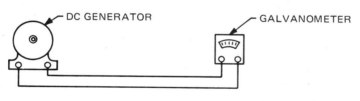

Fig. 10-43

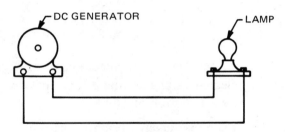

Fig. 10-44

6. Connect the miniature lamp to the generator as shown in figure 10-44.

7. Rotate the generator at different speeds and observe the amount of light in the lamp.

8. Disassemble the generator kit.

9. Record your observations.

QUESTIONS

1. Did the galvanometer read the same when the generator was operated at various speeds? Explain.

2. How did the speed of the generator affect the amount of light from the lamp?

3. Did you notice any difference in the meter reading when the generator was rotated counterclockwise? If it was different, how do you account for the difference?

4. Name the parts of the generator.

5. Explain the purpose of each part.

Experiment 2 CHANGING MECHANICAL ENERGY TO ELECTRICAL ENERGY

OBJECT

To study the construction of an automotive-type generator, to operate the generator, and measure the current output.

APPARATUS

1 - DC automobile generator
1 - 115-V AC motor, 1/2-hp. (370-watt)
1 - Mounting board, fastening devices, and belt
1 - Wattmeter
1 - DC ammeter (0-10 A)
1 - DC voltmeter (0-150 V)
3 - Lamp receptacles
    Variety of lamps
1 - DPST switch

Note: An alternator may be substituted for the generator. It would be advisable to have each experiment performed.

PROCEDURE

1. Disassemble generator, examine parts, and reassemble the generator.
2. Mount generator on mounting board with motor, attach belt, and hand test operation.
3. Assemble other equipment according to figure 10-45.

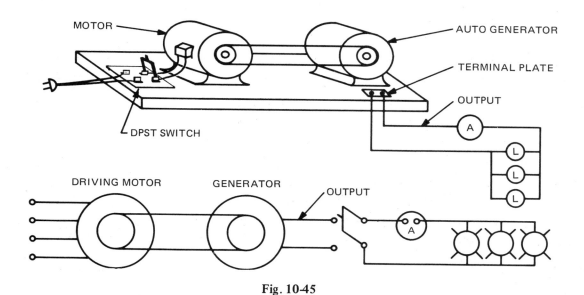

Fig. 10-45

4. Check the wiring.
5. Run the generator.
6. Measure and record the voltage and current when generator is delivering full load current and partial load.
7. Use wattmeter to measure power going into the motor when generator is producing about 5 amps and when it is producing about twice that much current. Record readings.
8. Find efficiency of motor-generator set.

OBSERVATIONS

Use a form similar to the one shown here to record your observations.

Generator Voltage . . . .   Full load   _____
                                                    Partial load   _____
Current . . . .   Full load   _____
                                                    Partial load   _____
Power consumed by motor under . . . .   Full load   _____
                                                    Partial load   _____
Efficiency . . . .   _____

QUESTIONS

1. Was this method of procuring dc an efficient use of electric power? Explain.
2. Did the motor carry as much current as the ampere rating on the nameplate?

Experiment 3 AC AND DC GENERATORS

OBJECT

To study the polarity characteristics of alternating- and direct-current generators.

APPARATUS

1 - AC-DC experimental generator kit
1 - Galvanometer

PROCEDURE

1. Connect the ac generator arrangement to the galvanometer as shown in figure 10-46.
2. Spin the generator shaft and observe the needle on the galvanometer.
3. Connect the dc generator arrangement to the galvanometer in the same manner.
4. Spin the generator shaft and observe the needle on the galvanometer.

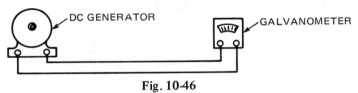

Fig. 10-46

QUESTIONS

1. Did the galvanometer needle always move in the same direction when attached to the dc generator?
2. Was the movement of the galvanometer needle the same when connected to the ac generator? If not, why not?

Experiment 4 AC AND DC GENERATOR WAVEFORMS

OBECT

To study the waveform of the output of a dc generator and of an ac generator.

APPARATUS

1 - DC generator
1 - AC generator
1 - Oscilloscope
1 - SPST switch

PROCEDURE

1. Connect the dc generator, oscilloscope, and switch as shown in figure 10-47.

2. Rotate the generator and observe the waveform.

3. Follow the same procedure with the ac generator.

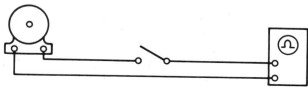

Fig. 10-47

QUESTIONS

1. Explain how the waveform for dc was different than the waveform for ac.

2. Make a labeled diagram of each.

PROJECTS

MAKE AN EXPERIMENTAL GENERATOR

The experimental generator, figure 10-48, can be assembled to produce a small amount of direct or alternating current. It was developed to be constructed in the school shop or at home.

For the production of dc it should be fitted with a segmented commutator, for ac, with a split-ring commutator.

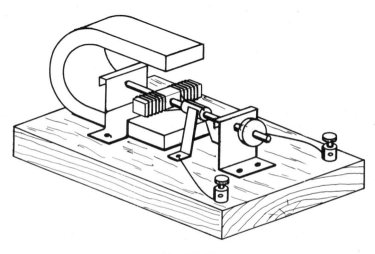

Fig. 10-48

For experimental and demonstration purposes, both type commutators can be placed on the shaft and two sets of brushes employed — one set for dc and one for ac.

MATERIALS

1 - Large horseshoe magnet
1 - Shaft, 1/8" (3 mm) welding rod
1 - Wooden base, size to fit individual plan
2 or 4 - Binding posts
1 - Shaft bearing
1 - Combination shaft bearing and magnet retainer
2 or 4 - Brushes, thin spring brass
1 - Armature, core iron or soft steel, 5/16" or 3/8" (8 mm or 9 mm) square
    Magnet wire, #26 or #28 gauge AWG
    Wood screws, #5 x 1/2" RH

PROCEDURE

Procure a horseshoe magnet. Design the parts according to the size of the magnet.

Make base, shaft bearing and magnet retainer, other shaft bearing, and brushes.

Next, make the armature assembly, figure 10-49. The armature must be a tight fit on the shaft.

To make the commutator, select a piece of brass or copper tubing. Cut a piece about 17/32" (13.5 mm) long. Square the ends and remove the burrs from the inside. Select a piece of hardwood dowel or plastic larger than the ID of the tubing. Place the material in the chuck on an engine lathe and turn it to fit the inside of the tubing. Drill a hole through the center, while it is in the lathe, to make a press fit on the armature shaft. Press the tubing over the dowel, face the end, and cut it off even with the other side of the tubing. Saw slots across each end as indicated on drawing, figure 10-50. The slots should not touch the hole.

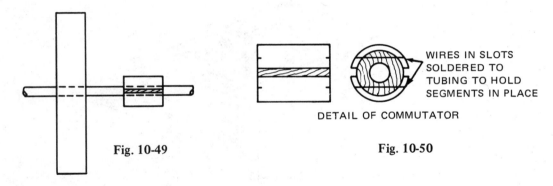

WIRES IN SLOTS
SOLDERED TO
TUBING TO HOLD
SEGMENTS IN PLACE

DETAIL OF COMMUTATOR

Fig. 10-49   **Fig. 10-50**

Press a piece of copper wire into each slot and solder ends to the tubing. Remove the excess solder and ends of the wires with a file. For a dc generator, saw a slot in the tubing on each side as indicated on the drawing. Carefully, remove the burrs. For an ac generator, make two commutators to be used as slip rings, figure 10-51.

If you wish to make a combination ac-dc generator, three are needed as shown in figure 10-52.

Notice how the wires from the armature are connected to the dc commutator and to each of the ac slip rings.

To complete the armature assembly, wind a layer of plastic tape on each side of the core where it will be covered with wire.

Wind two layers of magnet wire on each side of the armature. Start in the middle and end in the middle, figure 10-53. Care must be taken to continue winding in the same direction when going from one pole to the other. Secure the ends with cotton tape. *Note:* Leave about 3" (75 mm) of wire at end of each coil to fasten to the commutator or slip rings.

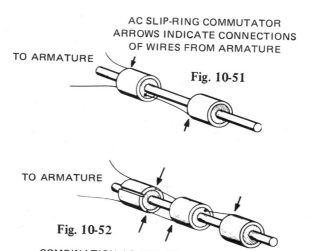

AC SLIP-RING COMMUTATOR ARROWS INDICATE CONNECTIONS OF WIRES FROM ARMATURE

TO ARMATURE

Fig. 10-51

TO ARMATURE

Fig. 10-52

COMBINATION AC AND DC COMMUTATOR ARROWS INDICATE CONNECTIONS OF WIRES FROM ARMATURE

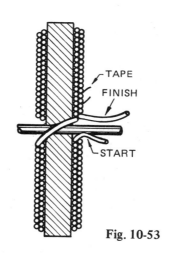

TAPE

FINISH

START

Fig. 10-53

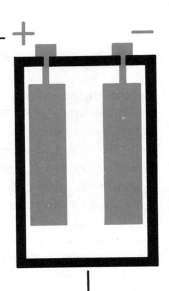

Chapter 11
Cells and batteries

Chapter 10 discussed how electrons could be given energy by the mechanical motion of magnets. A few years before that discovery was made, steady and controllable electric currents were produced by taking advantage of the potential energy that electrons in metal possess.

In 1800 Alessandro Volta, a physics professor, found that the source of energy existed in the combination of two metals and a liquid conductor. Volta built electric cells using various combinations of metals with salt water. He found out how to connect cells in series to get increased voltage and showed that the cells produced effects similar to electrostatic discharges.

A simple, but impractical, cell can be made by putting a strip of zinc and a strip of copper in water that has a little hydrochloric acid dissolved in it, figure 11-1. When a voltmeter is connected to the strips, it shows that about 0.75-volt emf is produced. The energy comes from the electrons in the zinc. More accurately, the energy is possessed by the zinc atoms and the hydrogen ions (H+) of the acid. All acids consist of positive-charged H+ ions and some kind of negative ion. The H+ ions attract electrons. Each zinc atom has two electrons that are energetic enough

to tear themselves away from the remainder of the zinc atom when they also subject to the attraction of the H+ ions.

Some results of transfer of electrons from zinc metal atoms to H+ ions can be seen when a piece of ordinary zinc is put into acid, whether any other metal is there or not. When H+ ions take electrons from the zinc, they become neutral H atoms. Being no longer positively charged, they are no longer attracted to negative ions in the water. They join in pairs, forming H_2 molecules. H_2 is ordinary gas that visibly bubbles away. The circle labeled Zn++ in figure 11-2 represents a zinc ion. A zinc *ion* is an atom that has lost 2 electrons. Zn++ ions are attracted into water by the negative ions of the acid.

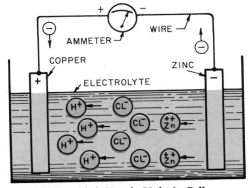

Fig. 11-1 Simple Voltaic Cell

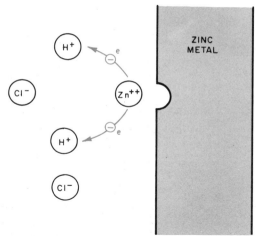

Fig. 11-2 Zinc Ion Leaving Metal

The electron transfer at the surface of the zinc, as described above, is not a useful electric current. In order to make some of those electrons travel a useful path, connect a wire from the zinc to the copper strip in the acid. The copper in the acid provides another place where the H+ ions pick up electrons, figure 11-3. The electrons handed off to the H+ ions at the copper surface are simultaneously replaced by electrons supplied by the zinc, the copper serving as a conductor to carry energetic electrons trying to get away from the Zn to the H+.

This simple cell is impractical because the useless transfer of electrons directly at the zinc surface is so rapid that after several minutes a small strip of zinc gives away all its energetic electrons and dissolves as zinc ions. The zinc metal, as such, is used up.

Leclanché improved this cell by using water and ammonium chloride instead of hydrochloric acid. Ammonium chloride consists of NH_4^+ and Cl^- ions. The ammonium ions, NH_4^+, do not attack the zinc directly and use it up rapidly, as HCl does. Leclanché also used, instead of copper, a carbon rod with manganese dioxide packed around it. Manganese dioxide (MnO_2) is a solid which does not dissolve in water. In manganese dioxide, the manganese atoms are strongly positive charged, making it a good electron attractor.

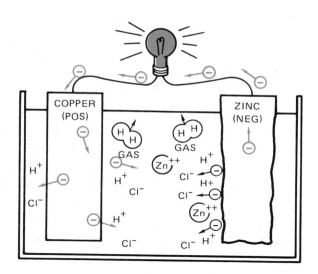

Fig. 11-3 Using Electric Current from a Simple Cell

When a wire is connected from the zinc to the carbon rod, energetic electrons from the zinc flow through the wire and carbon rod, attracted to the positive manganese. At the same time, zinc ions dissolve, attracted into the ammonium chloride solution in the wet blotting-paper lining, figure 11-4.

The carbon powder mixed in with the manganese dioxide, along with the carbon rod acts as an inert conductor to carry electrons to the manganese dioxide in the cell. (The Mn^{++++} atoms pick up two electrons each, becoming Mn^{++}. Mn^{++} ions do not have a strong enough attraction for electrons to have any further use in the cell.) The ordinary dry cell used in flashlights, portable radios, etc., is a sealed Leclanché cell. The inside of it is not dry, but a moist paste, when new. Often a zinc can is used, so the container is also the negative electrode. Dry cells cannot be effectively recharged.

PRIMARY CELLS

The name *primary cell* is given to those cells in which the electron transfer is not readily reversible. Primary cells are not intended for recharging. Cells that can be

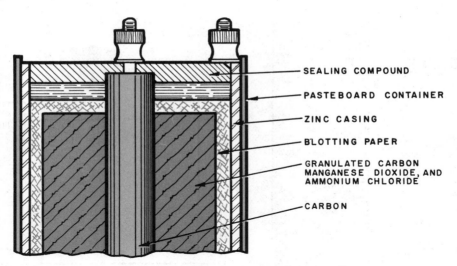

Fig. 11-4 The Leclanché Primary Cell

recharged, such as car batteries, are called *secondary cells.* The most common primary cell, the carbon-zinc, has been described. Some others are the alkaline dry cell, mercury cell, silver-zinc cell, and Lalande cell.

The alkaline dry cell is basically similar to the ordinary dry cell. However, it uses potassium hydroxide solution instead of ammonium chloride. Comparing alkaline cells with ordinary cells of the same size, the alkaline cell can produce higher continuous current because it has less internal resistance.

The *mercury cell* has two special advantages: (1) a high energy content in proportion to its size and weight, and (2) maintenance of voltage 1.35 over relatively long periods of time. These characteristics make it valuable for hearing aids, walkie-talkie sets, and other portable electronic equipment. As usually constructed, the center terminal is negative, connecting to a zinc cylinder or pellet, while the positive case connects to a mixture of mercuric oxide and carbon. The electrolyte is potassium hydroxide. The above three cells are the most widely used dry cells.

The *silver-zinc cell* is similar in construction to the mercury cell, but uses silver oxide as the positive plate. Voltage (1.5)

is quite constant throughout its life, and it has a higher energy-to-weight ratio than any other cell.

A *2-volt magnesium cell* uses magnesium bromide electrolyte and manganese dioxide at the positive terminal.

For special duties, primary cells which become active only when water or electrolyte is added are available. The Lalande cell (0.8 volts) uses zinc and copper oxide; the electrolyte is sodium hydroxide solution. The air cell (1.3 volts) also uses zinc with sodium hydroxide as an electrolyte, but oxygen from air diffusing into a porous carbon electrode is the active material of the positive plate. Other water-activated cells use a magnesium negative and copper chloride or silver chloride as the positive electrode.

Battery energy is expensive. Under favorable conditions, a standard flashlight cell might produce 5 watt-hours of energy. If the cell costs 39 cents, the buyer is paying $78.00 per kilowatt hour for energy. From mercury cells, energy costs over $300.00 per kilowatt hour.

Despite the high costs of energy, compared to 3 cents per kilowatt-hour for power-line energy, batteries must be used in mobile equipment such as submarines, automobiles, aircraft, and bicycle-mounted flashlights. When power line service is interrupted by storms, batteries provide emergency lighting and communication service.

Anodes and Cathodes

The terminal where electrons enter a device is called the *cathode;* the terminal where they leave the device is called the *anode.* Looking at the energy *user,* which may be a vacuum tube, or an electroplating bath, electrons are forced into the cathode by some other energy source, figure 11-5. Electrons in the device are repelled from the cathode toward the anode, from which point they are attracted away by the external energy source. The cathode is negative (well-supplied with electrons) and the anode is positive. Electrons flow through the device from negative to positive.

Looking at the energy producer, the anode is the electron-rich metal, zinc, the source of electrons supplied to an external energy user such as a lamp. Electrons enter the cell at the cathode, attracted by a relatively positive electrode. The anode is negative, the cathode is positive, energy in the cell pushes electrons out at the anode and pulls them in at the cathode.

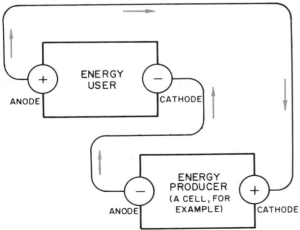

Fig. 11-5 Relationship of Anode and Cathode When Producing and Using Energy

The Fuel Cell

All the previously described cells use as an energy source the tendency of energetic electrons from a metal to transfer to some kind of electron-grabbing particle, usually a positively charged ion. During this process, the anode material itself is "used up" in the sense of being converted to some useless low-energy form. The fuel cell also uses this same type of electron transfer, with the important difference that the electron-giving material and the electron-taking material are fed in from outside the cell. The solid electrodes of the cell are not used up.

During the burning of any fuel — coal, oil, or gas — atoms of the fuel give electrons to oxygen of the air. Oxygen is the electron-grabber. The energy of the electrons immediately appears as heat. It is correct to think of the using-up of zinc in a flashlight battery as a slow oxidation, equivalent to slowed-down burning.

For many years, fairly successful attempts have been made at fuel-cell construction. The purpose is to control the electrons of ordinary cheap fuels, making them perform useful electrical work as they leave the fuel atoms and go to atoms of oxygen or some other electron-taker. Many problems require additional research. One problem is the requirement of high-purity fuel gas. Another problem is the development of cheap and reliable catalyst surfaces on which the essential reactions take place. High electrical efficiency has already been claimed for some experimental cells. The present necessity of auxiliary equipment, such as gas containers and pressure controls, is another problem. Electrical automobiles and tractors powered by fuel cells and a home-sized power plant are possibilities.

The hydrogen-oxygen fuel cell, figure 11-6, page 270, has hollow, porous, carbon electrodes immersed in potassium hydroxide

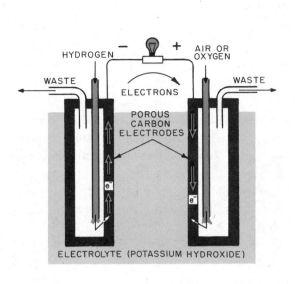

Fig. 11-6 Basic Hydrogen-Oxygen Fuel Cell

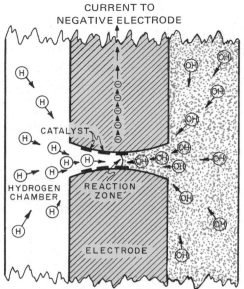

Fig. 11-7 Reaction in Pores of Electrode Between Hydrogen and Hydroxide Produce Electron Flow

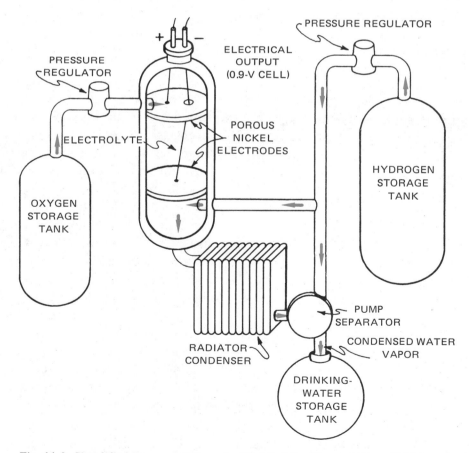

Fig. 11-8 Simplified Representation of a Fuel Cell for Manned Space Vehicles

solution. Hydrogen gas is pumped into one electrode and oxygen into the other. The porous carbon also contains certain metals or metal oxides called *catalysts*. A catalyst promotes a chemical reaction. In this case, the catalyst aids hydrogen molecules, which are pairs of atoms, to separate into single atoms so they can combine with the negative-charged hydroxide ions (OH^-) of the electrolyte, figure 11-7. This combining of H and OH^- forms a molecule of water, H_2O, with one loose electron left over. These surplus electrons are attracted to the oxygen-supplied electrode. On that electrode, electrons are taken up in the combining of oxygen + H_2O + electrons to form hydroxide ions, OH^-.

Hence, hydroxide ions are reformed at the cathode as fast as they are used up at the anode. The important resultant change is the conversion of hydrogen and oxygen to water and the production of electric energy.

SECONDARY CELLS

In a few types of cells, the electrochemical action is reversible. That is, if we connect the discharged cell to another electron source and push electrons back into the negative terminal of the cell, the materials in the cell will be restored to their original energetic form. Such cells are called *storage cells*, or *secondary cells*. In table 11-9, five types are listed in order of increasing cost.

STORAGE CELLS (Secondary Cells)

Type	Anode	Cathode	Electrolyte	Advantages	Uses
Lead-acid	Lead	Lead dioxide	Dilute sulfuric acid	Cheapest	Autos, submarines, aircraft
Edison (nickel-iron)	Iron	Nickel oxides	Potassium hydroxide solution	Long life without careful maintenance	Electrical traction, auxiliaries, marine, lighting
Nickel-cadmium	Cadmium	Nickel hydroxide	Potassium hydroxide solution		Engine-starting, controls and communications
Silver-zinc	Zinc	Silver oxide	Potassium hydroxide solution	High energy-to-weight ratio	Military, satellites, communications
Silver-cadmium	Cadmium	Silver oxide	Potassium hydroxide solution		

Fig. 11-9

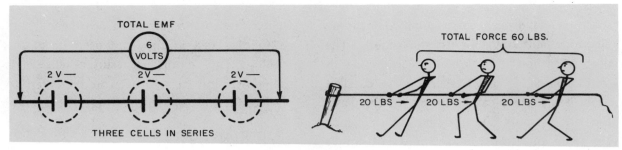

Fig. 11-10 Cells Connected in Series

Cells Connected in Series

As has been suggested already, cells are often connected in series in order to obtain higher voltages. The voltage produced by several cells in series is the total of the individual cell voltages. For example, three 2-volt lead storage cells in series produce 6 volts; six such cells in series produce 12 volts; four 1.5-volt dry cells in series produce 6 volts. To produce 90 volts from dry cells, sixty cells in series are required.

Cells Connected in Parallel

When the negative terminals of several cells are connected together, as in figure 11-11, one large negative plate is formed, and the connected positive plates act like one large positive plate. No increase in emf is obtained with this arrangement, but if the load resistor requires a total current of 15 amps, it can be provided by 5 amps through each cell.

This may be illustrated by the water-pump comparison. Three pumps in parallel pump more water into the open tank on top than one pump does. But assuming the water level is kept steady, the pressure is the same as that produced by one pump. If each is pumping 5 gallons (19 liters) per second, 15 gallons (57 liters) per second can run out through the waterwheel. This corresponds to the resistance load of the electrical circuit.

LEAD STORAGE BATTERY

A *battery* is a number of cells connected in series or in series-parallel.

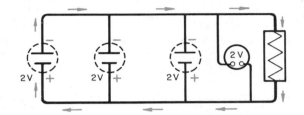

HEIGHT OF TANK SUFFICIENT TO MAKE 20 LBS. PER SQ. INCH (138 KILOPASCALS) PRESSURE

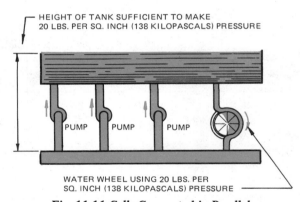

WATER WHEEL USING 20 LBS. PER SQ. INCH (138 KILOPASCALS) PRESSURE

Fig. 11-11 Cells Connected in Parallel

Each lead cell produces 2 volts. Six-volt automobile batteries have 3 cells in series; 12-volt batteries have 6 cells in series, figure 11-12.

The negative plate consists of metallic lead. When the cell is producing current, lead atoms on the surface of the plate lose two electrons each, becoming Pb^{++} ions. These Pb^{++} ions do not dissolve into the liquid, but remain on the plate and attract SO_4^{--} ions from the sulfuric acid solution, thus forming an invisibly-thin layer of $PbSO_4$ on the negative lead plate.

The positive plate consists of lead dioxide, PbO_2. Each lead particle is lacking four electrons, which were given to the

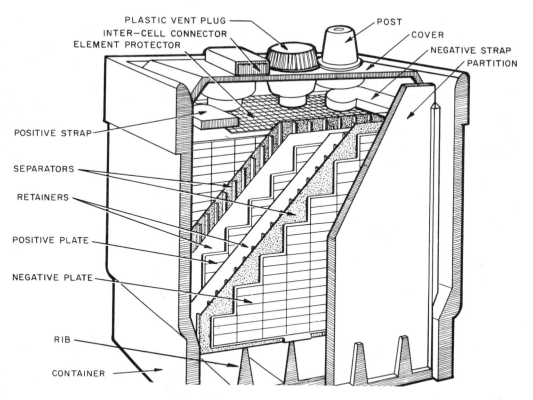

Fig. 11-12 A Lead Storage Cell

oxygen when the plate was formed. Each Pb^{++++} ion takes two electrons from the external circuit, becoming Pb^{++}.

The energy is obtained from the tendency of neutral lead atoms to give 2 electrons each to Pb^{++++} ions, both becoming Pb^{++} as a result of the transfer.

When the Pb^{++++} ions of the lead dioxide pick up the two electrons, they can no longer hold the oxygen, which goes into the acid solution and combines with hydrogen ions of the acid, forming water molecules. The Pb^{++} remains on the plate and picks up SO_4^{--} from the sulfuric acid solution, forming lead sulfate. These actions are shown in the diagram, figure 11-13, page 274. This chemical action can go on only where the plates are in contact with the sulfuric acid solution. In order to produce a large current, the plates are made so that a great deal of surface area is in contact with the

solution. In a cell, the plates are arranged as shown in figure 11-14, page 274. The negative plate is made of lead sponge, and the positive lead dioxide plate is also a porous structure permitting a large area of material to be wet by the electrolyte. Separators of wood, glass fibers, or similar porous material keep the plates from touching each other. To provide mechanical strength, both plates consist of an open framework of a lead-antimony alloy, into which the active material is pressed. The electrolyte is dilute sulfuric acid of specific gravity 1.28.

Batteries may be shipped wet, filled with the electrolyte, or dry, with the electrolyte in a separate container.

When charging a storage battery, place it in a well-ventilated room. Follow the manufacturer's instructions to connect the proper leads from the charger to

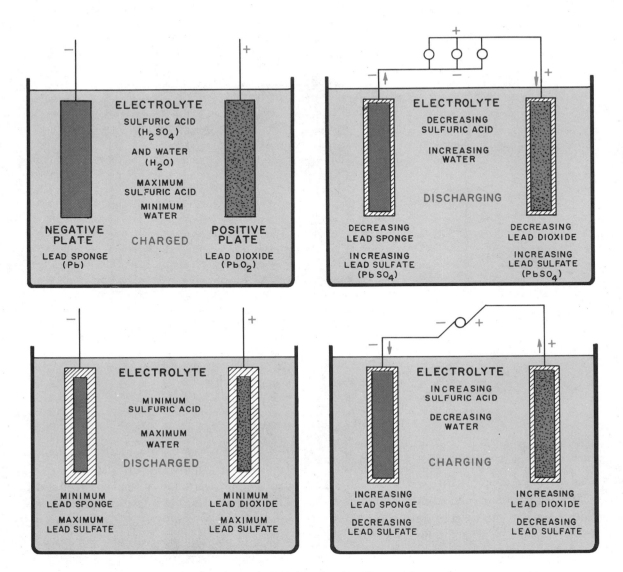

Fig. 11-13 Chemical Action in a Storage Battery

the battery terminals before power is applied. Always disconnect power before removing charger leads from battery terminals, since a spark could ignite the highly explosive hydrogen gas which is usually present during the charging process.

BATTERY CHARGING

A rectifier or a dc generator is a necessity for battery charging. The battery is charged by forcing electrons through it opposite in

Fig. 11-14   Top View of Plate Arrangement in a Storage Cell

direction to the current the battery normally produces.   In the lead cells, and in other storage cells, this reversal of current reverses the chemical changes that took place when the cell was furnishing energy.

Electrons forced onto the negative plate attract H^+ ions from the acid solution in the battery. The H^+ ions combine with sulfate ions (SO_4^{--}) from the plate, forming sulfuric acid again. The electrons put onto the negative plate change the Pb^{++} ions back to plain lead.

At the positive plate, H_2O decomposes, the hydrogen combining with the SO_4^{--} on the plate to make more sulfuric acid, and the oxygen combining with the lead to form lead dioxide again. The Pb^{++} of the discharged positive plate is converted to Pb^{++++} as the generator pulls electrons away, figure 11-15(B).

Ohm's Law in the Battery Charging Circuit

In a charging circuit such as shown in figures 11-15(A) & (B), the generator voltage has two duties: (1) it must equalize and overcome the battery emf, and (2) it must also produce enough volts (=IR) to produce current through the resistance in the circuit.

Example: If the rheostat is set for 2 ohms resistance, and a 15-volt generator is charging a 6-volt battery, find the current. First, subtract the 6 volts from the 15 volts, which leaves 9 volts to produce current through the 2 ohms resistance in the circuit (assuming that the ohms resistance of the rest of the circuit is small enough to neglect).

$$I = \frac{9 \text{ volts}}{2 \text{ ohms}} = 4.5 \text{ amps.}$$

A similar calculation would enable estimation of the ohms resistance necessary in the circuit to limit the charging current to a reasonable rate. Assume a generator produces 24 volts, and the battery and wiring has 0.4-ohms resistance.   A current of 6 amps is desired to charge a 6-volt battery. Calculate the necessary resistance to be added:

1. 24 − 6 = 18 volts to be used in the resistance of the circuit.

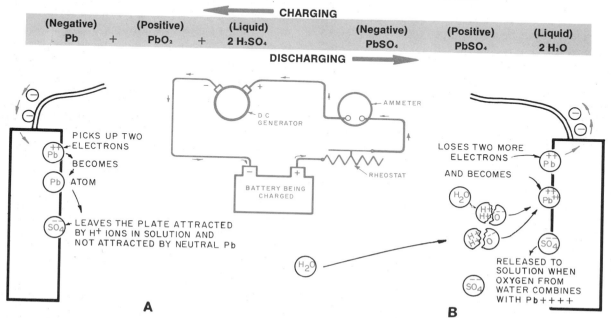

Fig. 11-15 (A & B) Chemical Action in a Cell During Charging Process

2. Total resistance of the circuit =

$$\frac{E}{I} = \frac{18}{6} = 3 \text{ ohms}$$

3. Circuit has 0.4 ohms already; 3 – 0.4 = 2.6 more ohms must be added in series in the circuit.

A too-high charging rate can damage a battery by (1) overheating and (2) gas-bubble formation inside the spongy plate material which forces active material to break away from the plate structure. The safest charging procedure is a rate of 10 amps or less, requiring about 24 hours. A battery can be charged on a constant-voltage circuit in 6 or 8 hours, starting with a 30- or 40-ampere rate which tapers down as the battery charge builds up. With this method, however, the battery should be checked to see that it is not overheating. 100°F (43°C) is the common accepted limit.

Battery Testing

A hydrometer measures the specific gravity of the electrolyte in each cell. A charged battery has enough sulfuric acid in the electrolyte so that its specific gravity is 1.25 – 1.28, (water = 1.00). As the battery discharges, SO_4^{--} of the acid is tied up on the plates and the electrolyte becomes more like plain water, its specific gravity approaching 1.1. Since occasionally a cell may not operate properly even when it has sufficient acid, a better method of battery testing measures the voltage of each cell when it is producing current. A good cell can produce 20 or 30 amps and maintain a 2-volt potential difference across the cell. A dead cell may read 2 volts when it does not have to produce current, but its voltage drops off as soon as a low resistance is placed across the cell.

> Be extremely careful when testing a storage battery with a hydrometer. Always hold the rubber tube on the hydrometer over the battery filler hole while taking the reading. Avoid getting drops of sulfuric acid on your hands or clothing. If acid comes in contact with your skin or clothing, neutralize it with baking soda or cold water.

The Ampere-Hour Rating

An *ampere-hour* (A.H.) is the amount of charge delivered by one ampere in one hour (1 ampere-hour = 3 600 coulombs). The ampere-hour rating of a battery is usually found from its ability to produce current for 20 hours, at 80°F. (26.7°C). A battery that can produce 6 amperes steadily for 20 hours deserves a 120 ampere-hour rating. A 120-ampere-hour battery can produce more than 120 ampere-hours if its discharge rate is 1 or 2 amps instead of 6 amps. It cannot produce 120 amps for one hour; the actual ampere-hour production depends on the current.

Battery Care

Particular care should be taken to avoid getting dirt into a cell. When the cap is removed, it should be set in a clean place, if it has to be set down at all. Dirt, especially a few flakes of iron rust, can spoil a cell permanently. Probably a great many automobile batteries are ruined because of the accidental entry of dirt into a cell.

A battery should not be allowed to remain in a discharged condition. If a battery is completely run down, it should be charged within a few hours at a slow rate. If allowed to stand discharged, the lead sulfate in the battery apparently hardens or crystallizes into a formation that is difficult to restore to lead and lead dioxide. The watery solution in a discharged battery can freeze in winter, breaking up the battery. That is another reason for keeping it charged.

The liquid should be maintained at a level that covers the plates. Distilled water is preferable, but faucet water is better than none at all. (Melted frost from the refrigerator is distilled water.) While water is lost from a battery mainly by evaporation, there is a slight loss from hydrogen and oxygen formed during charging. Since acid is lost only by cracking the case or tipping it over, acid seldom is needed.

EDISON AND NICKEL-CADMIUM STORAGE CELLS

The Edison nickel-iron battery and the nickel-cadmium battery (Junger & Berg, Sweden, 1898) are truly long-life batteries. Both are structurally stronger than the lead battery and not as subject to damage by extreme heat or cold. Both are often called alkaline cells, referring to the nature of the electrolyte, a 21 percent solution of potassium hydroxide, which is chemically a base or alkali rather than an acid.

The Edison cell was originally developed to serve as energy source for electric motor driven vehicles. It has less volume and less weight than a lead cell of the same ampere-hour rating and can withstand charging and discharging daily or more often for years. Its ampere-hour production is not harmed by over-charge or over-discharge and, unlike a lead cell, it is not damaged by remaining in a discharged condition for a long time.

The negative plates of the Edison cell consist of a nickeled-steel grid containing powdered iron, with some FeO and $Fe(OH)_2$. The iron is the source of electrons, which are attracted through the external circuit toward nickel ions, N_i^{++} and N_i^{+++}, on the positive plate. The positive plates are nickel-plated tubes containing a mixture of nickel oxides and hydroxides, with flakes of pure nickel for increased conductivity.

The disadvantages of the Edison cell are high initial cost and high internal resistance which limits maximum current, especially when the cell is cold. These disadvantages are enough to prevent its use in many situations. It is used in some portable lighting equipment and a few marine installations, where it neither gets nor needs the attention that lead cells require. It is appropriate for running electrical traction equipment, such as mine locomotives and forklift trucks, but not appropriate for starting engines.

The nickel-cadmium battery is a general-purpose battery, able to provide high current without excessive drop of voltage. It is used where ruggedness, long life, and low maintenance requirements outweigh the fact that the original cost of nickel-cadmium is several times the cost of lead cells. Expensive manufacturing processes are required for Ni-Cd cells, and the unprocessed nickel and cadmium are more expensive than lead. These batteries are used in railway signal systems, fire alarm systems, relay and switchgear operation, missile controls, and for starting aircraft engines and diesel engines in locomotives, oil-well pumps, etc.

Both sets of plates of the Ni-Cd battery are mechanically alike, the active materials are held in finely perforated thin flat steel pockets that lock into a steel frame, figure 11-16. The active material in the positive plate is nickel hydroxide mixed with graphite to improve conductivity; cadmium oxide is put into the negative plates. When the cell is charged, electrons forced onto the negative plate combine with Cd^{++} ions of the cadmium

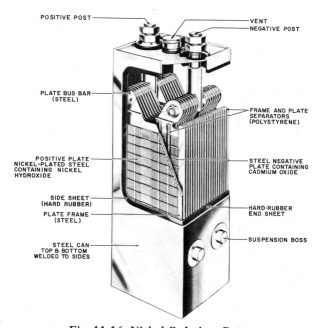

Fig. 11-16 Nickel-Cadmium Battery

oxide (CdO) converting them to plain cadmium metal atoms. On the positive plate, the removal of electrons from Ni^{++} ions of $Ni(OH)_2$ changes them to more strongly positive charged Ni^{+++} ions.

The density of the electrolyte in both the nickel-iron and the nickel-cadmium cells remains constant, and a hydrometer will not indicate the amount of charge.

INTERNAL RESISTANCE

Cells themselves have an internal resistance, which enters into a circuit calculation. In a circuit like figure 11-16 when electrons flow from the negative plate to the positive plate, there is a movement of ions in the electrolyte in the cell. This movement in the cell, like any current, does not go on with perfect ease; there is some resistance in the internal material of the cell.

A new large dry cell may have about 0.035 ohms internal resistance, and a high resistance voltmeter will indicate that emf is 1.5 volts. Assume we connect it, as in figure 11-17, to an external resistance of 0.715 ohms. To calculate the current: the total resistance in the circuit is 0.715 + 0.035

= 0.75 ohms (both resistances must be added for they are effectively in series; there is only one current path) and

$$I = \frac{E}{R} = \frac{1.5}{.75} = 2 \text{ amps}$$

How much would a voltmeter read if connected across the cell terminals while the 0.715-ohm resistor is also there? There are two ways of calculating this voltage:

1. The voltmeter is connected across a 0.715-ohm resistor that has 2 amps through it. E = IR = 2 x 0.715 = 1.43 volts.

2. The 0.035-ohm internal resistance of the cell has a 2-amp current in it. Of the 1.5-volt emf that the cell produces, some is used inside the cell, overcoming this internal resistance.

The voltage used inside the cell is E = IR = 2 x 0.035 = 0.07 volt. This 0.07 volt, subtracted from the 1.5 volts, leaves 1.43 volts for the external circuit. Depending on the internal and external resistances, the cell may or may not always produce 1.43 volts instead of 1.5.

With the same cell as before (1.5-volt emf and 0.035-ohm internal resistance) use an external load resistor of 0.115 ohm. Calculate the total resistance of the circuit.

$$0.035 + 0.115 = 0.15 \text{ ohms}$$

Calculate the current in the circuit.

$$I = \frac{1.5}{0.15} = 10 \text{ amps}$$

How much voltage exists at the terminals of the cell?

Like calculation (1) above:

$$E = 10 \times 0.115 = 1.15 \text{ volts}$$

In this case, a 1.5-volt dry cell puts out only 1.15 volts. Where does the rest of it go? The rest of the voltage was used inside the cell where there is a voltage drop. IR = 10 amps x 0.035 ohms = 0.35 volts used inside the cell.

$$1.5 \text{ volts} - 0.35 \text{ volts} = 1.15 \text{ volts}$$

The name *terminal voltage* is applied to this voltage at the terminals of the cell

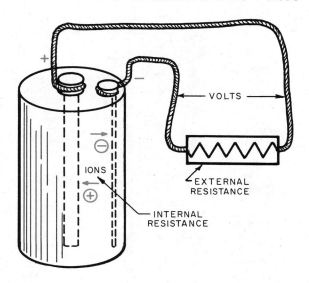

Fig. 11-17 Internal Resistance of Cell Affects Current

when it is in use to distinguish it from the emf, which is constant for a particular cell.

Maximum Current from a Cell

A cell uselessly produces its greatest current when it is short-circuited. Assume a wire of practically zero-ohms resistance is connected across the terminals of the 1.5-volt, 0.35-ohm cell. The amount of current is limited only by the internal resistance of the cell:

$$I = 1.5 \div 0.035 = 42.8 \text{ amps}$$

The terminal voltage is now zero because all the cell emf is used inside the cell. If this condition exists for more than a few seconds, the cell overheats, gases form in it, and the electrolyte starts boiling out of the top of the cell. As a result, the cell will be ruined.

If a dry cell is used to produce a moderate current for 10 or 15 minutes, there may be a noticeable drop in terminal voltage and current by the end of this time. The reason for it is a temporary increase in internal resistance due to the formation of a very small amount of hydrogen around the positive plate. When the cell remains on open circuit, this hydrogen is reconverted to H_2O, and the cell is restored to its original low internal resistance.

Dry cells generally fail because they become dry internally. An unused dry cell on a warm shelf for two years may lose its moisture by evaporation, despite the manufacturer's attempt to seal the top. A cell in use will lose its moisture when a hole is finally dissolved in the zinc. Drying out causes a great increase in internal resistance. A cell with one- or two-ohm internal resistance is of no value. It may produce enough current to make a voltmeter read 1.5 volts, but not enough current to light a flashlight bulb.

OHM'S LAW FOR CELLS IN SERIES

Three cells, each with 1.5-volt emf and 0.05-ohm internal resistance are connected in series to a 2-ohm lamp, figure 11-18. Calculate current in lamp and voltage at the lamp:

The current will be found from the total emf and the total resistance. With three 1.5-volt cells in series, the emf is 4.5 volts. The total resistance in the circuit is 0.05 + 0.05 + 0.05 + 2 = 2.15 ohms.

$$I = 4.5 \div 2.15 = 2.09 \text{ amps}$$

The voltage across the lamp is: 2.09 x 2 = 4.18 volts. Cells in series produce more voltage and more current.

OHM'S LAW FOR CELLS IN PARALLEL

One might assume that three or four cells in parallel are no better than a single cell because there is no increase in emf. It is true that in many situations there is no advantage to placing several cells in parallel. When the load resistance is rather small, there is an advantage, as shown in figure 11-19, page 280.

When one cell is in use, the total resistance is 0.2 + 0.06 = 0.26 ohms, and the current is 1.5/0.26 = 5.77 amps, which is a fairly large current for a single dry cell to produce for any length of time.

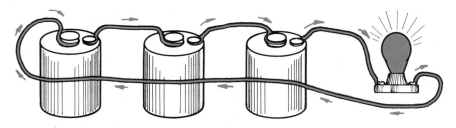

Fig. 11-18 Resistance of Cells Connected in Series

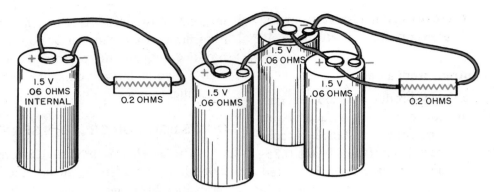

Fig. 11-19 Resistance of Cells Connected in Parallel

When three cells in parallel are used, the combined resistance of the three cells is 0.02 ohm (like three equal resistors in parallel, the combination has 1/3 the resistance of a single resistor) and the total resistance of the circuit is 0.02 internal plus 0.2 external = 0.22 ohm.

The current is 1.5/0.22 = 6.82 amps, which is somewhat more than the 5.77. But more importantly, this current is divided among three cells, each one producing 6.82/3 = 2.27 amps.

This sharing of the current load among several cells delays the temporary increase of internal resistance. It also increases the life of each cell, which is an advantage in that it is more convenient to replace the group of 3 cells every 30 days than to replace one cell every 10 days. Since the internal resistance of the parallel group is less than that of a single cell, the terminal voltage will be closer to the cell emf.

POINTS TO REMEMBER

- If electrons have a tendency to leave atoms of one material and combine with another material, the pair of materials are said to possess chemical energy.

- In an electromotive cell, materials are arranged so that electrons from a metal can flow usefully through an external circuit on their way to positive-charged ions in the cell.

- Terminal voltage = emf minus voltage drop inside cell.

- In series, cell voltages add. A parallel group has the same emf as one cell.

- Cell voltages: dry cell 1.5, lead storage cell 2.

- Keep dirt out of a car battery, keep it charged, and add water when needed.

- A secondary cell is one that can be recharged. A primary cell is not ordinarily intended for recharging.

REVIEW QUESTIONS

1. Where does the electrical energy in a simple cell come from?

2. Describe a Leclanché cell.

3. What is the difference between the Leclanché cell and an alkaline cell? A Leclanché cell and a mercury cell? What are the advantages of each?

4. What is the principal difference between a primary cell and a secondary cell?

5. Figure 11-14 shows a 13-plate lead storage cell. How are the plates connected — series or parallel?

6. A suggestion has been made that an old dry cell could be restored by punching holes in the container and soaking it in salt water. Is this likely to work?

7. Name the active materials in a lead storage battery.

8. What is an ampere-hour?

9. When a hard-rubber comb is negatively charged with static electricity, the electrons can be discharged quickly in a spark. Why don't the electrons all jump off the zinc plate of a dry cell just as fast?

10. How often should an ordinary car battery require charging?

11. The sketches in figure 11-20 represent various possible ways of connecting two dry cells and a lamp. The positive and negative are the terminals of the cells, shown in top view. The resistor represents the lamp. For each: State voltage at the lamp (0, 1.5, or 3 volts). Classify circuits as good or poor.

12. Is a car battery damaged from being run down by leaving ignition or lights on all night?

13. State items of main importance in care of a lead storage battery.

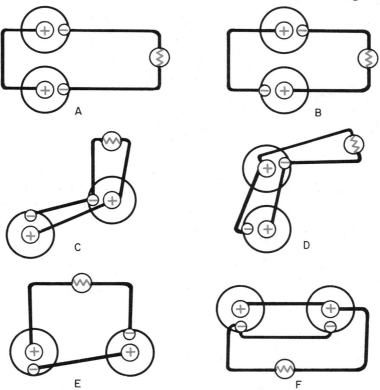

Fig. 11-20

14. Eight lead storage cells are arranged as in figure 11-21. (Top view of cell connections.) The emf of this battery is how much?

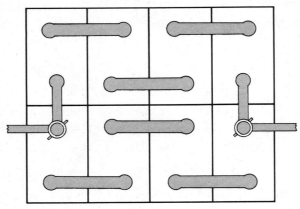

Fig. 11-21

15. Two 24-volt batteries, in series, are being charged by a 60-volt generator. Each battery has 0.02-ohms internal resistance. Calculate how much additional resistance is needed in the circuit to limit the current to 6 amps.

16. Describe two methods of testing the electrical charge in a lead plate storage battery.

17. Define terminal voltage as related to a dry cell.

18. In what ways does the fuel cell resemble a storage battery? How does it differ?

RESEARCH AND DEVELOPMENT

Experiments and Projects on Cells and Batteries

INTRODUCTION

Cells and batteries were the first source of current electricity. The history and development of cells and batteries is covered in Chapter 11.

The experiments which follow will show the relationship existing between cells and batteries and how they function.

The projects provide an opportunity for you to study the construction of cells and batteries and to learn how to use and maintain them.

EXPERIMENTS

1. To show different voltage and current values which can be obtained by various cell connections.

2. To observe the electromotive force set up between two electrical conductors of different materials immersed in a solution.

3. To test a storage battery.
4. To test the voltage of a cell and to calculate the current, with different resistors connected in the circuit.

PROJECTS

1. Study the construction of a lead storage battery and of a nickel-cadmium battery.
2. Make a simple cell.
3. Charge a storage battery.
4. Make a cutaway of a dry cell.
5. Design and build a solar battery.
6. Make a secondary cell.

EXPERIMENTS

Experiment 1 CELLS CONNECTED IN SERIES AND PARALLEL

OBJECT

To show different voltage and current values which can be obtained by various cell connections.

APPARATUS

6 - No. 6 dry cells
1 - DC voltmeter (0-10 or 0-15 V)

PROCEDURE

1. Connect the cells as shown in figure 11-22 and read the voltage of the group. Determine from the rating of the cells, the current capacity of this connection. Tabulate the type of connection, the voltage, and current.
2. Repeat Step 1 for figures 11-23 through 11-25.
   Note: Be careful that the proper scale range is used on the dc voltmeter.

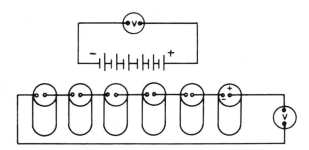

Fig. 11-22 Series

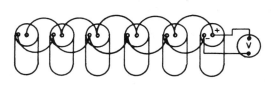

Fig. 11-23 Parallel

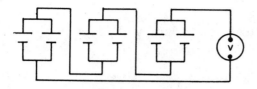

Fig. 11-24 Parallel-Series

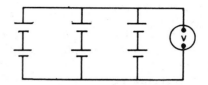

Fig. 11-25 Series-Parallel

OBSERVATIONS

Use a table similar to the one shown here to record your observations.

OBS. NO.	CELL CONNECTIONS	VOLTAGE OUTPUT	CURRENT OUTPUT
1	6 CELLS IN SERIES		
2	6 CELLS IN PARALLEL		
3	PARALLEL SERIES		
4	SERIES PARALLEL		

QUESTIONS

1. What arrangement of cells would give the greatest current when the external resistance to be overcome is small?

2. Which arrangement of cells would be most satisfactory to use to give the highest voltage possible with a minimum current?

3. For what kind of work are dry cells best adapted?

Experiment 2 SIMPLE ELECTRIC CELLS

OBJECT

To observe the electromotive force set up between two electrical conductors of different materials immersed in a solution.

APPARATUS

1 - Glass or plastic jar containing solution of sulfuric acid with specific gravity of 1.250
1 - Glass or plastic jar containing solution of copper sulfate or common salt
1 - DC voltmeter (0-2 V), low-reading
1 - DC ammeter (0-10 A)
1 - Strip of copper
1 - Strip of zinc
2 - Strips of lead
1 - Strip of carbon
1 - Low-voltage dc power supply
1 - SPST switch

> When mixing acid and water, always pour the acid into the water. Wear safety glasses and avoid getting acid on your hands and clothing.

PROCEDURE

1. Fill one container approximately 2/3 to 3/4 full with the sulfuric acid solution. Place the copper and zinc strips in the solution, being certain that the strips do not touch each other in the jar. Connect a voltmeter across the cell, figure 11-26. Read and record the instrument reading on a chart similar to the one below.

OBS. NO.	POSITIVE ELECTRODE	NEGATIVE ELECTRODE	CELL VOLTAGE
1			
2			
3			
4			

2. Repeat this test with combinations of different materials suggested by the instructor.

3. Repeat this group of four observations, using the same material combinations in the copper sulfate solution.

4. Put two clean strips of lead in the container of sulfuric acid solution. Find out if they produce any voltage.

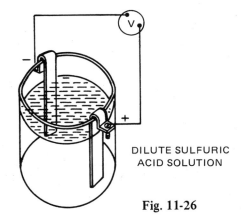

DILUTE SULFURIC ACID SOLUTION

Fig. 11-26

5. Connect a dc source (generator or low-voltage power supply) to the two lead strips so that the current is about two amperes, figure 11-27. Leave the current on for a few minutes. Notice a dark coating that appears on one of the strips. Which strip is (+) or (−)? This dark coating is lead peroxide.

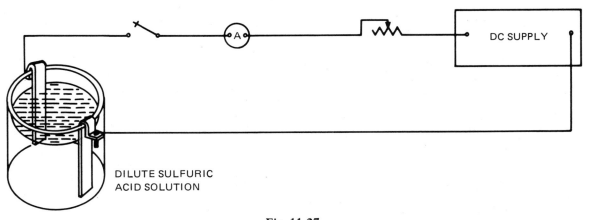

DILUTE SULFURIC ACID SOLUTION

Fig. 11-27

6. Disconnect the dc supply from the lead strips and connect a voltmeter to them, figure 11-28. How many volts does the meter read?

You have now made a lead storage cell. The lead and lead-peroxide combination is used in automobile storage batteries. See if your cell will light a flashlight bulb.

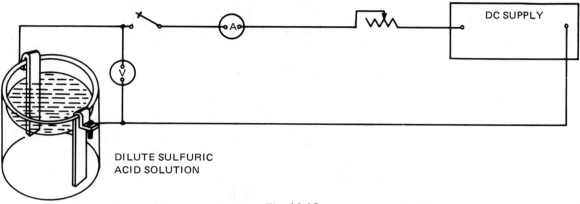

DILUTE SULFURIC
ACID SOLUTION

Fig. 11-28

Experiment 3 STORAGE BATTERIES

OBJECT

To test a storage battery.

APPARATUS

1 - DC ammeter (0-30 A)
1 - DC ammeter (0-150 A)
1 - DC voltmeter (0-10 V)
1 - SPST switch
1 - Hydrometer
1 - Lamp bank or rheostat
1 - Storage battery, 6- or 12-volt

Be extremely careful when testing a storage battery with a hydrometer. Always hold the rubber tube on the hydrometer over the battery filler hole while taking the reading. Avoid getting drops of sulfuric acid on your hands or clothing. If acid comes in contact with your skin or clothing, neutralize it with baking soda or cold water.

PROCEDURE

1. Check and record the specific gravity of each cell and the battery terminal voltage.
2. Connect apparatus as shown in figure 11-29.
3. Close the line switch and adjust the charging current to a value of about five amperes. Record line volts, volts across lamp bank, volts across battery and the charging current in amperes.

   Note: Be sure the screw plug caps are removed from each of the cells during the charging period.

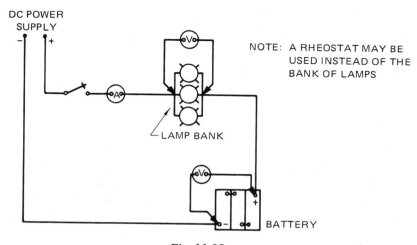

Fig. 11-29

4. Continue to charge the battery for a period of approximately one hour, then record the same readings as in Step 3.

5. Open the line switch and check and record the specific gravity of each cell and the battery terminal voltage.

OBSERVATIONS

Use a table similar to the one shown here to record your observations.

Obs. No.	Line Volts	Volts Drop R	Volts Drop Battery	Charging Rate	Hydrometer Readings I II III	Battery Volts Open Circuit	Battery Condition
1							Before Charging
2							After Charging

Sample Calculations: Resistance of lamp bank $R = \dfrac{E}{I}$

Watts loss - lamp bank $W = EI$

Watts consumed $W = EI$

QUESTIONS

1. What determines the proper charging rate of a lead storage battery?

2. How do you determine the state of charge of the cells of a storage battery?

3. Give some care and precautions to observe when using lead storage batteries.

PROJECTS

MAKE A SIMPLE CELL

The simple cell, figure 11-30, is one of the earliest methods of producing electric current. Electrons are carried by the movement of ions through the solution. The flow of electrons through the liquid is called electrolysis.

This cell is known as a primary cell because it cannot be effectively recharged unless the zinc is replaced.

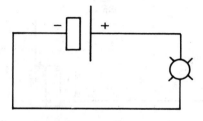

Fig. 11-30

MATERIALS

1 - Pyrex glass or plastic jar
1 - Strip zinc, 1" x 6" (25 mm x 150 mm)
1 - Piece copper, 1" x 1 1/4" (25 mm x 32 mm)
1 - Piece copper, 1/2" (12 mm) wide length determined by size of carbon
3 - Brass machine screws, #6-32 x 3/8" (M3.5 x 10 mm L) RH
3 - Brass nuts, #6-32 (M3.5)
1 - Miniature lamp receptacle
2 - Miniature lamps, 2- and 6-volt
1 - DC voltmeter (0-10 V)
    Sulfuric acid or powdered sal ammoniac
    Distilled water

PROCEDURE

Make the zinc element.
Make carbon-rod holder.
Attach the holder to the carbon rod.

Attach hookup wire to elements, figure 11-31.

Fill the jar three quarters full of a solution of 10 parts water and 1 part sulfuric acid. *Caution:* When mixing acid and water, always pour acid into the water. Wear safety glasses and avoid getting acid on your hands or clothing.

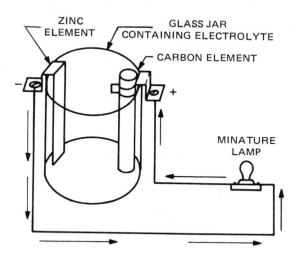

Fig. 11-31

Place carbon and zinc elements in container, figure 11-32.
Test voltage.
Attach wires to lamp receptacles and insert lamp of proper voltage.
Evaluate.

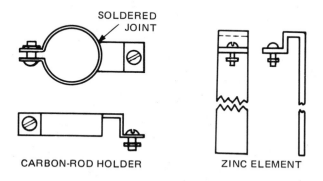

Fig. 11-32

DESIGN AND BUILD A SOLAR BATTERY

A solar battery converts the sun's rays into electrical energy. The cells are usually made of thin wafers of selenium with a metal backing. When light falls upon the selenium layer, electrons are emitted. The selenium has a positive charge, while the metal back plate is negative.

MATERIALS

  1 - Selenium rectifier
  1 - Wood base
  8 - Brass wood screws, #4 x 3/8" RH
12 - Terminal lugs

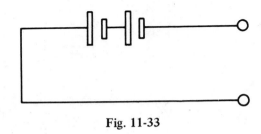

Fig. 11-33

PROCEDURE

Obtain and disassemble a selenium rectifier.
Select the plates you need and remove the coating with lacquer thinner.
Wash each plate thoroughly with soap and water.
Rinse and dry the plates.
Arrange the plates on a piece of graph paper and plan the base.
Mount the plates on the base according to figure 11-34.
Connect the cells and test the current and emf with a microammeter and a low-range voltmeter. Do this with cells connected in series and in parallel.
Evaluate.

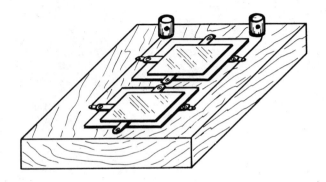

Fig. 11-34

MAKE A SECONDARY CELL

A secondary cell contains elements which are rechargeable. If a discharged cell is connected to another electron source and electrons are pushed back into the negative terminal of the cell, the elements in the cell will be restored to their original energetic form.

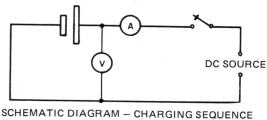

SCHEMATIC DIAGRAM – CHARGING SEQUENCE

Fig. 11-35

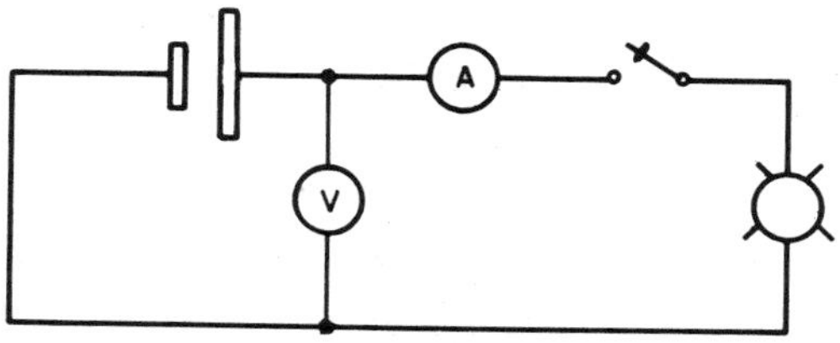

SCHEMATIC DIAGRAM – TESTING SEQUENCE

Fig. 11-36

MATERIALS

1 - Pyrex glass or plastic jar
2 - Strips lead, 1″ x 6″ (25 mm x 150 mm)
2 - Brass machine screws, #6-32 x 1/2″ (M3.5 x 12 mm) RH
2 - Brass nuts, #6-32 (M3.5)
1 - DC voltmeter (0-10 V)
1 - DC ammeter (0-10 A)
1 - Miniature lamp receptacle
2 - Miniature lamps, 2- and 6-volt
1 - Battery charger, 6-volt
1 - SPST switch
    Sulfuric acid
    Distilled water

PROCEDURE

Obtain a jar.
Make lead elements, figure 11-37.

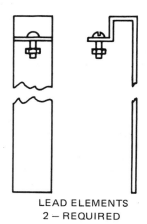

LEAD ELEMENTS
2 – REQUIRED

Fig. 11-37

Attach hookup wire to elements, figure 11-38.

Fill the jar three-quarters full of a solution of 10 parts water and 1 part sulfuric acid.

> When mixing acid and water, always pour the acid into the water. Wear safety goggles and avoid getting acid on your hands or clothing.

Place lead elements in container.

Arrange the parts on the bench and wire the circuit as shown in figure 11-35.

Adjust the charging current to 2 amps.

Charge the cell for about 30 minutes.

Remove the charging device and wire the circuit as shown in figure 11-36.

Evaluate.

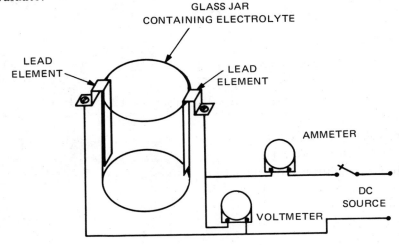

Fig. 11-38

Chapter 12
Small-scale sources of EMF

The greatest amount of industrial electrical energy is produced by rotating generators turned by mechanical energy of steam or water turbines. There are, however, various methods for the direct conversion of heat, light, mechanical motion, and nuclear energy into electrical energy without using water wheels or steam boilers and generators. So far, these methods do not produce important amounts of energy. Yet, they are highly useful because the small voltages they produce are used to control the operation of other equipment. Thermocouples, photocells, and pressure-sensitive (piezoelectric) materials are the *sensing elements* of the electrical brains that control furnaces, regulate speed of machines, play phonograph records, and perform a host of automatically controlled operations.

Furthermore, continuing investigation of these materials is producing new knowledge and suggesting new uses. Also, there is always the possibility that new methods of large-scale production of electrical energy may be developed.

THERMOCOUPLES

Direct Conversion of Heat to Electrical Energy

In Berlin in 1822, T.J. Seebeck reported the discovery that a circuit similar to the one in figure 12-1 would produce a steady electric current if the two junctions are at different temperatures. A and B are any two different metals. This direct production of an emf from heat is called the *Seebeck effect*.

This circuit is widely used for the measurement of temperatures, especially those beyond the range of liquid-in-glass thermometers. Voltage measurements are made instead of current measurements because the emf depends only on materials and temperature, while current would be influenced by many factors that determine the resistance of the entire circuit.

For accurate work, the voltage is measured by a potentiometer, figure 12-2, rather than by a millivoltmeter. A conveniently located potentiometer may be switched from one thermocouple circuit to another, permitting successive readings of temperatures at several different remote locations.

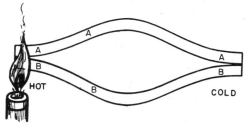

Fig. 12-1 The Seebeck Effect

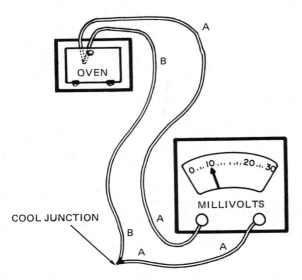

Fig. 12-2 Pyrometer (Thermocouple and Potentiometer)

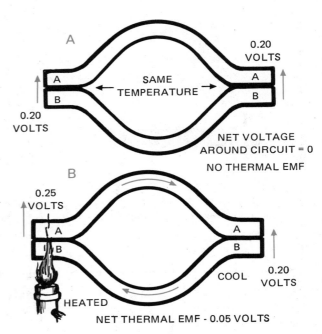

Fig. 12-3 Thermocouple Unit

The production of this emf is explained by a study of electron energies in the metals. When any two different metals are in contact, there is a tendency for a slight excess of electrons to drift from one metal and accumulate on the other. This slight accumulation of electrons causes a *contact potential difference* between the metals. It is a small voltage, very difficult to measure and usually noticed only as a nuisance in delicate measurements.

Application of heat changes the contact voltage at the heated junction, producing the useful *thermal emf.* The materials most often used for thermocouple junctions are: (1) copper and constantan for -300°F. to 750°F. (-184°C to 399°C); (2) chromel and alumel, 200°F. to 2 500°F. (93°C to 1 215°C); (3) platinum and platinum-rhodium alloy, 200°F. to 3 150°F. (93°C to 1 697°C). Voltage output can be increased by placing several thermocouples in series, figure 12-4. Such a device is called a *thermopile.*

In many situations, thermistors are replacing thermocouples for temperature measurement. A *thermistor* consists of an alloy whose resistance changes greatly with changes in temperature. It does not produce an emf and is used with a voltage source.

Thermocouple voltages may be used to operate relays or electronic controls. A commonly used circuit in gas-fired heating equipment is shown in figure 12-5.

Ordinarily the room thermostat acts as a switch, actuating a solenoid which opens the gas valve to the furnace when the room cools. The gas coming into the furnace is ignited by a pilot light. The pilot light also heats a thermocouple, producing a voltage that operates the coil of a relay. This relay holds the switch (S) closed. If the pilot light goes out, switch S is opened so that the main gas valve cannot be opened by the thermostat. This acts as a safety device, preventing an accumulation of unburned gas in the furnace, chimney, and other places.

The ordinary thermostat is not a thermocouple. It is a switch, operated by the bending of a compound bar as the temperature changes.

Invar is an iron-nickel alloy that does not change much in size when heated or cooled. Brass expands when heated and contracts when cooled. Strips of the two metals,

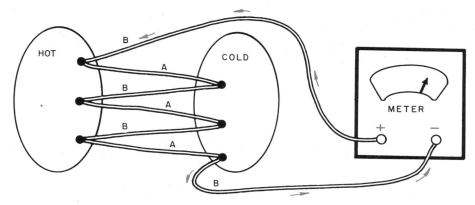

Fig. 12-4 Thermocouples in Series

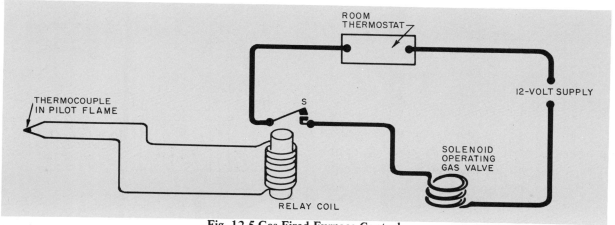

Fig. 12-5 Gas-Fired Furnace Control

brazed together, form the compound bar. Lengthening of the brass by heating forces the bar to bend as in figure 12-6, cooling bends it back again.

In the thermostat pictured, the brass-invar strip acts as the moving part of a switch, bending as it cools, to meet the contact at the right. A contact could be placed below the bar, so that a circuit can be closed by a temperature rise. Setting the temperature on the thermostat is a matter of adjusting the relative position of the bar support and the contacts.

Semiconductor Thermocouples

The thermoelectric properties of semiconductor alloys have been investigated with promising results. In a piece of N-type

(electron-rich) semiconductor heated at one end, electrons are forced toward the cold end. In a P-type (electron-poor) semiconductor, electrons tend to accumulate at the hot end. If in figure 12-3, A is an N-type semiconductor and B is a P-type semiconductor, the thermal voltages combine, helping to drive electrons in the same direction around the circuit with a useful emf of a few tenths of a volt. A thermopile consisting of a few hundred such junctions can produce power in useful amounts. Gas-fueled lead telluride thermoelectric generators now reliably energize communications and navigation equipment in remote locations. Three hundred gallons (1 136 liters) of liquid propane can keep a 50-watt unit operating for six months.

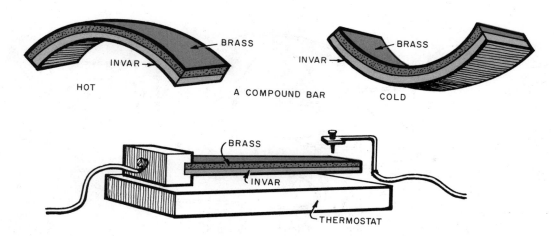

Fig. 12-6 Bimetal Thermostat

Small thermoelectric generators powered by the heat of a kerosene lamp can produce about 15 watts.

Energy Transfer in the Thermocouple

If a thermocouple circuit is allowed to produce current, heat is absorbed from the hot junction to produce the electrical energy. At the cool junction, some absorbed energy is reproduced as heat. The fact that this heat-to-electricity conversion can be reversed was discovered in 1834 by Peltier, figure 12-7. If in figure 12-3 we removed the heat source but inserted a battery in the circuit so it would maintain the same current, heat would still be absorbed at the left junction and produced at the right.

If a temperature difference is used to produce electrical energy in the circuit, it is called a thermocouple. If the same materials are used so that electrical energy produces a temperature difference, the result is called a *Peltier circuit.* The development of semiconducting materials has made the commercial manufacture of Peltier refrigerators practical. Lack of moving parts and noiseless operation are advantages of the Peltier refrigerators. To date, however, they are not as efficient as mechanical refrigerators and cost more to operate.

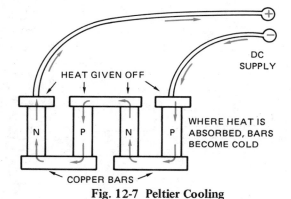

Fig. 12-7 Peltier Cooling

PIEZOELECTRICITY

EMF From Mechanical Pressure

The term *piezo* means pressure. When compressed, twisted, bent, or stretched, some materials develop electric charges on opposite sides of the material. The most common application of this effect is the use of crystals of Rochelle salt (sodium potassium tartrate) in crystal microphones and in crystal pickups for record players. The voltage produced is used as the input signal voltage for an amplifier. Sideways vibration of the needle twists the end of the crystal back and forth, producing alternating voltage on the sides of the crystal.

Many minerals show this piezoelectric effect. The best known mineral of this type is quartz. As with other pressure-sensitive materials, the slight bending of a properly cut slice of quartz develops opposite charges on its faces. This effect is reversible. That is, the application of a voltage (opposite charges) to the faces of the slice of quartz causes the crystalline quartz to bend slightly. By applying an alternating voltage of a frequency close to the natural mechanical vibration frequency of the crystal, the crystal is caused to vibrate, bending rapidly back and forth. This mechanical oscillation, in turn, sustains the continued production of alternating charges on the face of the crystal. Since a crystal will oscillate at only one definite frequency, the quartz crystal is used to control frequency of rapidly alternating voltages used in radio transmitters.

Some ceramic materials, such as barium titanate, show this piezoelectric property and can be used in record-player pickups. A piece of barium titanate tapped with a hammer produces enough voltage to flash a small neon lamp (NE-2, 0.04-watt). Wires from the lamp contact each end of the piezoelectric material, with a scrap of hard plastic as a cushion between hammer and the brittle barium titanate, figure 12-8. Ultrasonic vibrations (approximately 50 000 cycles) are produced by applying alternating voltage of that frequency to properly shaped pieces of barium titanate. These materials also can be used in pressure indicators and in indicators of mechanical vibration of machine parts.

PHOTOCELLS

EMF from the Energy of Light

The term *photocell,* as broadly used, includes three different types of light-sensitive devices:

1. A widely used photoelectric device is a *phototube,* figure 12-9. Light striking the curved metal cathode gives some electrons enough energy to escape from the metal. The free electrons can travel through the vacuum or low-pressure gas to the anode wire in the center if an external voltage is applied to make the anode more positive than the cathode. This phototube is not a voltage producer. Rather, it is a device that conducts when light falls on it. The so-called electric eye used in door openers and counting or sorting operations is a phototube.

2. Another photoconductive device consisting of a thin layer of a semiconductor (germanium, silicon, lead sulfide, cuprous oxide, cadmium sulfide, for example) can be used in a circuit. Light falling on these semiconducting materials gives enough energy to some of the electrons in the solid to make them free-moving inside the solid, so that the solid semiconductor is a better conductor when illuminated. Photo conductive cells are not intended to produce

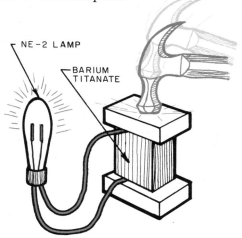

Fig. 12-8 Piezoelectrical Material

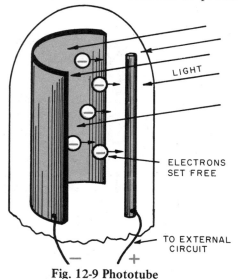

Fig. 12-9 Phototube

voltage, but are used along with voltage sources to take advantage of the change in resistance of the photo conductor when it is illuminated. For instance, they are useful in flame detectors and photographic meters, figure 12-10.

3. The term photocell should only be applied to devices that produce emf when illuminated. This *photovoltaic effect* is produced when the junction or boundary surface between a metal and a semiconductor is illuminated, figure 12-11. There are two ways of arranging the materials: (a) a very thin transparent metal film, sometimes silver or gold, is deposited on a sheet of semiconductor such as selenium, or (b) a very thin transparent layer of semiconductor is deposited on a metal. In both cases, the junction of metal and semiconductor can be illuminated. The direction of electron movement may be from metal to semiconductor, or from semiconductor to metal, depending on the nature of the impurities in the semiconductor. An improved photovoltaic cell called the *solar cell* is described in Chapter 17.

ELECTRICAL ENERGY FROM NUCLEAR ENERGY

Instead of burning coal or oil, some commercial power plants use energy from the breakup of uranium atoms as a source of heat. This heat is used to form steam, which powers an ordinary steam-turbine generator unit.

A method of efficiently directing nuclear energy so that it will immediately produce an electric current without the necessity of going through the customary steam-driven generator process has not yet been developed. In some experiments, charged particles that shoot out from a radioactive material are allowed to collect on a metal electrode. The tiny amount of charge produced does not seem to offer industrial possibilities.

Semiconductor junctions, as in a germanium transistor, can produce a small voltage when radiated. This is the principle of an experimental "atomic battery," sketched in figure 12-12. "A" is the radioactive source; a mixture containing the isotope Strontium-90 has been used. "B" is the semiconductor. Radiation from A releases electrons in the semiconductor. These freed electrons spill over into an electron-receptive alloy, C, in contact with the semiconductor. Note the similarity of this process to that which takes place in a photovoltaic cell. The voltage produced is small.

MAGNETOHYDRODYNAMIC (MHD) GENERATION

In the future, application of old principles in a new device may result in the efficient production of electrical energy from the heat of flames or nuclear reactors.

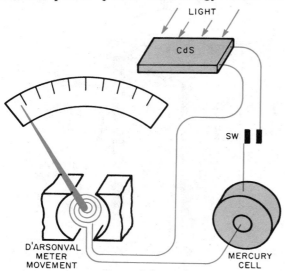

Fig. 12-10 Cadmium Sulfide Light Meter

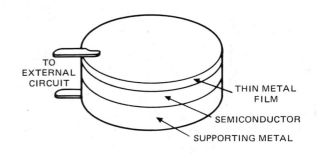

Fig. 12-11 Photocell

Chapter 10 described the generation of electric current when a conductor, such as a copper wire, is moved through a magnetic field. In an MHD converter, the conductor is an ionized gas instead of a wire. In figure 12-13, think of the superheater as a place for heating a gas, either by burning it or by applying heat from some external source. The superheated gas is at high pressure and at a temperature so high that gas molecules ionize, forming a cloud of electrons and positive ions.

The ionized gas blows through a nozzle, entering the magnetic field area at a temperature about 2 000°C (3 632°F.) and velocity over 1 500 mph (2 400 km/h). As shown in figure 12-13, the magnetic lines of force are directed at right angles to the page, as from a north pole over the page to a south pole behind the paper. Just as electrons in a wire in an ordinary generator are forced to move in a direction at right angles to both the magnetic field and their original direction of motion, in this device electrons are forced upward toward the anode. The anode may already be negative-charged due to previous accumulation of

electrons. But as more electrons come roaring by, the magnetic field deflects them and their high kinetic energy slams them on to the anode. This builds up the negative charge still more. Positive ions are deflected in the opposite direction. Therefore, they collide with the cathode and pick up electrons from the cathode. This maintains a positive charge (electron deficiency) at the cathode. The anode and cathode connect to the useful external circuit.

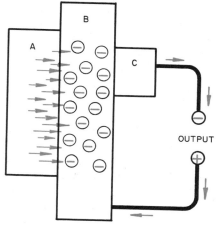

Fig. 12-12 Atomic Battery

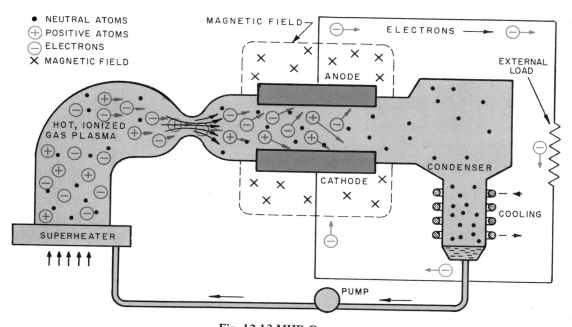

Fig. 12-13 MHD Generator

When and if MHD converters can be built to produce huge amounts of electrical energy, efficiency much higher than that for conventional steam-turbine generator plants is predicted. The size of the converter may be relatively small. One company has built an MHD generator that produced 2.5 kilowatts from a unit of about the same volume as this book.

Serious problems remain to be overcome. Extremely high temperatures are needed to make the gases ionize sufficiently and stay ionized instead of recombining. Heat sources capable of producing the extremely high temperature must be developed; an accompanying problem is either finding new materials to resist critically high temperatures or of making better use of materials in use already.

POINTS TO REMEMBER

- Light falling on the junction of a metal and a semiconductor displaces electrons from one to the other. This emf is used in photocells.

- Piezoelectricity (pressure-emf) is a voltage developed by mechanically distorting certain crystals and ceramics. This emf is used in pickups and control elements, not as a power producer.

- Heat applied to the junction of two metals will displace electrons from one metal to the other. This emf is used in thermocouples for high temperature measurement.

- The above energy conversions are reversible. Voltage applied to a piezoelectric material will change its shape. Current through a pair of thermocouple junctions releases heat at one end and absorbs heat at the other.

- An MHD generator charges its terminals by using a magnetic field to separate the charged particles in a high temperature ionized gas.

REVIEW QUESTIONS

1. What is Seebeck effect? Explain how the principle is used in industry.

2. Explain how thermal emf is produced.

3. What is a Peltier circuit?

4. What is piezoelectricity? Give an example of how it is used.

5. What is the difference between a photocell and a phototube?

6. Could a piezoelectric material be used to produce sound in a headphone or loudspeaker?

7. Name several semiconductors.

8. What is the difference between a thermocouple and a thermostat? Can one substitute for the other?

9. Name three piezoelectric materials.

10. Why are voltage measurements rather than electric current measurements made in the measurement of temperature with a thermocouple?

11. What difficulties are encountered in the operation of MHD generators?

12. Why is quartz crystal employed to control the voltages used in radio transmitters?

13. What is meant by the term photoconductor?

14. What is ionized gas?

15. Where have you seen ionized gas?

RESEARCH AND DEVELOPMENT

Experiments and Projects on Small Scale Sources of EMF

INTRODUCTION

Chapter 12 discussed small-scale emf devices, the principle upon which each operates, and how they are used in the home and elsewhere. Several experiments are suggested to provide an opportunity to become familiar with some of these devices and to learn how they operate. Because of the nature of this source of emf, the projects are of the assembly-disassembly, and examination of the component-part variety.

EXPERIMENTS

1. To plan and wire a circuit in which a bell or buzzer is operated by a photocell.

2. To produce electricity directly from heat.

3. To plan a device incorporating the use of a photocell and a solenoid to simulate a process used in industry for measurement or to produce mechanical movement.

4. To produce an electric current with a beam of light.

PROJECTS

1. Disassemble an old phonograph pickup and examine the parts.

2. Arrange a circuit consisting of a piezoelectric pickup, a high-gain AF amplifier, and an ac voltmeter for the purpose of measuring vibration.

3. Connect a circuit which employs a photoelectric cell to open a door or to count items on a moving belt.

EXPERIMENTS

Experiment 1 PHOTOELECTRIC RELAY CIRCUIT WITH SENSITIVITY CONTROL

OBJECT

To plan and wire a circuit in which a photocell is used to operate a bell or buzzer.

APPARATUS

2 - Transistors, P-N-P (2 N301, 2 N555) T_1, T_2
1 - Relay, 5 000-ohm
1 - Solar cell (International Rectifier B2M) PC_1
1 - Potentiometer, 5 000-ohm (R_1)
1 - Battery, 9- to 12-volt
1 - Terminal strip
1 - Base

PROCEDURE

1. Arrange parts on a wooden base or vector board.

2. Wire according to schematic diagram, figure 12-14.

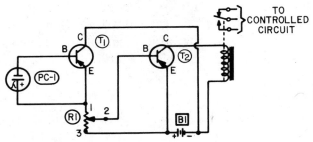

SCHEMATIC DIAGRAM FOR A VARIABLE-SENSITIVITY
PHOTOELECTRIC RELAY

(Courtesy of Howard W. Sams & Co., Inc.)
Fig. 12-14

3. Connect relay to bell circuit.

4. Provide light source to actuate solar cell.

5. Adjust potentiometer until relay closes.

6. Move the light away from the phototube until the relay opens.

7. Adjust potentiometer again and note effect.

8. Record your observations.

QUESTIONS

1. Explain how the light source striking the solar cell causes electrons to flow.

2. What is the purpose of the relay?

3. How could this circuit be used in a home?

Experiment 2 THE THERMOCOUPLE

OBJECT

To produce electricity directly from heat.

APPARATUS

1 - Galvanometer or millivoltmeter
1 - Piece copper wire, 18″ (460 mm) long
1 - Piece iron or nichrome wire, 18″ (460 mm) long
    Candle, match, torch flame, or boiling water for source of heat

PROCEDURE

1. Clean one end of each piece of wire.
2. Twist together one end of copper wire and one end of iron or nichrome wire. Make a tight joint about 1/2″ (12 mm) long, figure 12-15.
3. Braze the wire together at the twisted joint.
4. Connect the free ends of the wires to the measuring instrument.
5. Apply heat to the junction of the wires.
6. Observe the reading on the galvanometer or millivoltmeter.

QUESTIONS

1. What are some commercial uses for a thermocouple?
2. What causes the current in the wires?
3. Do you have a thermocouple in your home? If so, what does it do?
4. Could the electrical energy produced by a thermocouple be increased? Explain your answer.

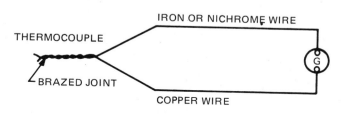

Fig. 12-15

Experiment 3 CONVERTING ELECTRICAL ENERGY TO MECHANICAL MOVEMENT

OBJECT

To plan a device incorporating the use of a photocell and solenoid to simulate a process used in industry for measurement or to produce mechanical movement.

APPARATUS

1 - Transistor, P-N-P 2N107 or 2N118
1 - Relay, 5 000-ohm, RL
1 - Solar cell (International Rectifier B2M) PC_1
1 - Potentiometer, 25 000-ohm (R_1)
1 - Battery, 9- to 12-volt
1 - Base
1 - Solenoid, spring-loaded
   Current source to actuate solenoid

PROCEDURE

1. Plan the arrangement of the parts.

2. Wire photoelectric relay circuit according to figure 12-16.

3. Wire solenoid circuit.

4. Attach solenoid circuit to relay terminals, figure 12-17.

5. Provide light source to actuate solar cell.

6. Adjust potentiometer until relay is energized.

7. Observe action of solenoid when relay closes.

8. Cut off light source and observe solenoid action.

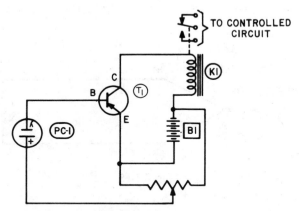

(Courtesy of Howard W. Sams & Co., Inc.)
Fig. 12-16

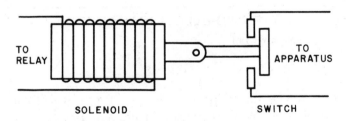

Fig. 12-17

QUESTIONS

1. How could this arrangement be used on model railroad equipment?

2. What other places could it be used?

3. How would it be used to turn on a night light?

PROJECTS

ARRANGE A PHOTORELAY AND ELECTRONIC CONTROL

This project is an unusual combination since it can be used to operate either as a photorelay or as an electronic control. It makes interesting use of a cold cathode-type tube and can be applied to operate many electrical devices such as burglar alarms, fire alarms, liquid level indicators, and mechanical measurements.

Photoelectric relays and similar electronic controls can be combined with other devices actually to control a whole sequence of machine operations in industry. For example, the many operations in making a motor block are all controlled in this way.

A punched tape with instructions on it can direct a whole series of machines to carry out the predetermined operations, including final assembly and inspection.

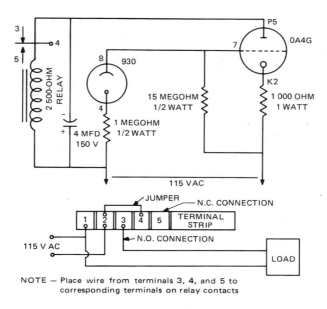

Fig. 12-18

MATERIALS

1 - Gas tube, #0A4G
1 - Phototube, #930
1 - Resistor, 1 000-ohm, 1-watt carbon, R-1
1 - Resistor, 15-megohm, 1/2-watt
1 - Resistor, 1-megohm, 1/2-watt
1 - Capacitor, 4-mfd., 150-volt
1 - Relay, 3 000- to 5 000-ohm plate sensitive (type LM5 or LB5 Potter and Brumfield)
2 - Tube sockets, 8-pin octal base
1 - Five post barrier screw terminal strip
1 - Line cord and plug
1 - Metal chassis, 1 1/4" x 3 1/4" x 3 1/4" (30 mm x 80 mm x 80 mm)

PROCEDURE

Plan the arrangements of the parts, figures 12-19 through 12-21.

Tubes should be positioned and tube sockets oriented so light can cross end of chassis and fall on concave side of #930 phototube.

Wire according to figure 12-18.

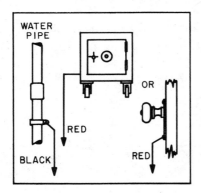

Fig. 12-19

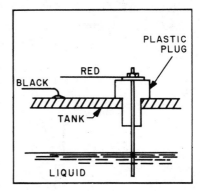

Fig. 12-20

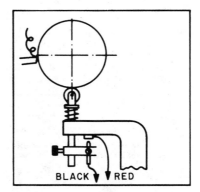

Fig. 12-21

Mount tubes on one side of the top of the chassis, and the relay on the other side. Place the relay on the same side as the terminal strip to make the leads the shortest. A relay of 3 000 ohms is preferred, and it should be able to operate on a small current, since 6- and 12-volt relays require too much current.

On the underside of the chassis, place the three resistors and the capacitor and make connections.

Connect the terminals of the relay to the corresponding terminals of the terminal strip. Be sure the wire is of sufficient size to carry the current the relay is capable of handling. Also, check to see that the device to be operated (load) operates on 115-volt ac and requires *less* current than the relay will handle.

Note: The jumper from terminal 2 to 4 on the terminal strip supplies 115 volts to the load. If the load is to operate on a different voltage, do not put in this jumper, but use hookup shown in figure 12-22. Notice that the return line from the load is not connected to terminal 1 on the terminal strip.

Get an old tube of the 8-prong type and break away the tube, retaining the base only. Solder a black lead to terminal 4 of the base, and a red lead to terminal of prong 2 of the base.

When all connections are made on the chassis, connect to 115 volts. Plug the tube base into the pin socket in place of the #930 tube. Cross the red and black leads and note that this changes the grid bias on the 0A4G tube, causing it to discharge through the relay. This will turn the current to the load on or off. Separate the red and black leads to return the circuit to normal. If the load is on when it should be off, remove the connection to the load from terminal 3 and fasten to terminal 5.

The device works on a current so small that it cannot be felt going through the body. Observe this by holding the red lead in one hand and the black lead in the other. *The relay will operate.*

Should the device fail to operate, here are some suggestions:

a. Check circuit and examine all connections.

b. If relay fails to close when leads are crossed, weaken the spring on the relay.

c. If relay chatters, substitute an 8 MFD in place of the 4 MFD.

d. If, after separating the leads, the relay fails to open, wrap the 0A4G tube in tin or copper foil and ground it to chassis.

To use the device as a photocell relay, place the #930 phototube in the octal socket. Light falling on the inner side of the curved plate will take the place of the crossed red and black leads and will permit a change in the grid voltage of the 0A4G tube. If the relay operates the instant the tube is plugged in, cover the tube, since the light intensity of the room is operating the tube.   A tube box with a hole cut in one side makes a good cover. In using the cover be sure that light entering the box will fall on the proper side of the tube. A flashlight makes a good light source, although a match or candle will work.

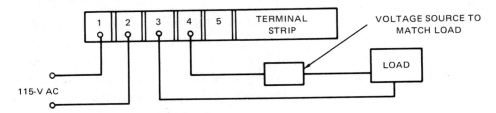

Fig. 12-22

Chapter 13

Motors

EFFECT OF MAGNETIC FIELD ON CURRENT-CARRYING WIRE

In Chapter 9 it was mentioned that when a wire carries electrons across a magnetic field the wire is pushed sideways.

The current in the wire has a magnetic field of its own. The magnetic field of the wire combines with the externally applied magnetic field, resulting in the field shown in figure 13-1.

To apply this effect in an electric motor, examine the effect of an external magnetic field on a rectangular loop of wire which is supplied with dc from a battery. For the present purpose, assume that the external magnetic field is supplied by a permanent magnet, figure 13-2, page 310.

The sections of the loop that lie parallel to the field are unaffected by the field. The side of the loop marked A will be pushed upward; the side marked B will be pushed downward. The following discussion explains the reason for this.

Figure 13-3, page 310, represents a vertical cross section of the loop in the magnetic field. At A, the heavy circle represents wire A of figure 13-2. The X in the circle represents electrons moving away from the observer (like the tail of an arrow flying away).

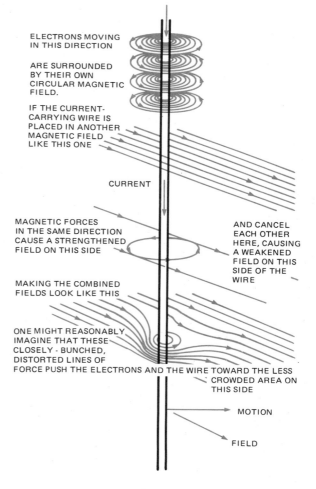

Fig. 13-1

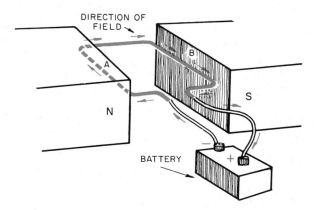

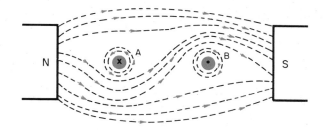

Fig. 13-3   Diagram of Magnetic Field in Fig. 13-2

Fig. 13-2   Current-Carrying Wire Loop in Magnetic Field

The dot in the circular section of wire B represents electrons moving toward the observer (the point of an approaching arrow). The circular pattern around A and B represents the magnetic field of the current in these wires.   (Using the left-hand rule, check the correctness of these directions.)

Underneath A, the north (N)→south (S) magnetic field combines with the field of the wire, making a strong field under the wire. Above A, the field of the magnet and the field of the wire are in opposite directions. They cancel, making the weak field represented by the less-concentrated lines above A. Wire A is lifted by the strong magnetic field beneath. To account for this lifting effect, think of the north→south lines of force as both repelling each other and attempting to straighten themselves.

A similar effect at B pushes wire B downward. B is pushed down by lines of force. This is similar to the way the round stick in figure 13-4 is pushed down by the string.

Note that the direction of motion of the wire can be quickly found by using the three-finger right-hand rule.   With first finger, middle finger and thumb at right angles to each other, forefinger = field, center finger = current, and thumb = motion.

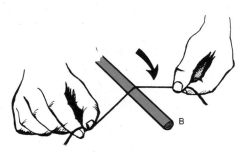

Fig. 13-4   Pressure of Magnetic Lines of Force on a Wire

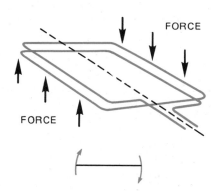

Fig. 13-5 Torque in an Electric Motor

TORQUE

The lifting of one side of the loop and the pushing down of the other side is a turning effect, or *torque,* on the loop of wire, figure 13-5. This combination of forces turns the armature of all electric motors. The same method explains the torque that turns the moving coil in a voltmeter or ammeter. The term torque, in general, means any attempt to cause mechanical rotation. Other examples of torque are the turning forces applied to steering wheels, driveshafts, wrenches, and screwdrivers.

The amount of a torque is calculated by multiplying the total applied force (pounds or newtons) by the radius of the turning circle. The unit of measure for torque is pound-feet (newton-meter).

As drawn in figure 13-2, the loop would rotate only about a 1/4 turn from the position shown until A and B were vertical. Continued lifting on A and pushing down on B are useless. In order to achieve continued rotation, the current direction in the loop must reverse when the loop reaches its vertical position.

This reversal can be accomplished for a single coil by a two-segment commutator, figure 13-6. At the instant shown, electrons flow from the negative brush through A and return through B to the positive brush. As A is lifted to the top of its rotation, the commutator segment that supplied electrons to A slides away from the negative brush and touches the positive brush. Thus the current in the loop is reversed. The momentum of the loop carries it past the vertical position. A is then pushed downward to the right, B is lifted to the left, and another half-turn of rotation continues, figure 13-7.

This impractical single-coil armature has a variety of faults, one of which is the irregularity of the torque it produces. When the loop is horizontal, the force on the wire is at its greatest. When the loop is vertical, there is no force on the wire.

A steadier torque is achieved by using the several coils of a drum-wound armature. The armature shown in figure 10-13 of Chapter 10 is as appropriate in a motor as it is in a generator.

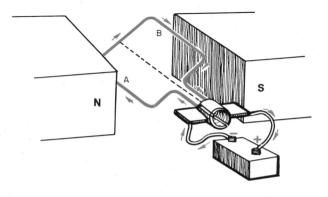

Fig. 13-6 How a Commutator Reverses Current to Produce Rotary Motion

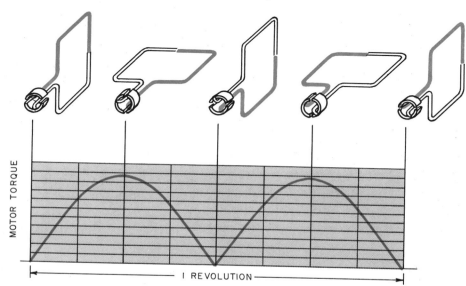

Fig. 13-7 Graph of Motor Torque Through One Revolution of Commutator

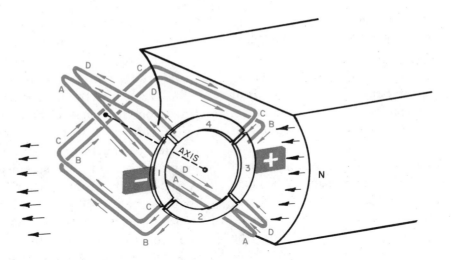

Fig. 13-8 Path Electrons Travel Through Coils, Brushes, and Commutator Segments

Figure 13-8 shows a similar, but simpler, drum winding. Assume that the south pole of the field magnet is at the left. It has been omitted from the diagram in order to show the windings clearly.

Trace incoming electrons from the negative brush through commutator segment 1, through coil A to segment 2, then through B to segment 3. There, the positive brush leads to the completed circuit (not shown) through a source of dc and back to the negative brush. From the negative brush, another circuit can be traced through armature coils C and D to the positive brush. Compare this drum-winding with that shown in figure 10-13 to note the similarity in principle.

The circuit of figure 13-8 is redrawn in more schematic fashion in figure 13-9. (The term *schematic diagram* means one that lays out the electrical circuit plainly, without regard to the actual appearance of the device. A photograph, or a *pictorial diagram* shows what the device looks like without showing electrical circuits.)

Two parallel circuits exist through this armature. All four of the coils contribute their torque to aid rotation of the armature.

In figure 13-8, a further rotation of 45° will bring coils B and C to a vertical position.

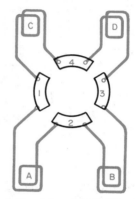

Fig. 13-9 Relationship of Coils, Brushes, and Commutator Segments

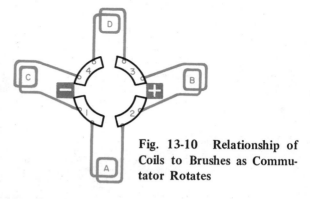

Fig. 13-10 Relationship of Coils to Brushes as Commutator Rotates

At that instant, the negative brush will touch both segments 1 and 4. Segments 2 and 3 contact the positive brush.

As shown in figure 13-10, there is no current in coil C because both ends of it touch the same brush. Likewise, there can be no current in coil B. This condition exists at the instant when coils B and C are in a horizontal position. Even if current was present in coils B and C, it would produce no torque. The absence of current in the coils during this short interval is no disadvantage. At this same instant, coils A and B are in a vertical position producing maximum torque.

In a more complicated drum winding (for example, one with twelve coils), there would be ten coils carrying current and producing torque during the short time that two of the coils were inactive, like B and C in figure 13-10.

Calculation of Torque

The amount of torque developed by a motor can be predicted before the motor is built. As you might expect, greater current, more turns of wire, and stronger magnetic field all contribute to greater turning effect.

In the following formula, ϕ (phi) means total flux, the number of magnetic lines of force passing through the armature. Z is the number of wires (not coils) along the armature. For example, eight wires in figure 13-8 would be represented by Z. I_a is the total armature current, and m is the number of parallel paths through the armature, for example, two in figure 13-8.

$$\text{Torque (Pound-feet)} = \frac{\phi \times Z \times I_a}{425\,000\,000 \times m}$$

To illustrate the use of this formula, assume an armature 6 inches long is placed between magnet poles as in figure 13-11. The armature carries 200 turns of wire; that is, 400 wires for Z in the formula. Flux density in the field magnet is 50 000 lines per square inch; 6″ x 4″ = 24 square inches, so 24 x 50 000 = 1 200 000 lines leaving the north pole and passing through the armature. The total armature current is 10 amps, which passes through two parallel paths in the armature, figure 13-10.

$$\text{Torque} = \frac{1\,200\,000 \times 400 \times 10}{425\,000\,000 \times 2}$$
$$= 5.65 \text{ pound-feet}$$

HORSEPOWER OUTPUT

The preceding calculation of torque does not itself tell the power of the motor; speed must be taken into account also. One horsepower (hp) is defined as a rate of doing work equal to 33 000 foot-pounds per minute, (Chapter 4). From this definition, horsepower can also be calculated from torque and rpm. This is true for all machines; it is not limited to electric motors.

$$\text{hp} = \frac{2\pi \times \text{torque} \times \text{rpm}}{33\,000}$$

The above formula may be rewritten as: hp = 0.000 19 x torque x rpm or

$$\text{Torque} = \frac{\text{hp}}{0.000\,19 \times \text{rpm}}$$

Assuming that the armature developing 5.65 pound-feet of torque in the previous example is operating at 1 250 rpm, horsepower can be found:

$$\text{hp} = \frac{2 \times 3.14\,(\pi) \times 5.65 \times 1\,250}{33\,000}$$
$$= 1.34 \text{ hp}$$

Incidental to this discussion of power, 746 watts = 1 horsepower, and 1.34 horsepower = 1 000 watts. It had been assumed

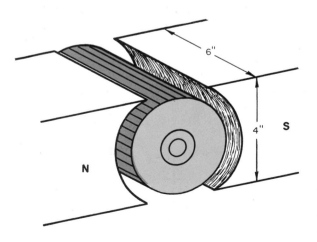

Fig. 13-11

that this armature has a current of 10 amps. Therefore, if it were 100 percent efficient, a power input of 100 volts x 10 amps = 1 000 watts would produce the 1.34 horsepower. In the metric system, power is measured in watts.

Experimentally, the torque of a motor is measured directly by a device called a *prony brake,* figure 13-12. Tightening the bolts makes the brake tend to turn along with the motor pulley. The brake arm is restrained by the stationary spring scale, so a torque load is placed on the motor. The torque is the product of the net force on the scale, F in pounds, times the effective length (L) of the torque arm in feet.

$$\text{Torque} = F \times L$$

Speed of the motor can be found by the use of a revolutions counter and stopwatch, a tachometer, or a calibrated stroboscopic light. Thus the true mechanical horsepower output (or *brake horsepower)* can be determined from measurements of torque and speed.

If the horsepower output is known originally from electrical data, torque can readily be calculated for any known value of rpm by using

$$T = \frac{hp}{0.000\ 19 \times rpm}$$

EMF GENERATED IN A MOTOR

Whenever an electric motor is running, its wires are cutting lines of force. Therefore, an emf is generated in the armature. Using members from the previous example, the emf can be calculated using the generator formula (Chapter 10):

$$emf = \frac{\text{turns x 4 x flux x rpm}}{\text{Paths x 60 x 100 000 000}}$$

$$= \frac{200 \times 4 \times 1\ 200\ 000 \times 1\ 250}{2 \times 60 \times 100\ 000\ 000}$$

$$= 100 \text{ volts}$$

This generated 100 volts is necessarily in a direction opposite to the direction of the current through the armature of the motor. This conclusion can be reached in either of two ways:

(1) With the help of the three-finger rules for motor and generator, it can be seen that the motion caused by current in a wire will cause an emf that opposes the current, figure 13-13.

(2) If the generated voltage were in the same direction as the current, a power company would not be needed; we would simply spin the armature, short-circuit the brushes, and let the motor run itself.

Importance of Generated EMF

It is sometimes difficult for a beginner to realize the value of the emf that a motor

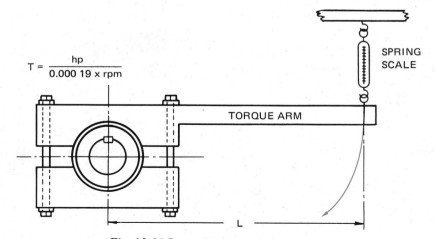

$$T = \frac{hp}{0.000\ 19 \times rpm}$$

SPRING SCALE

TORQUE ARM

L

Fig. 13-12 Prony Brake to Measure Torque

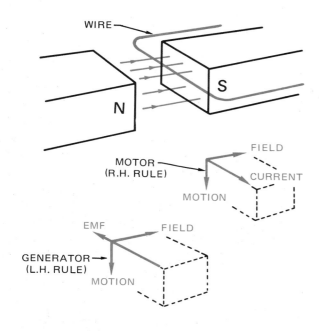

Fig. 13-13 Left-Hand Rule and Right-Hand Rule Applied to Rotation of an Electric Motor

generates. The generated emf is a measure of the useful mechanical energy obtained from the electrons passing through the armature.

Sometimes this emf is called *back emf,* which emphasizes the fact that it opposes the voltage applied to the motor. This is a useful opposition. The generated back emf hinders the movement of the incoming electrons in the same way that the weight of a bag of groceries opposes the efforts of a boy carrying the bag up a flight of stairs. If there were no opposition, no useful work would be done. If the boy left the groceries on the sidewalk, he could run upstairs much faster, but that would not deliver the groceries. An electric heater-element generates no back emf; therefore the electrons may run through it rapidly. However, they produce no mechanical work.

If the armature of an electric motor is fastened so that it cannot turn, it will generate no back emf. Then the electrons can flow through the windings more easily. However, the motor will greatly overheat.

Returning to the numerical example used previously, there is an armature rotating at 1 250 rpm, generating 100 volts, and taking a current of 10 amps from the power line. Assume that the armature wiring has 2-ohms resistance and determine the total (line) voltage needed to operate the motor.

The power line has two jobs: (1) it must supply 100 volts, which is converted to mechanical energy, and (2) it must supply enough additional voltage to force the 10-amp current through the 2-ohm wire resistance. It takes 20 volts to put 10 amps through 2 ohms, so the line voltage must be 100 + 20 = 120 volts.

Total voltage applied to the armature equals back emf generated in the armature plus voltage (IR) used to overcome wire resistance.

To carry this calculation one step further multiply the total applied voltage (120 volts in the example) by the current (10 amps). The total power input is 1 200 watts.

Multiplying the generated emf (100 volts) by the 10 amps results in 1 000 watts, which is the useful mechanical power output. Note that this same figure has already been found once before. It was found from the 1.34 hp output that was found from torque and rpm.

The power input is 1 200 watts; the output is 1 000 watts. Where does the other 200 watts go? By multiplying the 20 volts used on resistance by the 10 amps, we get 200 watts. This represents the rate at which energy is converted into heat in the armature. Two hundred watts is the rate of heat production.

Total input power (Applied volts x amps)	=	Useful mechanical power output (Back emf x amps)	+	Heating rate (IR x amps)

The useful power produced can be calculated from the back emf and the current.

This is exactly like saying that the useful power (working rate) accomplished by the grocery boy can be calculated from the weight of the bag and how fast he lifts it. The weight is not doing the work, but the boy does more work if the bag weighs more.

Here is another sample calculation. Given a 5-ohm armature that takes 6 amps on a 115-volt line when operating at its normal rating, find: (a) back emf generated, (b) power input, (c) useful power output, (d) heating rate, (e) efficiency, (f) current and heating rate if the motor is stalled.

(a) 6 amp x 5 ohm = 30 volts used on resistance

115 - 30 = 85-volts back emf

(b) Power input = 115 x 6 = 690 watts

(c) Power output = 85 x 6 = 510 watts

$$\frac{510 \text{ watts}}{746 \text{ w/hp}} = 0.68 \text{ hp}$$

Fig. 13-14 12 000-HP DC Motor for Steel Rolling-Mill Drive

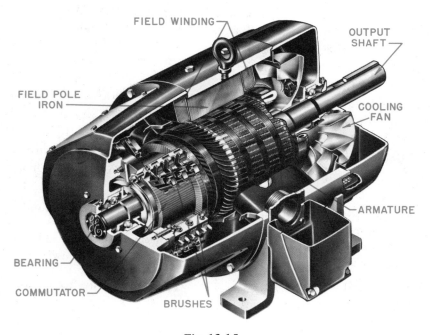

FIELD WINDING

OUTPUT SHAFT

FIELD POLE IRON

COOLING FAN

ARMATURE

BEARING

COMMUTATOR

BRUSHES

Fig. 13-15

Fig. 13-16 Armature for DC Motor

(d) Heating rate = 30 volts x 6 amp
= 180 watts

(e) Efficiency = $\dfrac{\text{power out}}{\text{power in}}$ =
$\dfrac{510}{690}$ = 0.74

(This is the *armature efficiency,* not the overall efficiency of the entire motor. The armature efficiency can also be found from back emf ÷ line volts, 85 ÷ 115 = 0.74.)

(f) If the motor stalls, the current becomes:
115 V ÷ 5 ohms = 23 amps
(Power input becomes 115 x 23 = 2 645 watts, all of which is the heating rate.)

> Always select a motor of sufficient power to carry the anticipated load. An overloaded motor will become hot and possibly burn out or cause a fire.

The magnetic field in which the armature rotates may be provided in either of three ways: a permanent magnet, an electromagnet in series with the armature, or an electromagnet in parallel with the armature. In these three types of motor, energy is delivered to the rotating armature through external wiring, brushes, and the commutator. In still another type of widely used motor, alternating current in the field coils induces

Fig. 13-17 Winding a DC Traction Motor Armature

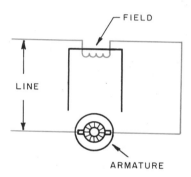

Fig. 13-18 Schematic Diagram of Series-Wound Motor

a current in the rotor, just as alternating current in the primary coils of a transformer induces current in the secondary. This ac induction motor will be described in a later chapter.

SERIES MOTORS

The *series motor*, figure 13-18, is used in home and shop to operate small high-speed tools and appliances, such as mixers, vacuum cleaners, drills and other portable electric tools, figure 13-19. Such devices benefit from one or both of these characteristics of the series motor: (1) high rpm on low-torque loads, and (2) ability to develop large torque when slowed down. The induction motor is basically simpler to build, but to get speeds over 3 600 rpm

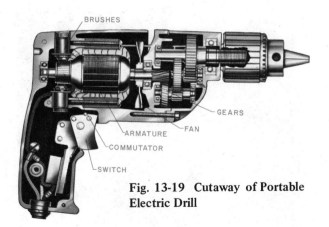

BRUSHES

GEARS

ARMATURE

FAN

COMMUTATOR

SWITCH

Fig. 13-19 Cutaway of Portable Electric Drill

requires gears or belt pulleys. If overloaded, the induction motor stalls at times when the series motor will keep running.

Small series motors operate on either ac or dc. If the current through the motor is reversed, armature current and field flux reverse simultaneously, maintaining torque in the same direction. To reverse rotation of the motor, either the connections at the brushes or at the field coil must be reversed, but not both.

In an automobile, the starter motor is a series motor because high torque from a small motor is required. Series motors also power fans in car heating systems and window raising-and-lowering devices for the handicapped. Industrially, large direct current series motors are used on cranes, hoists, subway cars, electric locomotives, or on any machine where extremely high torque is required for starting a load or continuing to move a suddenly applied overload during operation.

However, large series motors cannot be used where a relatively constant speed is required from no-load to full-load, or where the motor may have to operate at no-load. For those applications, either ac induction motors or dc shunt (parallel field) motors may be used.

The following series of steps explains why the series motor develops high torque at low speeds.

(1) Assume the motor has been operating at moderate speed and a mechanical load is suddenly applied to the motor. The first effect of this is slowing of the motor.

(2) Back emf in the motor is proportional to speed, so slowing the motor reduces the back emf generated.

(3) Reduced back emf lets the line voltage push more current through the motor.

(4) More current in the field coils makes a stronger magnetic field.

(5) Torque depends on the product of field flux and current, so if current is multiplied by 3, and flux is multiplied by 3, the torque will be 9 times stronger than before.

A sudden reduction of load makes the armature turn more easily. Its first response is to speed up slightly. The increased speed generates a little more back emf, reducing the current coming into the motor. Yet, the existing lowered current and flux create enough torque to accelerate the motor. If the motor still carries some load, eventually current and flux becomes small enough to maintain the reduced torque requirement without further acceleration. However, a very substantial increase of rpm occurs before current and flux are reduced enough.

If the load is completely removed, current and flux continue to make torque that accelerates the motor. The only retarding load is friction. Speed continues to increase; back emf gradually increases. Hence, the current becomes smaller and smaller until finally the torque is very small, as small as friction, figure 13-20. The great speed increase is harmless in a small motor where 10 000 rpm may even be desirable. In larger motors, the tendency of things to travel in a straight line becomes more important. At 10 000 rpm, the surface of an 8 inch (203 mm) diameter armature is traveling at about 4 miles (6.5 km) per minute. Each ounce (28.3 g) of copper wire in the slots requires a force of 700 pounds (317 kg) to hold it in place. So, the big armature might explode before it reaches 10 000 rpm.

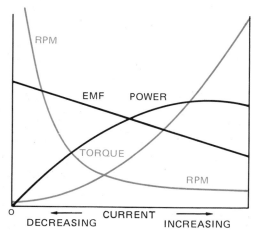

Fig. 13-20 Series Motor Characteristics

SHUNT MOTORS

The magnetic field of the shunt motor is supplied by an electromagnet in parallel with the armature, figure 13-21. Compared with the armature, the field winding has more turns of smaller wire, so the field current is small. As long as the line voltage is constant, the unchanging field current in this simple motor maintains a steady magnetic field regardless of variations in armature current.

Compared with the series motor, the shunt motor maintains a much steadier speed and does not develop the huge increase in torque when slowed down by overload.

Torque, as always, is proportional to armature current and field flux. If the motor is slowed by applying more mechanical load, reduction in rpm reduces the back emf. This allows more current through the armature, and the torque increases. But if the current is multiplied by 3, the torque is multiplied by 3 (not the 9 suggested for series) because the field flux stays constant. If the motor is overloaded, it slows to a stop. In the stalled motor there is no back emf, current is abnormally high, and the $I^2 R$ heating rate is extremely high. The motor should be equipped with a bimetal thermostat that will disconnect it from the line before it is damaged by excessive heat.

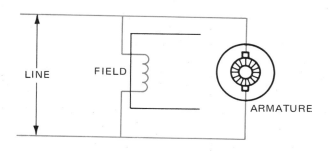

Fig. 13-21 Schematic Diagram of Shunt Motor

The shunt motor will not "run away" when the load is removed, as the series motor may do. Suppose the mechanical load is entirely removed. The first response will be to increase its speed slightly. Back emf depends on rpm and field flux; flux is constant in this motor. A small increase in rpm promptly raises the back emf enough to reduce the armature current so there is no excess torque to continue accelerating the armature. (In the series motor, the increase in rpm was accompanied by decreased flux, so back emf did not rise as much for a small speed change.)

Simple shunt motors are useful industrially where a relatively constant speed is desired from no-load to full-load with severe overload unlikely, and where dc is readily available. (Shunt motors are not operated on alternating current because the current in the field and armature will not reverse at the same instant.) Shunt motors can be used on fans and blowers, as the motor unit for a motor-generator set, and for operating metalworking machines, textile machines, woodworking machines, and similar machines.

MOTOR CONTROL

By adding some means of externally controlling the field current and armature current independently in the shunt motor, now both speed and torque can be adjusted to meet the momentary needs of the driven machine, figure 13-22.

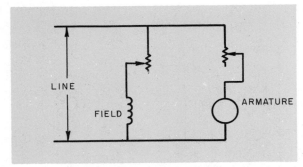

Fig. 13-22 Schematic Diagram of Speed Control for Shunt Motor

Field Control

The following discusses what happens to the speed of the motor when the magnetic field is weakened by adding resistance in the field circuit, thus reducing the field current. (Assume the motor has been running and delivering constant torque and power.) The first effect of field weakening is reduction of the back emf generated. That is proportional to field flux and rpm. Reduction of back emf allows considerably more current through the armature; the torque immediately increases. Torque does depend on both armature current and flux, but the armature current increases much more than the flux decreases. The increased torque accelerates the motor to a higher speed, causing back emf to start increasing. Therefore, armature current is gradually reduced again. Reduced armature current finally reduces the torque to a point where the motor no longer accelerates and delivers the previous value of torque at higher speed. This result might at first be unexpected. Reducing the field strength causes the motor to drive the machine faster with a higher armature current, slightly reduced back emf, and higher power output. Increasing the field strength reverses all these effects.

Armature Control

If the current in the armature of the shunt motor is reduced, the results may not be surprising. (Assume again that the motor

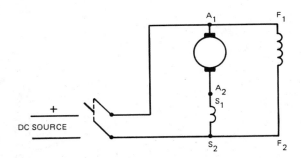

Fig. 13-23 Schematic Diagram of Cumulative Compound Motor

has been driving a constant load and field strength is constant.) The first effect of reduced armature current is reduced torque, so the motor slows. But slowing the motor reduces the back emf which lets armature current build up again to its previous value. The motor continues to deliver the same torque, but at reduced speed and reduced power output.

Control of current, especially armature current, by rheostats, is wasteful of energy. In a great variety of modern machines, the two currents in the motor are controlled by electronic devices. Usually these devices are controlled rectifiers which deliver dc to the motor from the ac power lines. At the same time they regulate the amount of current to field and armature. Another electronic device controls the rectifier. A change in any imaginable measurable quantity can be used to generate a voltage that will control electronic rectifiers. For example, a change of temperature, position, tension, speed, or pressure can initiate events that cause a motor to restore conditions to some desired preset value.

Compound Motors

Reviewing the characteristics of series and shunt motors, we found the shunt motor has a more constant speed, but a series motor of the same power rating can exert much greater torque when required. These two desirable features can be achieved by putting both a series winding and a shunt winding on the field-magnet iron of a motor. Usually, connections are arranged so that the series-winding current is in the same direction around the magnets as the shunt-winding current, allowing both windings to aid each other in producing flux. This connection forms a cumulative-compound motor, figure 13-23. At heavy loads, when the motor slows down, the increased current in the series field boosts the field strength, giving added torque. At light loads, the constant-flux shunt field limits the top speed, just as it does in a shunt motor. This avoids the runaway behavior of the series motor. Some uses for this type of motor are rolling-mill drives, metal shears, metal stamping presses, and elevators.

Reversal of Rotation

Reversal of a motor is usually accomplished by a reversing switch placed so as to control only the armature current,

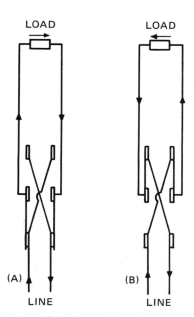

Fig. 13-24 Double-Pole Double-Throw Switch Used as a Reversing Switch

figure 13-24. Reversing current to the entire motor (field and armature) accomplishes nothing. With both field and armature current reversed, rotation continues in the same direction. In connection with some types of motor-control systems that require a quick stop, the armature current may be reversed momentarily to provide reverse-torque to stop the motor. This method of achieving a sudden stop is called *plugging.*

POINTS TO REMEMBER

- Electrons moving through a magnetic field are pushed sideways in a direction at right angles to the field and at right angles to their original direction.

- The above fact explains the motion of armatures of dc and ac motors and the controlled movement of electron streams in electric arcs or in vacuum tubes.

- The drum armature is an effective way of arranging coils to produce a continuous torque in a motor.

- Torque means turning effect. It is calculated as force x radius and is measured in pound-feet or newton-meters.

- Torque of a motor depends on flux, number of wires, and current according to the formula:
$$\text{torque} = \frac{\phi \times Z \times I_a}{425\ 000\ 000 \times m}$$

- One horsepower is defined as a rate of working. It equals 33 000 foot-pounds per minute:
$$\text{hp} = \frac{2\pi \times \text{torque} \times \text{rpm}}{33\ 000}$$

- The total voltage applied to the armature of a motor does two things:
  1. Part of it (=IR) produces heat by causing current through resistance.
  2. The other part, which is equal to the generated emf, runs the motor, producing mechanical energy.

- Applied voltage = emf + IR

- Power input = useful power output + $I^2 R$ heating rate.

- Torque is proportional to flux and armature current. Increase of either increases torque.

- Counter emf is proportional to flux and rpm. Increase of either increases emf.

- The shunt motor has its field coils in parallel with the armature. Its constant field strength limits input current and speed at no-load.

- The series motor has its field coil in series with the armature. At no-load, reduced field permits enough current to produce torque that accelerates motor greatly. At heavy-load, torque is high, speed low.

- Weakening the field speeds up the shunt motor; reduced armature current slows the shunt motor.

- Comparing advantages of series and shunt motors: The shunt motor has the more constant speed, but a series motor of the same power rating can exert a much greater torque when necessary without a very large increase in current.

- Compound motors have both series and shunt fields. Relative ampere-turns of the two fields determine speed and torque characteristics of the motor.

REVIEW QUESTIONS

1. How should magnet poles, figure 13-25, be placed so that the wire will move upward when the switch is closed?

2. In figure 13-26, a brass strip is fastened to the left terminal, and rests in loose contact with the right-hand terminal. What happens when a battery is connected to the terminals (a) so that the left one is negative? (b) reversed, with right negative?

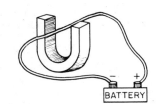

Fig. 13-25

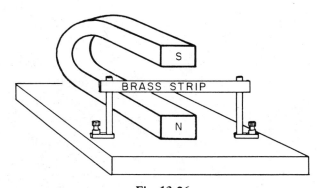

Fig. 13-26

3. What happens to the wire referred to in question 1 if an alternating current (60 hertz) is sent through the wire?

4. Determine the direction of rotation of the armature of figure 13-8.

5. a. What is torque?
   b. What factors would determine the amount of torque on a coil placed in a magnetic field?

6. What is the generated emf in a motor?

7. The magnetic field in which the armature of a motor rotates may be provided in either of three ways. What are these three ways?

8. What are the advantages of a series motor? Name some disadvantages.

9. Explain why a series motor develops a high torque at low speeds.

10. How does a shunt motor differ from a series motor in construction and operating characteristics?

11. What factors determine the torque in a motor?

12. What happens when a shunt motor is overloaded? How can this condition be avoided?

13. What is the chief advantage of a shunt motor?

14. What happens to the speed of a motor when the strength of the magnetic field is reduced by adding a resistance in the field circuit? What happens if the current in the armature is reduced?

15. How is a compound motor constructed? Name some advantages of such a motor.

16. Explain how the direction of rotation of a motor may be reversed.

17. List several industrial applications of a series motor, a shunt motor, and a compound motor.

18. Reduced line voltage may cause a shunt motor to overheat. Why?

RESEARCH AND DEVELOPMENT

Experiments and Projects on Motors

INTRODUCTION

In Chapter 13, you have learned about torque, horsepower, importance of emf generated in a motor, and how series, shunt, and compound motors operate. In the experiments which follow, you will have an opportunity to test your knowledge of the operation of the various motors and to make some calculations. In the projects, suggestions are made regarding the designing and constructing of simple low-voltage motors, the connecting and operating of motors, and the study of the various parts of a motor.

EXPERIMENTS

1. To understand the operating principle of series-wound, shunt-wound, and compound-wound dc motors.

2. To study the operating characteristics of a commercial-type, series-wound dc motor.

3. To study the operating characteristics of a commercial-type, shunt-wound dc motor.

4. To study the field windings of a compound dc motor and its operating characteristics.

5. To learn how to reverse the direction of rotation of a dc motor by changing the armature coil connections, and by changing the field coil connections.

PROJECTS

1. Design and construct a simple series-wound dc motor with a switch in the circuit arranged to reverse the rotation of the motor.

2. Design and construct a simple shunt-wound dc motor with a switch in the circuit arranged to reverse the rotation of the motor.

3. Connect and operate a fractional horsepower dc motor with a starting rheostat.

4. Disassemble a fractional horsepower dc motor and examine the parts. Reassemble and operate.

EXPERIMENTS

Experiment 1 SERIES-WOUND, SHUNT-WOUND, AND COMPOUND-WOUND DC MOTORS

OBJECT

To understand the operating principles of series-wound, shunt-wound, and compound-wound dc motors.

APPARATUS

1 - Experimental motor kit
1 - Speed indicator
1 - DPST switch

PROCEDURE

1. Assemble parts of a series-wound motor.

2. Connect to current with a DPST switch.

3. Operate motor and check rpm with speed indicator (motor running free and under load).

4. Find torque and record.

5. Make a circuit diagram of the electron flow through the motor.

6. Plan a circuit diagram to reverse the direction of rotation.

7. Remove the necessary wires and make the new connections to reverse the motor.

8. Disassemble motor and assemble as a shunt-wound motor and make the same observations.

9. Follow the same procedure to produce and observe a compound-wound motor.

OBSERVATIONS

Use a table similar to the one shown here to record your observations.

MOTOR TYPE	RPM RUNNING FREE	RPM UNDER LOAD
SERIES		
SHUNT		
COMPOUND		

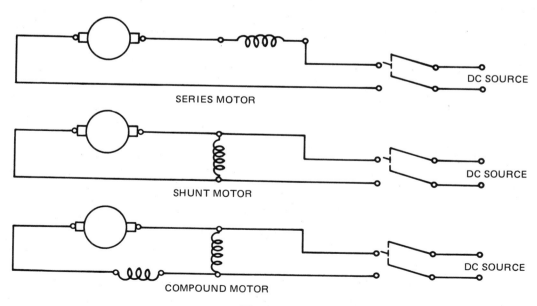

Fig. 13-27

QUESTIONS

1. Which motor had the greatest rpm? How do you account for the answer?

2. Which motor had the most torque?

3. Was the torque on the series-wound motor the same as the others? If it was different, what was the reason?

4. What are the most important features of each motor?

5. What is a significant feature of the series-wound motor that the others do not have?

6. Could you plan a circuit in which a DPDT switch would be incorporated to reverse the direction of rotation on a motor without removing wires each time? Draw a wiring diagram.

Experiment 2 SERIES-WOUND DC MOTOR

OBJECT

To study the operating characteristics of a commercial-type, series-wound dc motor.

APPARATUS

1 - Series-wound dc motor, commercial-type
1 - Speed indicator
1 - Starting rheostat
1 - DPST switch
1 - Prony brake
1 - Spring scale

PROCEDURE

1. Connect motor according to figure 13-28.

2. Operate motor and check rpm with speed indicator (motor running free and under load).

3. Find torque and record.

4. Make a circuit diagram of the electron flow through the motor.

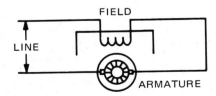

Fig. 13-28

OBSERVATIONS

Use a table similar to the one shown here to record your observations.

MOTOR TYPE	RPM RUNNING FREE	RPM UNDER LOAD	TORQUE
SERIES			

QUESTIONS

1. What are the advantages of a series motor?

2. What was the nature of the torque at low speed? Explain the answer.

3. Suggest several applications of series motors.

Experiment 3 SHUNT-WOUND DC MOTOR

OBJECT

To study the operating characteristics of a commercial-type, shunt-wound dc motor.

APPARATUS

1 - Shunt-wound dc motor, commercial-type
1 - Speed indicator
1 - Starting rheostat
1 - DPST switch
1 - Prony brake
1 - Spring scale

PROCEDURE

1. Connect motor according to figure 13-29.

2. Operate motor and check rpm with speed indicator (motor running free and under load).

3. Find torque and record.

4. Make a circuit diagram of the electron flow through the motor.

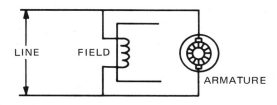

Fig. 13-29

OBSERVATIONS

Use a table similar to the one shown here to record your observations.

MOTOR TYPE	RPM RUNNING FREE	RPM UNDER LOAD	TORQUE
SHUNT			

QUESTIONS

1. What are the chief advantages of a shunt motor?

2. Did the torque of the motor change under load? Explain your answer.

3. What conclusion did you reach concerning the most important operating feature of a shunt-wound motor?

4. On what type equipment would a shunt motor be used?

Experiment 4 COMPOUND DC MOTOR

OBJECT

To study the field windings of a compound dc motor and its operating characteristics.

APPARATUS

1 - Compound dc motor, commercial-type
1 - Speed indicator
1 - Starting rheostat
1 - SPST switch
1 - Prony brake
1 - Spring scale

PROCEDURE

1. Connect motor according to figure 13-30.

2. Operate motor and check rpm with speed indicator (motor running free and under load).

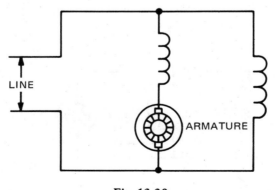

Fig. 13-30

3. Find torque and record.

4. Make a circuit diagram of the electron flow through the motor.

OBSERVATIONS

Use a table similar to the one shown here to record your observations.

MOTOR TYPE	RPM RUNNING FREE	RPM UNDER LOAD	TORQUE
COMPOUND			

QUESTIONS

1. How is a compound motor constructed?

2. Describe the operating characteristics of a compound motor.

Experiment 5 REVERSING ROTATION OF A SERIES MOTOR

OBJECT

To learn to reverse the rotation of a dc motor by changing the armature coil connections and by changing the field coil connections.

APPARATUS

1 - DC motor

1 - DPDT switch

PROCEDURE

1. Connect motor and operate.

2. Operate and check rotation.

3. Locate wires leading to the motor brushes, disconnect at brushes and reconnect through a DPDT switch wired as a reversing switch, figure 13-31.

   Note: If the motor has a single field winding, it may also be possible to reverse rotation by reversing field current instead of armature current.

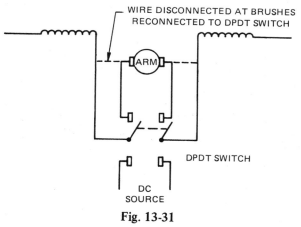

Fig. 13-31

QUESTIONS

1. What causes a motor to run in the opposite direction when the wires are reversed?

2. Could the direction of rotation of the motor be reversed by reversing the leads from the dc power source? Explain why or why not.

PROJECTS

DESIGN AND CONSTRUCT A SIMPLE EXPERIMENTAL MOTOR

This motor, figure 13-32, is easily constructed. It can be made as a simple series-wound type or, by arranging separate terminals from the armature and field coil, it can be operated either as a series-wound or shunt-wound motor. If a second field coil is wound over the primary field coil and separate leads brought out to terminals, it can be operated as a compound-wound motor. If the commutator segments are set in the same relationship to the armature as they appear on the drawing, figure 13-33, the motor will be self-starting.

MATERIALS

1 - Pc. soft steel, 3/8" x 2 3/4" (19 mm x 70 mm)
1 - Pc. soft steel, 1/8" x 3/4" x 9 5/8" (3 mm x 19 mm x 244 mm)
1 - Pc. steel rod, 1/8" x 3" (3 mm x 76 mm)
1 - Pc. brass or copper tubing, 3/8" OD x 1/2" (9 mm OD x 12 mm)
1 - Pc. dowel or plastic to fit inside diameter of tubing
2 - Pcs. sheet iron, #20 or #22 gauge x 1" x 2 7/8" (25 mm x 73 mm)
2 - Pcs. spring brass, #30 gauge x 1/4" x 2 1/4" (6 mm x 57 mm)
2 - Wood screws, #6 x 3/4" RH
6 - Wood screws, #4 x 1/2" RH
6 - Binding posts or Fahnestock clips
1 - Wooden base, 3/4" x 3 1/2" x 6" (20 mm x 90 mm x 150 mm)
   Magnet wire, #22 AWG Formvar

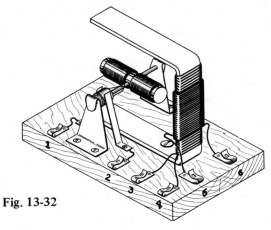

Fig. 13-32

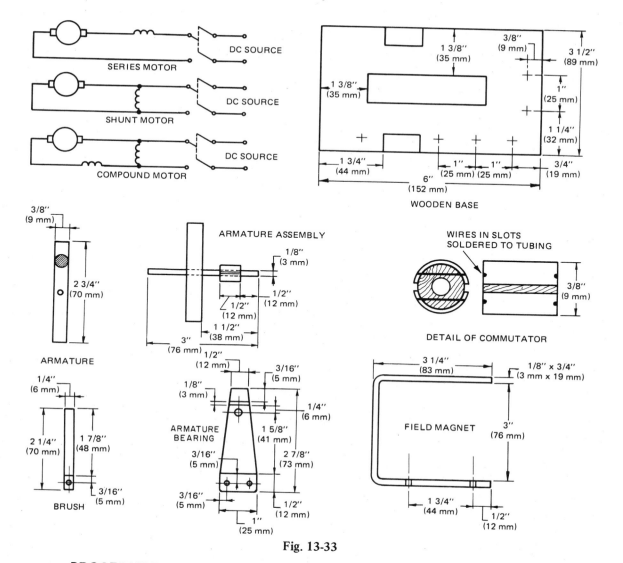

Fig. 13-33

PROCEDURE

Make the field magnet, armature bearings, brushes, and base.

Next, make the armature assembly. The armature must be a tight fit on the shaft. To make the commutator, select a piece of brass or copper tubing. Cut a piece about 17/32" (13.5 mm) long. Square the ends and remove the burrs from the inside. Select a piece of hardwood dowel or plastic larger than the ID of the tubing. Place the material in the chuck on an engine lathe and turn it to fit the inside of the dowel. Drill a hole through the center while it is in the lathe, to make a press fit on the armature shaft. Press the tubing over the dowel, face the end, and cut it off even with the other side of the tubing. Saw slots across each end as indicated on the drawing. Press a piece of copper wire into each slot and solder the ends to the tubing. Remove the excess solder and the ends of the wires with a file.

Saw a slot in the tubing on each side as indicated on the drawing. Carefully remove the burrs.

Wind a layer of plastic tape on the field magnet where the coil will be wound and on each side of the armature.

Wind two layers of #22 magnet wire on each side of the armature. Start in the middle and end in the middle, figure 13-34. Care must be taken to continue winding in the same direction when going from one pole to the other. Secure the ends with cotton tape. *Note:* Leave about three inches of wire at the end of each coil.

Starting at the bottom, wind two layers of #22 gauge magnet wire on the field magnet. The first layer should be 2 3/4" (70 mm) long. Starting 1/2" (12 mm) from the bottom and ending 1/2" (12 mm) from the top, wind two layers of #22 gauge magnet wire over the first coil and in the same direction, figure 13-35. Secure ends of each coil with cotton tape.

Coat the coils on the field magnet and armature with shellac or varnish.

Assemble, test, and evaluate.

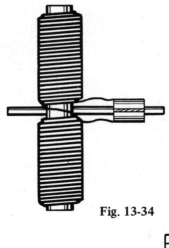

Fig. 13-34

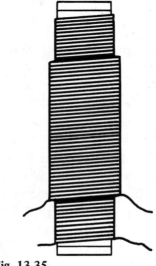

Fig. 13-35

EXPERIMENTAL MOTOR NO. 2

This motor, figure 13-36, with a three-pole armature, is self-starting and has considerable torque for its size. It may be connected to operate as a series motor or as a shunt motor on 6-volt direct current.

MATERIALS

48 - Pcs. #22 gauge sheet iron
1 - Pc. CRS 1/8″ x 3 1/2″ (3 mm x 89 mm)
1 - Bolt, 1/2″ x 3″ (M12 x 75 mm L)
1 - Nut, 1/2″ (M12)
1 - Machine screw, #10-32 x 1″ (M5 x 24 mmL)
2 - Pcs. sheet iron, #22 gauge, 1″ x 2 3/4″ (25 mm x 70 mm)
1 - Pc. Copper-clad phenolic board, 1 1/4″ (32 mm) diameter
2 - Pcs. Spring brass wire, #16 gauge x 4″ (100 mm)
4 - Pcs. soft steel, 1/8″ x 1″ x 4 1/4″ (3 mm x 25 mm x 108 mm)
2 - Fiber washers, 1/16″ x 1 1/2″ diameter (1.5 mm x 38 mm diameter)
1 - Wooden base, 3/4″ x 3 3/4″ x 6″ (20 mm x 95 mm x 150 mm)
2 - Fahnestock clips
1 - Washer, #6 (M3.5)
4 - Wood screws, #4 x 1/2″, RH
3 - Wood screws, #6 x 1/2″, RH
1 - Pc. tubing, 1/8″ ID x 1/4″ (3 mm ID x 6 mm)
1 - Pc. tubing, 1/8″ ID x 3/8″ (3 mm ID x 9 mm)
2 - Nuts, #5-44 (M3)
1 - Fiber washer, 1/2″ (12 mm) diameter with 1/8″ (3 mm) hole
1 - Pulley
Magnet wire, #22 gauge AWG Formvar

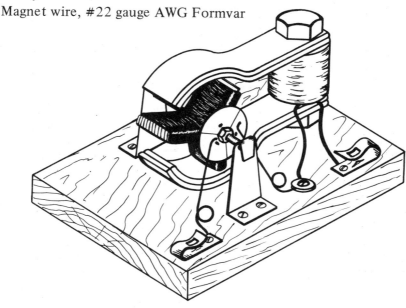

Fig. 13-36

PROCEDURE

Study the drawings, figures 13-37 and 13-38, and the suggested procedure. Procure the materials and make the parts.

When cutting the armature laminations, be careful to keep the metal flat.

Remove all burrs, stack the pieces evenly, clamp them together, and drill the hole through the center. Select a drill which will produce a tight fit between the laminations and the shaft.

Insert the shaft and balance the rotor.

Before winding the armature coils, round the corners of each rotor pole with a fine file. Next, wrap two layers of plastic tape around each one. Starting at the center each time, wind four layers of #22 magnet wire on each pole. Wind each coil in the same direction and with the same number of turns.

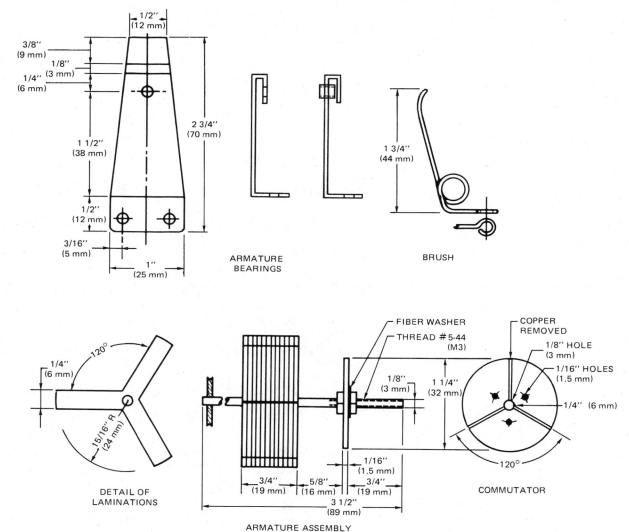

Fig. 13-37

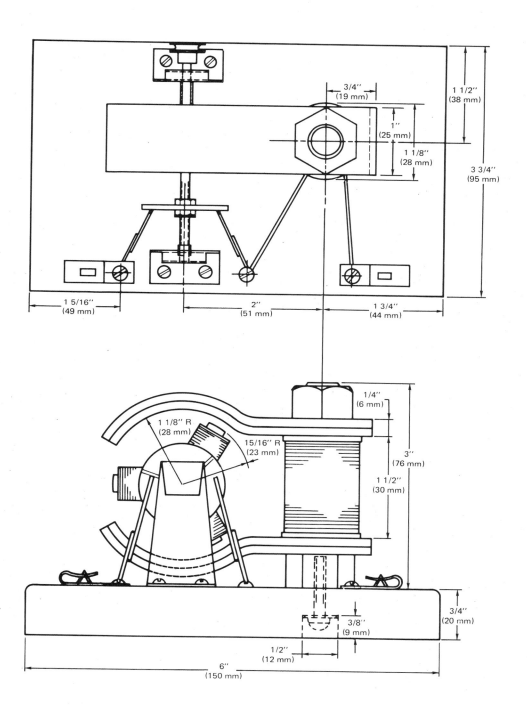

Fig. 13-38

Leave sufficient wire on each end to reach the commutator. Secure the ends with cotton tape.

Check each coil for ground and coat with shellac or varnish.

The field coil is wound on the bolt over insulating material. It consists of 8 layers of #22 magnet wire. When winding the coil, start from the nut end. Leave enough wire on each end to make connections.

The bolt is drilled and tapped so the frame and coil can be fastened to the base with a machine screw.

The field pieces may be made from either 1/16" (1.5 mm) or 1/8" (3 mm) x 1" (25 mm) band iron or soft steel. If 1/16" (1.5 mm) material is used, eight pieces will be required.

Since the shaft is threaded on the commutator side, the armature bearing is fitted with a short piece of tubing to prevent the threads from binding.

The commutator is made of copper-clad phenolic board and secured to the shaft with two nuts. A fiber washer is placed between the disc and the nut on the copper-clad side to prevent a short. Enough copper should be removed from each segment at the edge of the hole to prevent it from touching the shaft.

The armature coils are connected so the beginning end of each is fastened to a commutator segment along with the finish end of the coil to the right of it. The wires are threaded through holes in the commutator and soldered to the copper face.

Note: A three-segment commutator may be made with tubing.

The brushes are formed from brass spring wire. They are mounted to the base with wood screws. Care should be taken to adjust them for the proper tension.

Assemble the parts and test. The position of the commutator segments in relation to the poles may need to be adjusted to provide maximum torque.

Evaluate.

EXPERIMENTAL MOTOR NO. 3

This motor, figure 13-39, is unique because of the type of field coils and the use of sections of iron pipe as part of the field metal. Another feature is the manner in which the commutator segments are held in place with fiber rings.

A three-pole armature similar to the one in motor No. 2 could be substituted for the two-pole armature suggested in this motor; a three-segment tubular-type commutator could be used instead of the disc.

MATERIALS

1 - Pc. soft steel, 3/8" (9 mm) diameter or 3/8" x 3/8" x 1 7/8" (9 mm x 9 mm x 48 mm)

1 - Pc. CRS, 1/8" x 3 1/4" (3 mm x 85 mm)

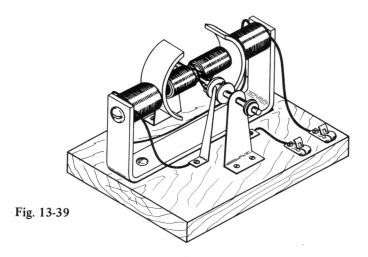

Fig. 13-39

1 - Pc. soft steel, 3/8" dia. or 3/8" x 3/8" x 1 7/8"
1 - Pc. maple dowel, 3/8" x 2" (9 mm x 50 mm)
1 - Pc. brass or copper tubing, 3/8" OD x 9/16" (9 mm OD x 14 mm)
2 - Pcs. sheet iron, #20 or #22 gauge, 1" x 3" (25 mm x 76 mm)
2 - Pcs. spring brass, #30 gauge, 1/4" x 2 1/4" (6 mm x 57 mm)
2 - Pcs. fiber, 1/16" x 3/4" (1.5 mm x 19 mm) square
4 - Pcs. fiber, 1/16" x 1" (1.5 mm x 25 mm) square
1 - Pc. iron pipe, 2" ID x 3/4" (50 mm ID x 20 mm)
1 - Pc. band iron, 3/16" x 3/4" x 9 1/2" (5 mm x 19 mm x 214 mm)
1 - Pc. wood, 3/4" x 3 3/4" x 6" (20 mm x 95 mm x 150 mm)
2 - Machine bolts, 1/4" x 1 1/2" (M6.3 x 40 mm L)
2 - Fahnestock clips
8 - Wood screws, #4 x 1/2" RH
2 - Wood screws, #6 x 3/4" RH
    Magnet wire, #22 gauge AWG Formvar

PROCEDURE

Study figures 13-40 through 13-42. Procure the required materials, and make the metal parts and the base.

Next, make the coils. Both coils should be wound in the same direction. For a series motor, the beginning of one coil should be connected to the end of the other.

The commutator is made similar to the one in previous experimental motor, except that fiber rings are used to hold the segments in place instead of wires soldered to the segments. This system has an advantage when more than two segments are needed.

The armature, except for the dimensions, is made exactly the same as in the first experimental motor.

When all the parts are finished, assemble and test. Adjust the tension of the brushes until the rotor spins easily.

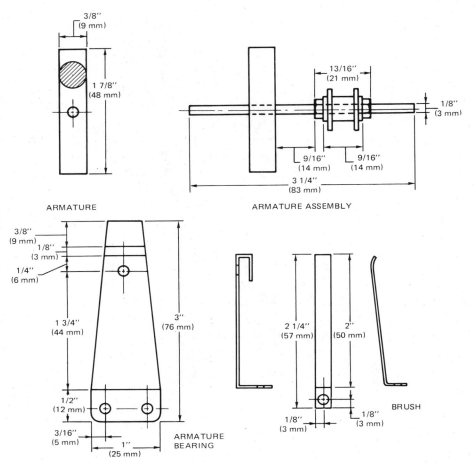

ARMATURE

ARMATURE ASSEMBLY

ARMATURE BEARING

BRUSH

Fig. 13-40

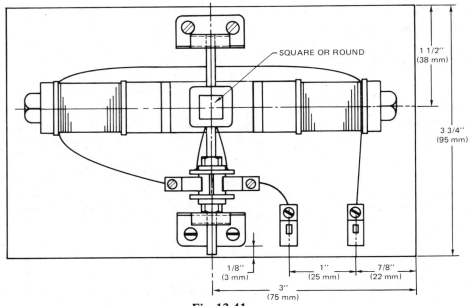

SQUARE OR ROUND

Fig. 13-41

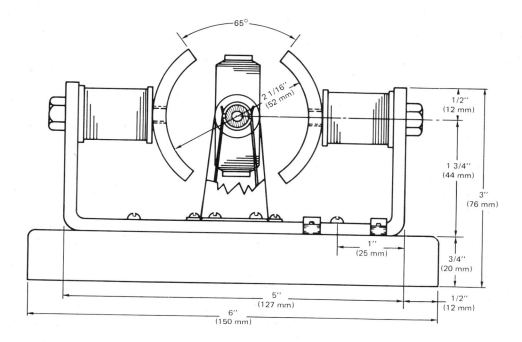

Fig. 13-42

Experiment with the position of the commutator segments in relation to the armature poles until the motor is self-starting and has a good torque. Evaluate.

Chapter 14
Alternating current

The section on ac generators in Chapter 10 should be reviewed, particularly the meaning of the sine wave graph, because that method of representing alternating voltages and currents will be used. Alternating current in coils of wire produces a special effect that should now be studied.

EMF INDUCED IN A COIL

Assume a coil is connected to a source as in figure 14-1. Whether the supply to the coil is dc or ac, electrons start flowing, and a magnetic field forms in and around the coil. This field is the total effect of all the little circular fields around each turn of wire in the coil. The electrons take a short time to build up speed, and the magnetic field increases in extent and in number of lines as the current increases, figure 14-2.

These lines of force originate from single wires. As the lines of force expand outward, they sweep across the other wires in the coil. This produces emf in all the turns of the coil. The direction of this emf can be found from Lenz's Law which states, "the induced emf opposes the change that causes it." In this case, the change is the starting up of a current, so the induced emf (volts) opposes the voltage of the source that is starting up the current. The induced emf

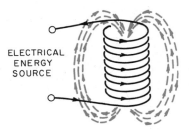

Fig. 14-1 Magnetic Field Around an Energized Coil

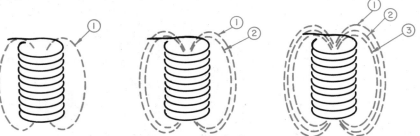

Fig. 14-2 Magnetic Field Expanding as Current Increases

produces no current itself. However, it does hinder the production of current by the source. It is because of the induced opposing emf that "the electrons take a short time to build up speed", as stated above, instead of immediately starting out at the value of current found from I = E/R.

Like any generated emf, the voltage induced in the coil itself depends on the rate of cutting of lines of force by wires, or of wires by flux lines. In equation form, emf = $\frac{\text{turns x flux change}}{\text{seconds x } 10^8}$ (10^8 is an abbreviation for 100 000 000). The term *flux change* means the number of lines of force that sweep through the coil during some small time interval. "Seconds" is written in the formula, but the time is generally a fraction of a second, especially when the coil is connected to an ac supply.

In a coil connected to 60 hertz, the current is allowed only 1/240 second to start from zero and increase, then it is time for it to start decreasing. During the 1/240 second, figure 14-3, the expanding magnetic field is inducing emf that opposes the increase of current. The opposition is so severe that the current never reaches the value that would be found from I = E/R. This means

that I = E/R does not work for alternating current in coils. This is especially true in coils wound on iron, because iron produces a large amount of magnetic flux through the coil. The following does work: ac amperes, I, are still proportional to applied volts, E. And in coils we have, along with a little resistance, the more important opposition of induced emf. This opposition is given the name *inductive reactance.* If you have a coil with zero resistance*, a new formula that looks like Ohm's Law can be written, I = $\frac{E}{\text{inductive reactance}}$. To make this still more like Ohm's Law, inductive reactance itself is measured in ohms. However, the letter symbol is X_L instead of R. Resistance, R, depends on the length, thickness, and material of the conductor. X_L, inductive reactance, is an entirely different opposition to current due to induced emf's when the current in a coil changes. Alternating current is always changing in amount and direction. As long as the current is alternating, reactance exists. Reactance depends on the frequency of the alternating current. At a higher frequency than pictured in figure 14-3, less time is allowed for each cycle of current change; it does not have enough time to rise far. At a frequency of 120 hertz, the maximum value of current is only half as high as at 60 hertz. In other words, the opposition is twice as much.

LAGGING CURRENT

In addition to limiting the amount of alternating current in a coil, inductive reactance causes another important effect. It causes the back-and-forth current surges to be out of step with alternating voltage impulses delivered to the coil by the ac power line.

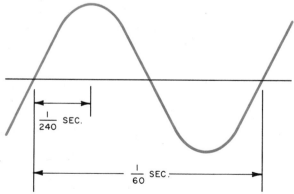

Fig. 14-3 Alternating-Current Wave Pattern of One Cycle

---

* At temperatures close to absolute zero, some metals and alloys do have zero resistance. Supercooled wire carries current without any I^2R power loss; current started in a supercooled lead ring will keep going for days without any battery in the circuit. The reactance of copper coils at normal temperature on an iron core is so high compared with resistance that it is not unreasonable to neglect R for a while in this discussion.

In figure 14-4(A) the black sine wave represents 60 hertz alternating voltage to be applied to a coil. Assume the reactance of the coil is so large that its resistance is unimportant. The coil is connected at an instant when line volts = 0. As volts increase (0 to 1 on horizontal time scale) current increases, as shown by the rising color line. During the time interval from 1 to 2, the current continues to increase in the forward direction because voltage is still in the forward direction. Between 1 and 2, I does not increase as fast as it did at first. This is because the forward voltage is decreasing, and induced emf in the coil is still opposing increase of current. Between 0 and 2, the behavior of I is comparable to speed developed by a heavy friction-free cart when suddenly pushed by someone (like V) who applies increasing force and continues to push forward, but not as hard as before. As long as V is pushing forward, the cart continues to pick up forward speed (I).

If the person changes his or her mind and decides to quickly bring the cart back again (times 2 to 3 on graph), difficulty will be encountered because the cart already has forward motion. This forward motion must be stopped before the cart can be rolled back, figure 14-4(B). Likewise, the current in the coil at time 2 will not suddenly stop when reverse voltage is applied from the power line. The magnetic field surrounding the coil at time 2 must collapse. While doing so, it is opposing change of current, driving electrons forward in the same direction they had already been traveling. During the time interval between 2 and 3, previous current is finally stopped, and line voltage in the reverse direction causes electrons to begin moving in the reverse direction during the 3-to-4 interval, figure 14-4(C).

At time 4, the situation is much as it was at time 2. The voltage impulse in the reverse direction has ceased, but reverse current still exists, having been brought up to a momentary

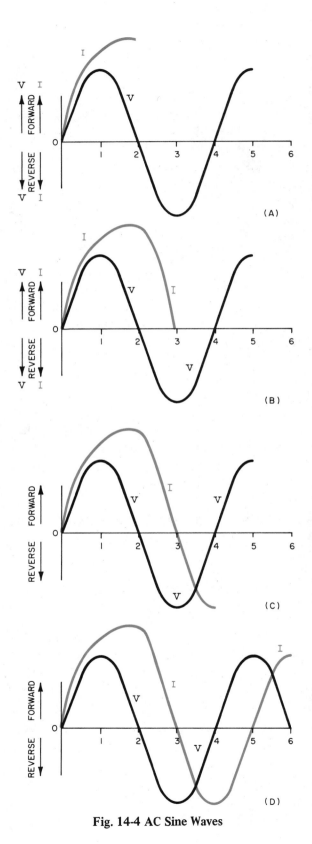

Fig. 14-4 AC Sine Waves

maximum. The time between 4 and 5 is used to slow that current to a stop, for induced emf opposes stopping it. At 5, reverse current has been brought to zero and forward voltage has reached a peak; forward current is built up again by forward voltage between times 5 and 6.

Disregard the time interval between 0 and 2, because what happened then will not be repeated. Examine the continuing relation of the voltage and current waves, as shown in figure 14-5. Along the horizontal time scale, early events are at the left, later events are toward the right. Notice that the peaks of current occur 1/4 cycle later than the peaks of voltage. Or, dividing one cycle into 360 electrical degrees, we say that the current lags the voltage by 90°. To *lag* means to be late, or delayed.

Power In A Reactance

Because current lags voltage by 90° in a purely inductive circuit, and because the magnetic field of a coil possesses energy, some very useful energy and power relationships arise. Figure 14-6 is a copy of figure 14-5, with another line, P, added. P is the power graph; each point on it can be calculated (Watts = Volts x Amps) by multiplying simultaneous values of V and I from the voltage and current waves. Whenever either V or I is zero, P = 0. Between points 3 and 4, V and I are both in the same (reversed) direction. There are two reasons for the appearance of the P graph above the axis in the forward direction. One is that algebraically, two quantities pictured as negative on the

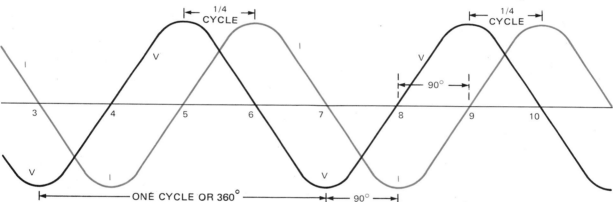

Fig. 14-5 Lag of Current Behind Voltage in Inductive Reactance

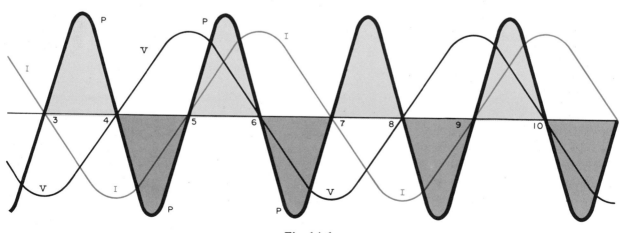

Fig. 14-6

343

graph multiply to make a positive quantity. Another more important reason is that volts and amps are in the same direction. When this occurs, energy flows from the source into the coil, building up current to a maximum at time 4. During the 3-4 interval the coil absorbs energy, just as it does in the 5-6 interval. Observe the 4-to-5 interval, when V is rising to a maximum. All that V is really doing is stopping the previously established current. At time 4 when I was at a peak, a large amount of magnetic flux existed in and around the coil. That magnetic field had energy. While the current slows to zero, the magnetic field collapses, dumping its energy back into the source. The power graph is drawn below the horizontal axis between points 4 and 5, not only because positive V times negative I makes negative watts, but also because *negative* watts indicate a reverse flow of energy back to the source, as compared with the forward flow of energy from the source to the coil. Keep in mind that the power graph shows instantaneous values of watts. This is the rate of energy transfer, not the amount of energy. Amount of energy is shown on the graph by the area of the shaded portions between the P wave and the axis. For example, at time 6, watts is zero; the coil is not giving or receiving energy. But at that instant, point 6, the magnetic field contains all the energy accumulated between times 5 and 6, while current and flux were building up. The gray area between 6 and 7 represents the amount of energy fed back to the source. If a wattmeter is connected into the circuit between the source and the coil, figure 14-7, it does not read 240 watts. If an ideal re-actance coil has zero resistance, such a coil takes zero watts. A real iron-core coil that takes 2 amps on 120 volts might, for example, have 3 ohms resistance; in that case the wattmeter reads 12 watts. With ac, "watts = volts x amps" will need much correction. This correction will be explained later. Watts = $I^2 R$, with R meaning resistance, is true still.

Reactance with Resistance

Abandon the ideal coil, or rather, think of it as having resistance in series with it to represent a real coil. In figure 14-8, there is an ideal coil with reactance but no resistance, connected in series with a resistor that is not a coil. If we put 3 amperes through the circuit, voltages are as shown in the sketch. Voltmeter reading = amps x ohms, over a period of time, whether the ohms represent resistance or reactance. Figure 14-9 shows graphically these two voltages, along with the 3-amp current. Remember this is a series circuit; the same 3 amps exist in X_L and R. Electrons must start and stop moving at the same instants in both X_L and R, for there is only one circuit path. But the voltages across X_L and R do not become zero at the same instant. In the resistor, the voltage becomes zero at the same time that the current becomes zero. Also, maximum voltage is accompanied by maximum current mainly because there is no reason for it to do otherwise. The energy of the current is always changing into heat in the resistor. There is no energy saved up as there is in the magnetic field of the iron-core coil. The

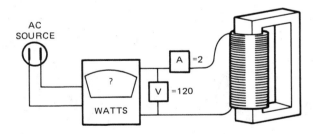

Fig. 14-7 Measuring Watts in an Iron-Core Coil

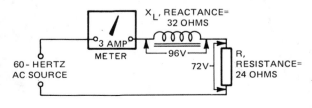

Fig. 14-8

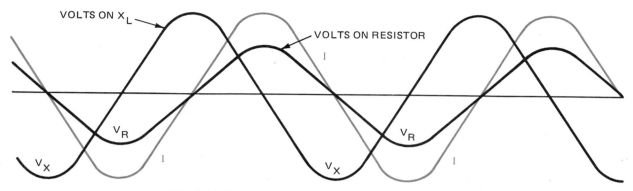

Fig. 14-9 Voltage and Current for X and R in Series

saved-up energy in the magnetic field of the coil puts the 96 volts of the coil out of step with changing current by 90° as explained previously.

Addition of Voltages

Using series-circuit reasoning, one might expect that the coil voltage and resistor voltage (96 volts and 72 volts) can be added, giving the line voltage supplied at the 60-hertz source of figure 14-8. They can be added, but the total is not 168 volts. According to the same voltmeter that measured the 96 and 72 volts, it is 120.

Figure 14-10 shows two sine waves, of relative heights 4 and 3, (like 96 and 72) and (in color) their total. Two dc voltages of 96 and 72 are constant and steady quantities, which add to 168. Alternating voltages of 96 and 72 add to 168 only if they are in step (*in phase*). To be in phase, both must reach their peak value at the same instant. They both become zero together. If they are out of step, they can be added taking instantaneous values and directions into account, as was done in figure 14-10. For example, place a ruler along the vertical dotted line near the middle of figure 14-10: At that time-instant on the graph, one sine wave in black has a value of +1.5, the other black sine wave has a reverse value of –3.46. Their total (color) is –3.46 +1.5 = 1.96 in the reverse direction. Try the ruler vertically at some other time-instant on the graph. Observe

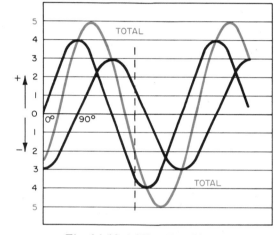

Fig. 14-10 Adding Two Sine Waves

that the values for the two black points, taking direction into account, add to give the value on the color sine wave. The graph shows that the total wave has height = 5, and lags the height = 4 wave by 37°.

If alternating voltages had to be added by the above laborious method, direct current would probably be used for everything. Fortunately, there is an easy way of adding sine wave alternating quantities. Examine the mechanical device in figure 14-11. A leaky bucket hangs on a peg stuck in a very slowly rotating wheel. The colored paint leaking out of the bucket drizzles straight down on to a sheet of paper. If the paper is stationary, the paint is spread back and forth, making a line on the paper like that between the thumbs at the left. If the

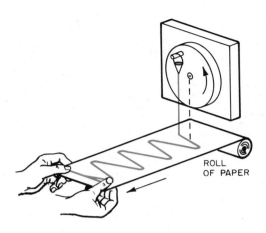

Fig. 14-11 Sine Wave Recording Device

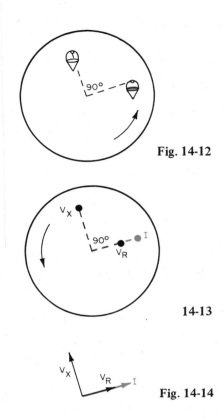

Fig. 14-12

14-13

Fig. 14-14

operator pulls the paper to our left at steady speed, the paint draws a sine wave on the paper. One rotation of the wheel produces one cycle of sine wave. Could such a machine make two sine waves, as shown in figure 14-5? All that is needed is to drive another peg in the wheel and hang another bucket, with black paint. The added peg could be at the same distance from the center as the first peg, making the waves of equal height. To make the color wave lag the black one by 90°, the pegs must be separated by 90° on the wheel, with the black leading and color lagging in the rotation.

Compare the three sine waves of figure 14-9 with the peg arrangement of figure 14-13. While a bucket of black paint on the V_X peg draws the V_X sine wave, a bucket closer to center on the V_R peg draws the V_R wave of figure 14-9. A color bucket at I will draw the color wave in phase with V_R, but with higher peaks than V_R. The peg arrangement of figure 14-13 tells everything that figure 14-9 does in much more convenient form. The distance of the pegs from the center gives the height of the sine wave. The angle between the radii drawn to the pegs gives the separation in electrical degrees between the waves in figure 14-9. Pegs, paint, and the wheel are no longer needed. Instead, simply use the three radius-lines shown in figure 14-14 to represent the three waves of

figure 14-9. The arrowheads merely emphasize where the end of the radius line is, for its length is important in representing the amount of volts or amps. The overlap of V_R and I shows their in-phase relationship. Their 90° lag behind V_X is evident in the 90° angle. These radius lines are usually called rotating *vectors*.

VECTOR DIAGRAMS

Look back at figure 14-10 and draw vectors as a simpler replacement for it. First, draw a line 4 units long to represent the wave of height 4. It makes no difference in what direction it is drawn. It is drawn horizontally in figure 14-15(A). Next, draw a vector for the wave of height 3, because its 90° phase difference is easy to show. But having established a vector for the first-chosen wave, this next vector must be located to show that it lags the first one by 90°. In accordance with the customary counterclockwise rotation of vectors, the vec-

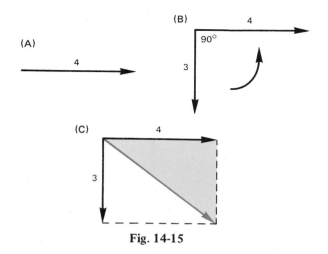

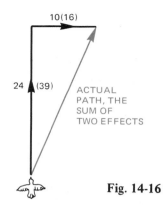

Fig. 14-16

(A)

(B)

(C)

Fig. 14-15

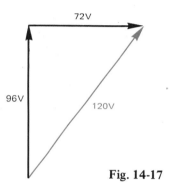

Fig. 14-17

tor for the second-chosen wave, height 3, is drawn downward. (Drawing the vector upward would indicate it is leading the first by 90° instead of lagging.) Next, locate the vector for the color total wave of height 5. Although figure 14-10 does not show it accurately, the color wave lags the first one by 37°. It is ahead of the second black wave by 53° (90 – 37 = 53). With a protractor it can be located between the two vectors already drawn in figure 14-15. It is time to look carefully at the results. Figure 14-10 was a tedious and complex way of adding two sine waves. Figure 14-15 is the easy way of adding two sine waves with a 90° separation — using vectors.

The addition of directional quantities might be confusing for beginners. Here is a nonelectrical example.

(1) See figure 14-16. A bird is trying to fly straight north at a speed of 24 miles per hour (39 kilometers per hour). The air through which the bird is flying is moving east at 10 miles per hour (16 kilometers per hour). After flying for one hour, where is the bird in comparison with the starting point? We have a right triangle diagram, in which

$$X^2 = 10^2 + 24^2$$
$$X = 26$$

The bird has traveled 26 miles (42 kilometers) in one hour. We have added two velocities at right angles, 10 (16) and 24 (39), and found 26 (42) for the answer.

(2) Back to an electrical example, figure 14-8. Problem: to add two alternating voltages, 90° apart. Figure 14-9 will remind us that the 72 volts lags the 96 volts. We can draw the voltage vectors as in figure 14-17.

$$X^2 = 96^2 + 72^2$$
$$X = 120 \text{ volts}$$

In the circuit of figure 14-8, with 32-ohm X_L and 24-ohm R, a 120-volt source put 3 amps through the circuit. Using the relationship, Ohms = Volts per Amps, 120/3 = 40 ohms. Are there 40 ohms in the circuit? The voltage vector relation $x^2 = 96^2 + 72^2$ can be used to prove something. The general relationship used was (Total line volts)² = (Volts on X_L)² + (Volts on R)². This was found from the triangular diagram of vectors. For each of the three expressions of volts above, make an appropriate V = amp x ohms substitution. The long equation changes to

$$(I \times \text{total ohms})^2 = (I\, X_L)^2 + (I\, R)^2$$

or

$$I^2 \times (\text{total ohms})^2 = I^2\, X_L^2 + I^2\, R^2.$$

Divide each term by I^2, and get

$$(\text{total ohms})^2 = X_L^2 + R^2.$$

In the example, X_L is 32-ohms reactance and R is 24-ohms resistance.

(Total ohms)^2 = $32^2 + 24^2$ = 1 024 + 576 = 1 600. (Total ohms)^2 = 1 600, total ohms = 40, which is the result obtained from $\dfrac{120 \text{ volts}}{3 \text{ amps}}$. But what is the nature of this total ohms quantity, reactance or resistance? It is both, and it gets a new name, *impedance*. When the current is divided into applied line volts, the ohms found are the impedance of the circuit. Impedance, given the letter symbol Z, can also be found from X and R as shown above.

> Impedance is $Z = \dfrac{\text{line E}}{I}$ and $Z^2 = X^2 + R^2$.

Inductive reactors are simply coils intended for use in circuits where their reactance is an advantage. Such coils have many uses in communication equipment. These will be described later. They also are useful in controlling the amount of ac. For example, they are used in dimming banks of theatrical lights, because reactance limits current without wasting energy and producing unwanted heat. As suggested before, the amount of ohms reactance depends on the amount of magnetic flux in a coil, which can be varied by moving an iron core into or out of the coil, figure 14-18.

A FLUORESCENT LAMP CIRCUIT

Iron-core reactance coils are installed in fluorescent lamp fixtures. These coils help start the operation of the lamp and control the current through the lamp. Such a current-limiting device is called a *ballast*. Figure 14-19 shows the simple arrangement used in fluorescent desk lights having push-button starting

switches. The 115 volts applied to the circuit causes no current through the tube at first, because the resistance of the tube is too high to permit an arc to start. Closing the starting button allows a current through the heater filaments at the ends of the tube and through the ballast. Heating of the end-filaments warms the mercury vapor. Electrons are emitted from the hot filament, making the tube ready to conduct. When the starting switch is opened, the collapsing magnetic field in the ballast coil induces a momentary high voltage in the circuit. The high voltage drives electrons through the mercury vapor. The vapor ionizes, and conduction is underway.

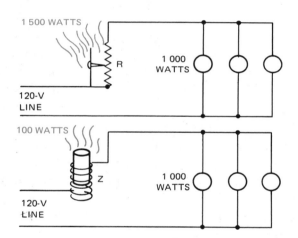

Fig. 14-18 Current Control by Resistance and by a Reactor

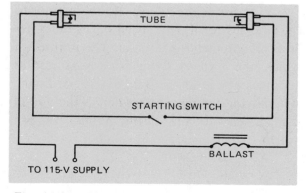

Fig. 14-19 Fluorescent Tube Circuit with Manual Starting Switch

During operation the voltage across the tube may be from 50 to 100 volts, depending on the size of tube. The inductive ballast limits the current without causing too much heat production, which would occur if a series resistor were used instead.

Some fluorescent fixtures have an automatic starting switch instead of the push-button type of figure 14-19.

The glow switch, figure 14-20, replaces the starting switch shown in figure 14-19. The glow tube is a neon-filled glass bulb containing a U-shaped bimetallic strip and a fixed contact which is normally open. When the circuit is connected to the 115-volt line, current is small due to the high resistance of the glow tube. Also, there is little voltage drop across the series reactor. Voltage across the glow tube contacts is enough to start a little arc discharge between the bimetal strip and the fixed contact. This arc heats the bimetal, which bends and touches the fixed contact.

The starting switch is now closed, and the tube filaments are heated. Closing the contact in the glow tube stops the glow tube arc, the bimetal cools, and the contacts open. At the instant of opening, the inductive voltage kick generated in the series reactor coil starts conduction in the fluorescent tube.

For longer (40$^+$-watt) tubes, a ballast as shown in figure 14-22 is advisable. The upper

section of the ballast, like the secondary of a transformer, has added voltage induced in it from the winding that is across the 120-volt line, yet its reactive effect still limits the current while applying more voltage to the tube during operation.

The instant-start fluorescent lamp circuit basically looks like figure 14-22 without the starter. Instant-start lamps are made with special cathodes to permit this method of starting. A higher voltage, transformer-type ballast starts electrons moving without the preheating.

The most satisfactory starting arrangement is the rapid-start ballast of figure 14-23. This is now used with most new installations. The ballast has low-voltage windings (few turns) that quickly heat the tube cathodes, causing enough ionization to start the arc from the voltage of the main ballast windings. The low-resistance cathodes remain

STARTER HEATS UP   CATHODES OF FLUORESCENT LAMPS HEATING   FLUORESCENT LAMPS ON

Fig. 14-21 A Glow Switch Starter

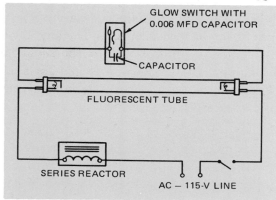

Fig. 14-20   Fluorescent Tube Circuit with Glow Switch

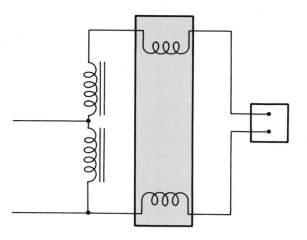

Fig. 14-22   Fluorescent Tube Circuit with Glow Switch and Ballast

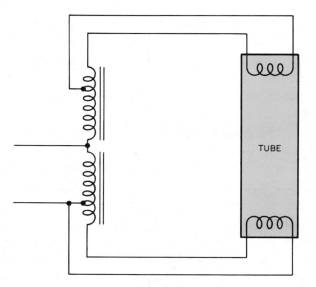

Fig. 14-23   Basic Rapid-Start Fluorescent Tube Circuit

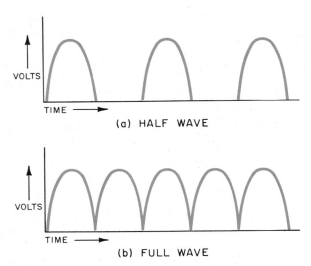

Fig. 14-24   Current Characteristics of a Half Wave and a Full Wave Rectifier

heated during lamp operation.   Rapid-start tubes can be used in the older (figure 14-22) preheat fixtures.

CHOKES

Choke coils are mentioned in connection with rectifiers or radio work.   Choke coil is another name for an inductive reactance coil. *Rectifiers* are devices that convert alternating voltage into pulses of voltage in one direction. (Rectifiers are discussed more fully in Chapter 17.)   A rectifier that simply stops voltage in one direction and lets the other through from the ac line is called a *half-wave rectifier*.   Its output is shown on a graph in figure 14-24. If reverse ac pulses are turned around so that all of them come out forward, the output is called *full-wave*.   120 pulses per second are produced from a 60 hertz ac line.   In either case, the output generally needs smoothing out to make it more steady and uniform (like battery voltage).   A choke (inductor, inductive reactor) in series with the output helps in the smoothing.   The reactor develops back emf to oppose increase of current.   It also opposes decrease of current by inducing forward emf when the current starts to fail.

INDUCTANCE

A replacement ballast for a 20-watt fluorescent is easy to buy.   Experimenters needing a reactor (ballast, inductor) have problems when they read the electrical-parts catalog.   One filter choke is rated "8 hys, 300 mA, 100 ohm", another is "3 h, 150 mA", with no mention of volts or reactance.   The 300 mA or 150 mA tells, in milliamperes, the maximum current intended for the coil.   The "100 ohm" is dc resistance, not reactance. Hys or h stands for henries, a quantity that is determined by the number of turns, their arrangement, and the core material.   Henries are a measure of inductance, which is the ability of the coil to induce voltage when the current in the coil changes.   The amount of inductance is proportional to the square of the number of turns.   It is greater for a coil of large diameter and also if a core of high magnetic permeability is used.

> By definition, a one-henry coil induces one volt when its current is changed in the rate of one ampere per second.

The last part of that definition, amperes per second, means a rate of change of current.

For example, if someone slowly and steadily increases the current (dc) through a heater from 2 amps to 5 amps, taking 10 seconds to do so, what is the rate of change of current?

The amount of change is 3 amps in 10 seconds. For one second, the amount of change is 3 amps divided by 10 seconds, or the rate of change is 0.3 amps per second.

When something is connected to a 60-hertz ac line, and the current is what is called a steady 5 amps (ac), it is not actually steady. The current changes constantly. The current changes from zero to maximum in one-fourth of a cycle, which is 1/240 second. Using as an example an alternating current that builds up from zero to a peak value of 2 amps in 1/240 second, the rate of change is the amount of change divided by time or 2 divided by 1/240 which equals 480 amps per second. In such a circuit, a one-henry coil would generate an average of 480 volts during that 1/240 second.

The following shows how to use henries of inductance:

> Inductance and frequency of the ac determine the ohms reactance of the inductor.

$$X_L = 2\, f\, L$$

In that formula, X_L is ohms reactance, f is frequency, and L is henries of inductance. The catalog could not state the reactance of a listed coil for it depends on the frequency in the circuit where one uses it.

Using the above formula to find the reactance of the above 8-henry choke coil in a half-wave rectifier circuit (60 pulses per second): X_L = 2 x 3.14 x 60 x 8 = about 3 000 ohms, which is 30 times its dc resistance. Thinking of the half-wave pulses of figure 14-24 as equivalent to ac and dc mixed together, the 8-henry coil has 30 times as much opposition to changing current (ac) as it has to dc, thus smoothing out the changes into a steadier dc.

Effective Values of AC

It was stated above that what is called a steady amount of ac is not really steady, it is always changing. When an ac ammeter reads 1 amp, that means that the current through the meter will heat a resistor just as fast as a steady 1-amp dc will heat it. If an ac reaches a maximum value of 1 amp during its cycle, it will not heat a resistor as much as 1-amp dc because the ac has instantaneous values lower than 1 amp most of the time.

The table in figure 14-25 shows values of I^2 for 10 degree intervals of a sine-wave current rising to a 1-amp maximum. The heating rate depends on I^2 (watts = $I^2 R$). The average value of I^2 for the 1/4 cycle, or for a whole cycle, is 0.5. The number 0.5 is important because in order to be able to use $I^2 R$ and other power formulas properly in ac calculations, ac meters should be scaled so that, if a current rises to a peak value of 1 amp, the meter will read a number for I. When squared, this number will equal 0.5. That number is $\sqrt{0.5} = 0.707$ amp. So, when an ac ammeter reads 0.707 the current has the same watts-producing effect in a resistor as 0.707 amps dc, figure 14-26.

The same relation applies to volts, for watts = V^2/R. A 120-volt ac line has the same power-producing effect in a resistor as 120 volts dc. In order to do so, the ac voltage starts from zero and rises to a sine-wave peak value of $\frac{120}{0.707} = 168$ volts.

Effective values of ac current and voltage are termed *root-mean-square* (rms) values. In the above calculation we squared, took an average (mean) and then a square root, in order to find the useful number to print on the meter scale.

The Saturable Reactor

Figure 14-18 showed how reactance can be varied by changing the position of the iron core. Complete removal of the iron would

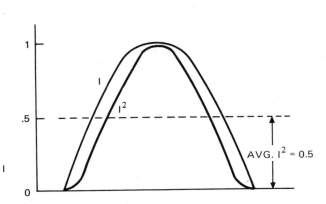

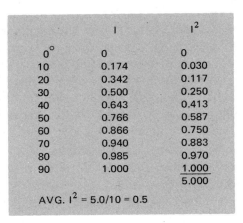

	I	I^2
0°	0	0
10	0.174	0.030
20	0.342	0.117
30	0.500	0.250
40	0.643	0.413
50	0.766	0.587
60	0.866	0.750
70	0.940	0.883
80	0.985	0.970
90	1.000	1.000
		5.000

AVG. I^2 = 5.0/10 = 0.5

Fig. 14-25

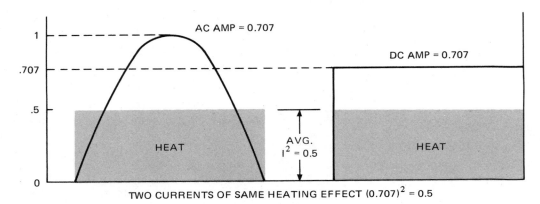

TWO CURRENTS OF SAME HEATING EFFECT $(0.707)^2 = 0.5$

Fig. 14-26 Two Currents Having the Same Heating Effect

reduce the reactance so much that nearly full line voltage would be applied to the load. Another method of varying reactance with the core stationary is to put dc through another many-turn coil on the same core. With sufficient dc ampere-turns, the iron of the core becomes *saturated*, figure 14-27. That is, it becomes so strongly magnetized that changes in the ac coil affect the total magnetic flux only slightly.   Reactance depends, not on amount of flux, but on the amount of change of flux.   Therefore, the addition of dc reduced the reactance.

POINTS TO REMEMBER

- Induced emf opposes the change that causes the induced emf.

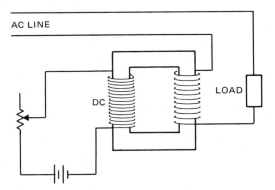

Fig. 14-27 Basic Saturable Reactor

- Induced volts equals:

$$\frac{\text{number of turns x flux change}}{\text{seconds x 100 000 000}}$$

- When coils conduct ac, extra ohms of opposition, called inductive reactance, appear in the coil.

- Inductive reactance ohms (X_L) equals $2\pi f$ L (f = frequency, L = henries)

$$X_L = \frac{E}{I}$$

where E = induced opposing voltage.

- Inductive reactance tends to make coil current lag coil voltage by 90°; during one complete ac cycle the net power used by the coils only = $I^2 R$.

- Ohms impedance (Z) = $\frac{E}{I}$

where E = applied voltage to circuit.

- For a coil, $Z^2 = X^2_L + R^2$

- The inductive ballast for vapor lamp circuits provides a momentary high starting voltage, then limits current through the lamp.

- Inductance, measured in henries, is the ability of the coil to induce voltage due to change of current. One henry induces 1 volt when the current changes at the rate of 1 amp per second.

- Effective ac volts and amperes are defined so that they produce the same heating rate as dc volts and amps.

- Maximum (peak) values of ac volts and amps during a cycle rise to 1.414 times as much as the effective value which appears on voltmeters and ammeters.

REVIEW QUESTIONS

1. What is meant by the term flux change?

2. Explain inductive reactance. On what does it depend? How is it measured?

3. What causes reactance to exist in a coil?

4. Define the standard unit of measurement of inductance.

5. Describe a sine wave.

6. Explain what is meant by phase angle. How is it expressed?

7. What is impedance?

8. What is a vector?

9. Explain what a cycle is in relation to electric current.

10. State the phase relationship of voltage and current in an inductive reactor.

11. If a coil has 50-ohms reactance and 50-ohms resistance, how much is its impedance?

12. In the coil referred to in question 11, which leads – current or voltage? By how much?

13. Explain why the lead-lag relationship of question 12 exists.

RESEARCH AND DEVELOPMENT

Experiments and Projects on Alternating Current

INTRODUCTION

An induced current opposes the motion that causes it. Another way of stating Lenz's Law is: Induced voltages and induced currents oppose change of magnetic field. In Chapter 14, the cause and effect of inductive reactance and inductance are explained. Inductance is the property of coils which is important in alternating-current devices.

EXPERIMENTS

1. To examine the effects of inductance in an alternating current.

2. To observe the induced electromotive force in a coil:
   a. With an air core
   b. With a laminated iron core
   c. With a solid iron core

3. To study the action of iron-core reactance coils in fluorescent lamp fixtures.

PROJECTS

1. Design and make the coil required to assemble the circuit illustrated in figure 14-18. Assemble, operate, and observe.

2. Design and construct the necessary apparatus to prove the following:
   a. Inductance in a coil is determined by number of turns of wire on the coil.
   b. Inductance may be changed by reducing or enlarging the diameter of the coil.
   c. Inductance will vary according to the permeability of the material from which the core is made.

EXPERIMENTS

Experiment 1 INDUCTANCE IN AN AC CIRCUIT

OBJECT

To observe the effects of inductance in an alternating-current circuit.

APPARATUS

1 - AC voltmeter (0-150 V)
1 - AC ammeter (0-5 A)
1 - Lamp bank
1 - DPST knife switch
1 - Reactor coil with removable iron core
Note: See Project 1 for reactor coil suggestions.

PROCEDURE

1. Connect apparatus as shown in figure 14-28.

2. Adjust the iron core until it is down as far as possible in the reactor coil.

3. Have the instructor check the circuit before closing the line switch.

4. Adjust the resistance value of the lamp bank until the voltage drops across the reactor coil and the lamp bank are about equal.

5. Record the line voltage, line current, resistance or lamp bank voltage, and the coil voltage.

6. Remove the iron core from the coil and repeat Step 5.

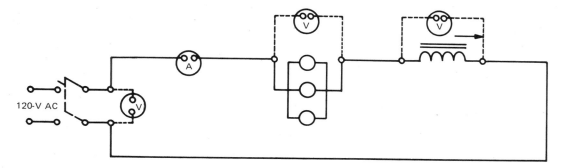

Fig. 14-28

OBSERVATIONS

Use a table similar to the one shown here to record your observations.

OBS. NO.	E—LINE	E_R	E_{X_L}	I—LINE
1				
2				

QUESTIONS

1. Explain the effect that the iron core in the coil had upon the inductance.

2. Explain why the algebraic sum of the line voltage drops in the circuit does not equal the line voltage.

3. Explain the reasons for any phase difference between the line voltage and line current.

Experiment 2 INDUCED ELECTROMOTIVE FORCE

OBJECT

To observe the induced electromotive force in a coil:
a. With an air core
b. With a laminated iron core
c. With a solid iron core

APPARATUS

1 - AC ammeter (0-10 A)

1 - AC voltmeter (0-150 V)

1 - Autotransformer

1 - SPST switch

1 - Air-core coil of 600 turns, #24 magnet wire, with 65 turns of same wire wound over 600. (Use standard coil bobbin with 1″ (25 mm) hole.)

PROCEDURE

1. Connect the apparatus as shown in figure 14-29.

2. Start with autotransformer set at zero and turn on the current.

3. Measure amperes and volts at a low voltage. Record meter readings.

4. Make the same observations at a higher voltage. Record meter readings.

    Note: Check the heat produced in the coil as the voltage is increased. If it becomes hot to the touch, turn off the current and allow it to cool.

5. Insert a laminated iron core in the coil and repeat Steps 2, 3, and 4.

6. Insert a solid iron core in the coil and repeat the same steps.

7. Calculate: Volts ÷ Amps and record. Is this answer to be called resistance or something else?

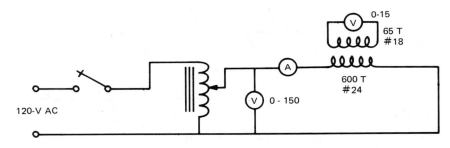

Fig. 14-29

OBSERVATIONS

Use a table similar to the one shown here to record your observations.

VOLTAGE	AMPS	VOLTS	TYPE OF CORE
LOW			AIR CORE
			LAMINATED CORE
			SOLID CORE
HIGH			AIR CORE
			LAMINATED CORE
			SOLID CORE

QUESTIONS

1. What happened to the meter readings when the iron core was inserted in the coil?

2. Was there any difference in the behavior of the meters with the different iron cores?

3. How can you account for the variation in each test?

Experiment 3 INDUCTIVE REACTORS

OBJECT

To study the action of iron-core reactance coils in fluorescent lamp fixtures.

APPARATUS

1 - 30-watt fluorescent tube
1 - Ballast
1 - End fixture caps
1 - Push-button switch
1 - AC voltmeter (0-150 V)
1 - Glow switch

> When experimenting with intense light, wear appropriate colored safety glasses.

PROCEDURE

1. Assemble the component parts as shown in figure 14-30.

2. Measure voltage across ballast (series reactor) and tube with starting switch open. Use 20-watt, 2-wire ballast for a single tube light.

3. Close switch and release after tube is operating. Measure volts across the tube and across the ballast. Should these voltages add arithmetically to 115? State your reasons why they should or why they should not.

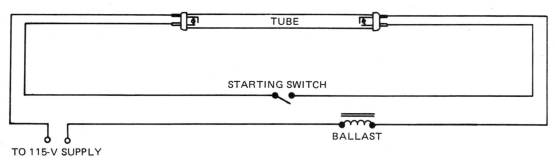

Fig. 14-30

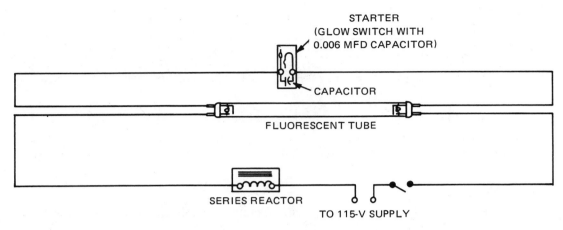

CIRCUIT FOR FLUORESCENT TUBE

Fig. 14-31

Use a table similar to the one shown below to record your observations.

SWITCH POSITION	VOLTAGE ACROSS BALLAST	VOLTAGE ACROSS TUBE
OPEN		
CLOSED		

4. Assemble the component parts as shown in figure 14-31.

5. Remove the starter after the tube is operating and observe the effect.

QUESTIONS

1. What is the function of the ballast?

2. What happened when an ordinary starter was substituted for the starting switch?

3. Will the tube continue to produce light if the starter is removed?

PROJECTS

DESIGN AND MAKE TAPPED INDUCTANCE COILS

These coils were planned for use in proving that:

1. The inductance in a coil is determined by the number of turns of wire on the coil.

2. Inductance in a coil may be changed by reducing or enlarging the diameter of the coil.

3. Inductance in a coil will vary according to the material from which the core is made.

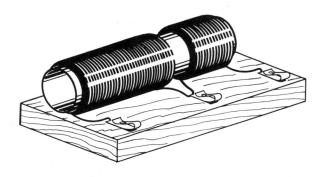

Fig. 14-32

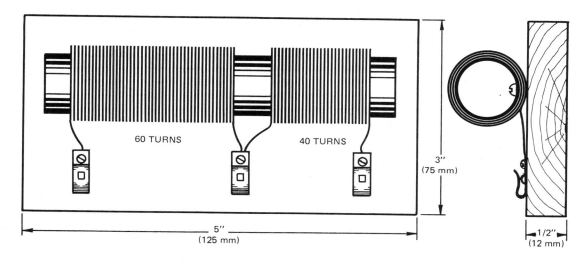

Fig. 14-33

MATERIALS

1 - Fiber tube, 1 1/2″ (38 mm) ID x required length
1 - Fiber tube, 3/4″ (20 mm) ID x required length
2 - Wooden bases, 1/2″ x 3″ x 5″ (12 mm x 75 mm x 125 mm)
6 - Fahnestock clips
6 - Wood screws, 1/2″ x #6 RH steel
    Magnet wire, #22 AWG

PROCEDURE

Calculate the length of the tube required to hold 100 turns of #22 wire. Allow 1/2″ (12 mm) on each end for fastening coil to base.

Wind two sets of coils.

Varnish the coils.

Make and finish the bases.

Make laminated iron cores; make aluminum and wooden cores.

Assemble and test as in figure 14-34 (A through C), page 360.

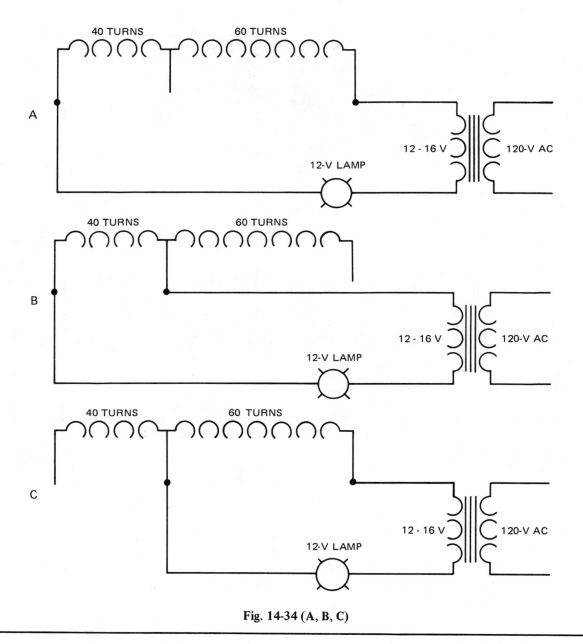

Fig. 14-34 (A, B, C)

PLAN AND ASSEMBLE A SATU-RABLE REACTOR

This saturable reactor, figure 14-35, is an interesting device that can be used to control the amount of ac to a load, such as a lamp.

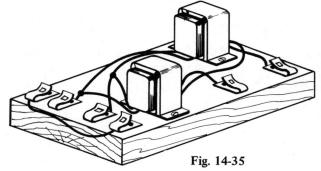

Fig. 14-35

MATERIALS

2 - Identical radio power transformers, center tapped with 250-volt or more secondary. Transformer may also have low-voltage windings, which will not be used.

1 - Base, type and size optional

6.- Binding posts

When working with transformers, remember that some produce high voltages. Check with the instructor before connecting transformers into a circuit or attempting voltage measurements. Always use extreme caution when working with or around high voltages. They are very dangerous.

PROCEDURE

Two identical radio transformers.

Make a suitable base and mount the transformers and binding posts. Make the circuit connections according to figure 14-36.

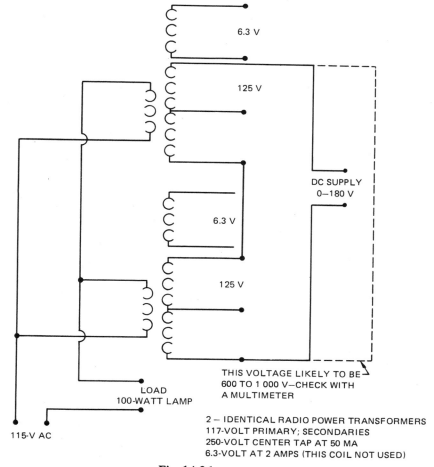

Fig. 14-36

Connect the device to the ac supply and test the dc supply connections with an ac voltmeter. If the meter reads 600 to 1 000 volts, reverse the connections on one coil of one transformer.

Test dc supply terminals again. The ac voltage should be close to zero. Your saturable reactor is now ready to use.

To test the unit, connect the dc terminals to a dc supply of 0 to 180 volts. Next, connect the load terminals to a 100-watt lamp and vary the dc voltage while you observe the amount of light from the lamp.

Evaluate.

Chapter 15

Capacitors and their uses

DESCRIPTION

Perhaps the first capacitor to attract one's attention is the variable capacitor seen in a radio receiver, figure 15-1. A discarded radio or TV set can supply a variety of capacitors. The internal construction of small cylindrical or flat capacitors that are torn apart is similar to that shown in figure 15-2. The two long strips of aluminum foil are separated from each other by strips of waxed paper and rolled and covered with wax or plastic. They illustrate well the general description of a capacitor:

> A *capacitor* consists of two layers of conducting material separated by an insulator.

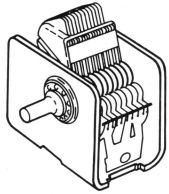

Fig. 15-1 Variable Air Capacitor

The tiny disk capacitors in figure 15-3 are ceramic capacitors. The insulation between the silver plates is a ceramic material, such as barium titanate, in disk form. Some small rectangular capacitors contain sheets of foil separated by rectangular thin pieces of mica. The mica serves as the dielectric. *Dielectric* means insulator.

When the larger cylindrical aluminum can-type capacitor is opened, the two conductors may not be recognized immediately. In the container is a roll of thin aluminum which is one conductor. The light gray coating on this aluminum foil is aluminum oxide. It is the insulator, or dielectric. The other conductor is a liquid, such as borax dissolved in water. This second conductor fills the space between the aluminum foil and the container which connects to the liquid electrolyte. The

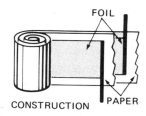

Fig. 15-2 Flat Paper and Foil Capacitor

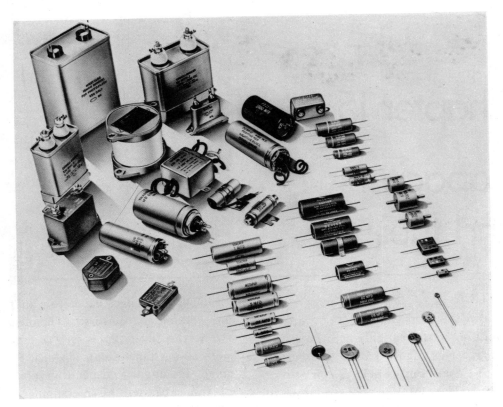

Fig. 15-3 Representative Commercial Capacitors

electrolyte is sometimes in a jellylike form, or it may saturate a cloth wound between the positively charged aluminum foil. This construction is given the name *electrolytic capacitor,* figure 15-4. Electrolytic capacitors are generally used in rectifiers.

CHARGING A CAPACITOR

When a capacitor is connected to a dc source, figure 15-5(A), electrons move onto one plate of the capacitor and stop there because they cannot get through the insulating material. While they are accumulating there, they repel electrons that are drawn to the positive terminal of the source from the other plate. The whole charging process is finished as quickly as perhaps 0.01 second if there is not much resistance in the circuit. If the capacitor remains connected to the battery, there is no current, and the capacitor remains charged. If the capacitor is disconnected, figure 15-5(B), there is still no current.

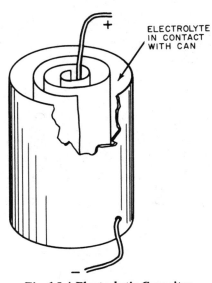

ELECTROLYTE IN CONTACT WITH CAN

Fig. 15-4 Electrolytic Capacitor

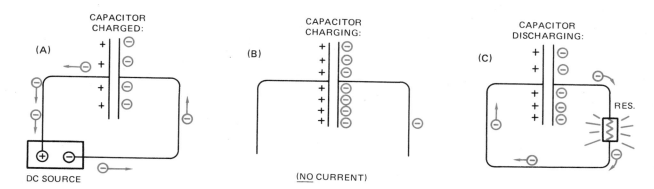

Fig. 15-5 Function of a Capacitor

The capacitor then remains charged, for there is no chance for electrons to move. Notice that a charged capacitor has its plates charged oppositely. The number of missing electrons on the positive plate equals the number of extras on the negative plate.

If the two loose wires from the charged capacitor are connected to a small lamp, figure 15-5(C), electrons can move. The extra electrons on the negative plate repel each other and are attracted by the positive charges in the other plate. The discharging process is usually over in a short time, depending on the size of the capacitor and the amount of ohms resistance in the circuit.

On a capacitor one may find some numbers, for example, "0.05 μf, 600 v dc". The "600 v" is a maximum voltage rating, meaning the capacitor can be used on circuits where the voltage is 600 or less. The abbreviation for *microfarads* is μf. Microfarads are a measurement of capacitance, which is not a measurement of how much the capacitor can hold. *Capacitance* is a ratio of amount of charge to voltage used. For example, a capacitor that collects and stores a charge of 200 coulombs when connected to a 100-volt dc source is a huge capacitor. Its charge-to-volts ratio is 200 coulombs per 100 volts = 2 coulombs per volt. If the same capacitor is connected to a 20-volt source, it would get a charge of 40 coulombs.

The 2 coulombs per volt ratio still holds. The term *farad* (fd) means coulombs per volt.

Another example is a capacitor that acquires 0.001 coulomb of charge when connected to a 1 000-volt dc supply. To find its capacitance, that is, to find coulombs per volt, divide $\frac{\text{coulombs}}{\text{volts}} = \frac{0.001}{1\,000} = 0.000001$ which is one-millionth of a coulomb per volt, or, one-millionth of a farad, called a microfarad, (μf). An even smaller unit, the *micro-microfarad* (μμf) is one-millionth of a millionth of a farad. A capacitor rated as 0.001 microfarad can also be called 1 000 μμf.

Most of the capacitors in radio and television are used with alternating voltages of various frequencies. When a capacitor is connected to an alternating voltage, each plate keeps changing its charge from positive to negative and back again. Electrons bounce back and forth (ac) in the wires leading to the capacitor, even though electrons cannot cross through the insulation in the capacitor. Some of the most important uses of a capacitor depend on facts to be developed by looking at some further details of capacitor behavior.

AC IN THE CAPACITOR

In figure 15-6, the sine wave shows the changing voltage applied to the capacitor.

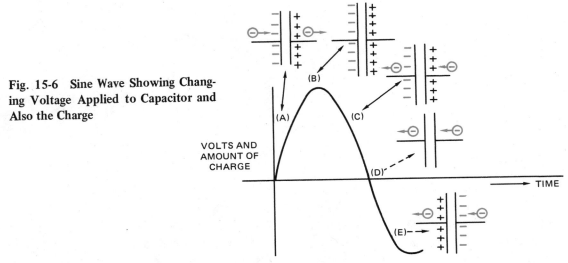

Fig. 15-6 Sine Wave Showing Changing Voltage Applied to Capacitor and Also the Charge

VOLTS AND AMOUNT OF CHARGE

TIME

It also shows the charge (coulombs) on the capacitor, since charge is always proportional to volts. As the voltage increases (slope A), electrons move on to the plates, and the capacitor charges to maximum at peak B. Slope C of the curve shows the forward voltage decreasing. The decrease permits electrons to start running backward, starting to discharge the capacitor. Compare this effect to blowing up a balloon. If you reduce the forward pressure, the balloon discharges air back into your mouth. At point D on the sine voltage graph, pressure has dropped rapidly to zero and fast-moving electrons have just discharged the capacitor. The reversal of voltage (slope E) forces electrons to keep moving, now charging the capacitor in the reverse direction. Notice a similarity in current direction at C and E, although the charge and voltage are reversed. At C electrons are moving off the negative plate, impelled by their own potential energy. They are permitted to move because the external voltage is decreasing. At E electrons are driven by the external energy source, piling up negative charges at the right. At C, the charge on the capacitor causes the flow. At E, the externally driven flow causes the capacitor to charge.

Figure 15-7 shows a graph of current, amps, along with the voltage sine wave. Review the above discussion in connection with figure 15-7, keeping in mind that the color current graph shows amount and direction of *flow rate = current.* Whenever the capacitor has accumulated maximum amount of charge, the motion of charge is zero. In figure 15-7 the peaks of the current waves occur early. The voltage peaks later on the time scale. After the first half-cycle, the two waves are out of step, or out of phase, by 90°. Figure 15-8 shows the same information as figure 15-7, but represented by rotating vectors.

While a capacitor is charging, it takes energy from the ac supply lines. When a capacitor discharges it dumps that energy back into the supply, so over a period of time it consumes no net energy. The power graph, P in figure 15-9, shows the rate at which energy is taken from or given back to the supply source.

Between points 1 and 2 on the horizontal time-axis, V and I are both in the forward direction, and the capacitor is charging. By time 2 when V is at its maximum, the capacitor has stored an amount of energy that is represented by the color-shaded area above the axis between 1 and 2. Between 2 and 3,

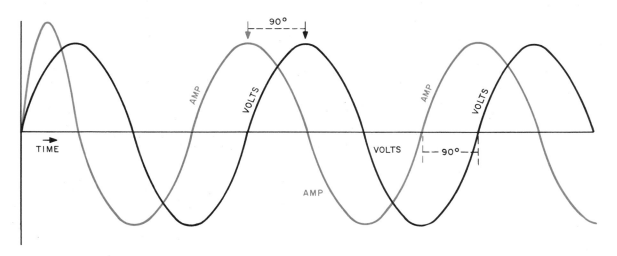

Fig. 15-7 Graph of Current Along with the Voltage Sine Wave. Current at Capacitor Leads Voltage by 90°.

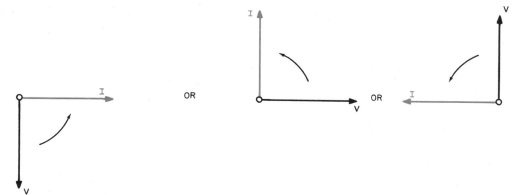

Fig. 15-8 Vectors for Current and Voltage at Capacitor

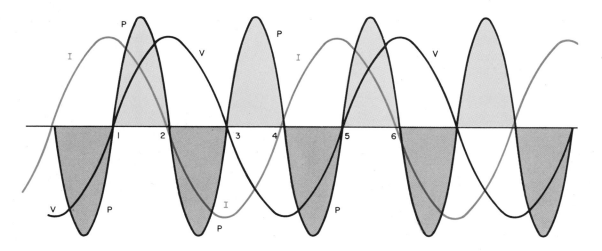

Fig. 15-9 Graph of Power in Capacitor Reactance

the discharging capacitor produces the reverse current I. The gray-shaded area below the axis between 2 and 3 represents energy given out by the capacitor. Between 3 and 4, V and I are in the same direction, taking energy from the supply again as the capacitor recharges.

> Radios, televisions, and other electronic equipment have large capacitors. These capacitors retain their charge after the power is disconnected. A charged capacitor can give a severe shock. To protect the worker and test equipment, they are discharged before work is begun. This is done by shorting the capacitor terminals to the chassis with a screwdriver that has an insulated handle.

CAPACITIVE REACTANCE

The fact that any capacitor has a definite charge-to-voltage ratio (capacitance) is one reason why the ac in a capacitor is limited in amount. The ability of a capacitor to limit current is called *capacitive reactance*, X_C. Capacitive reactance is measured in ohms, like the current-limiting ability of resistors and inductive reactors. The formula $X_C = \frac{1}{2\pi f C}$ gives the relation of capacitive reactance X_C (ohms) to frequency (f) and capacitance. Frequency is hertz, or cycles per second, and C is farads in the above formula. Because C is more often expressed in microfarads than in farads, it is more convenient to use $X_C = \frac{1\,000\,000}{2\pi f C}$, with C now meaning microfarads.

As an example, suppose we want to find the current in a 10-μf capacitor when it is connected to the 120-volt, 60-hertz line: First, using $X_C = \frac{1\,000\,000}{2\pi f C}$ to find ohms:

$$X_C = \frac{1\,000\,000}{2 \times 3.14 \times 60 \times 10}$$
$$= \frac{1\,000\,000}{3768}$$
$$= 265 \text{ ohms.}$$

Next, Ohm's Law, $I = \frac{\text{Volts}}{\text{Ohms}}$ will tell the current:

$$I = \frac{120}{265} = 0.45 \text{ amp.}$$

Actually, if a 10-μf capacitor is connected in series with an ammeter to the 120-volt ac line, the current might be any amount from 0.4 to 0.5 amp. This is because capacitors are often built to a 10 percent tolerance.

Capacitor and Resistor In Series

Capacitors are often used with resistors in a circuit. Figure 15-10 represents a capacitor with X_C = 40 ohm connected in series with a 30-ohm resistor. As in any series circuit, there is only one current, but more than one voltage to consider. The relation of the alternating voltages to the current is best

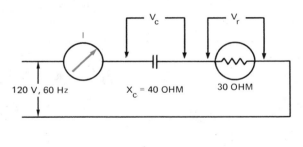

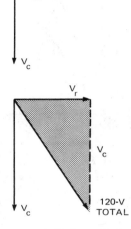

Fig. 15-10   Current and Voltage Relationship for Capacitor in Series with a Resistor

shown by a vector diagram, as was used in Chapter 14. Start the diagram by drawing a vector for the current I in any position. The vector representing the voltage across the resistor V_R is drawn in the same direction as I, overlapping it. At the resistor the voltage is in phase with the current. At the capacitor, current leads voltage by 90°, figure 15-8, therefore draw the voltage of the capacitor V_c 90° behind the current, which puts I 90° ahead of V_c.

In series circuits, individual voltages add up to equal the applied line voltage, but alternating voltages do not add as simply as dc voltages. V_c and V_R could be added by the long process used in figure 14-10, but it is easier to get the total by using vectors. From the triangle of the last vector diagram in figure 15-10, (Total volts)^2 = $(V_c)^2 + (V_R)^2$, or total volts = $\sqrt{V_c{}^2 + V_R{}^2}$.

The term impedance, Z, has already been used to mean the total ohms opposition in a circuit. To find the relation between total ohms impedance and the individual values of ohms resistance and ohms capacitive reactance, these forms of Ohm's Law,

Total volts = I Z; V_c = I X_c and V_R = I R will be inserted into the expression (total volts)^2 = $V_c{}^2 + V_R{}^2$, so it becomes

$$I^2 Z^2 = I^2 X_c{}^2 + I^2 R^2$$

Dividing through by I^2 gives

$$Z^2 = X_c{}^2 + R^2$$

The above is a useful formula because it enables the prediction of the behavior of a capacitor and resistor in limiting current. Figure 15-10 pictured a capacitor of X_c = 40 ohm in series with R = 30 ohm. The total ohms of this combination is 50 ohms, found from Z^2 = $40^2 + 30^2$ = 1 600 + 900 = 2 500. Z^2 = 2 500; Z = 50.

If the voltage applied to the circuit is 120, and Z = 50 ohms, then I = E/Z = 120/50 = 2.4 amps. The voltage across the capacitor in the circuit = I X_c = 2.4 x 40 = 96 volts.

The voltage across the resistor = IR = 2.4 x 30 = 72 volts. These voltages agree with the vector relationship shown in figure 15-10, because $120^2 = 96^2 + 72^2$.

Capacitor and Resistor In Parallel

Rearrange the above capacitor and resistor, placing them in parallel. In figure 15-11, the 30-ohm resistor is connected directly to the 120-volt source, and the current in it is 120/30 = 4 amps. Closing the switch at the capacitor will connect the capacitor in parallel with the resistor. After this is done, the resistor is unaffected, since its connection to the 120-volt supply is still there. The ammeter in the capacitor branch will read I = 120/40 = 3 amps. How much

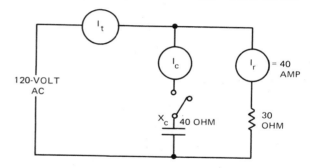

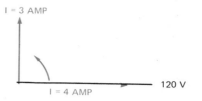

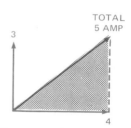

Fig. 15-11 Current and Voltage Relationship for Capacitor in Parallel with a Resistor

will an ammeter in the line read as the value of the total current, I_t? A vector diagram is necessary to find out the total of 3 amps and 4 amps in this case.

In this parallel circuit, there is only one voltage, which is shown by the 120-volt vector. In a resistor, current and voltage are in phase because the resistor does nothing to create any phase difference. The 4-amp current in the resistor is in phase with the 120 volt, so it is drawn as a vector in the same direction as the 120. The 3-amp current in the capacitor leads the voltage (the same voltage) by 90°, so its vector is drawn 90° ahead of the 120-volt line. Because the currents, 3 amps and 4 amps, are out of step, they add to make 5 amps, as shown in the vector triangle relationship. Total $I^2 = 3^2 + 4^2$, $I^2 = 9 + 16 = 25$, $I = 5$ amps.

The most valuable uses of capacitors are in circuits where they are connected with coils (inductors) to enable control of voltages and currents without energy loss, to produce alternating currents of desired frequencies, or to sort out one frequency of several that may exist in a circuit.

RESONANCE

If a capacitor and a coil are connected in series to the 120-volt, 60-hertz line, some surprising voltages may develop. Figure 15-12 suggests connecting a capacitor, 16.5 μf, $X_c = 160$ ohms, in series with a 0.58 henry inductor with no resistance, but with inductive reactance $X_L = 220$ ohms. There will be a current in the series circuit, shown in the vector diagram by the colored arrow, I. In that diagram, V_L is shown 90° ahead of I, because the current in the inductor lags voltage by 90°. In a series circuit there can be only one current, so the same current vector applies for the capacitor as well as for the coil. At a capacitor, current leads voltage by 90°, shown in the vector diagram by placing V_c, capacitor

voltage, 90° behind I. This puts these two voltages, V_L and V_c, 180° apart. "Two voltages are 180° out of phase" is a way of saying "two voltages are always opposite in direction." So that two opposing voltages can add to equal 120 volts, one must be 120 more than the other ($V_L - V_c = 120$). $V_L = I\ X_L$, and since X_L was given as 220, $V_L = (220 \times I)$. $V_c = I\ X_c$, X_c was given as 160 ohms, so $V_c = (160 \times I)$.

Putting these terms, 220 I and 160 I, in place of V_L and V_c in $V_L - V_c = 120$, results in 220 I - 160 I = 120. The subtraction gives 60 I = 120, therefore I = 2 amps. Two amps in the 160-ohm X_c produces 320 volts at C. Two amps in the 220-ohm X_L produces 440 volts across the coil. Since the 320 volts is in opposite direction to the 440 volts, it subtracts from the 440. This leaves 120, which equals line voltage, or total volts.

The impedance of this circuit can be found from the fact that 120 volts puts 2 amps through it, 120/2 = 60 ohms, which is the difference of X_L and X_c, 220-160. Because of the *oppositely-directed* voltages, X_L and X_c subtract in determining their

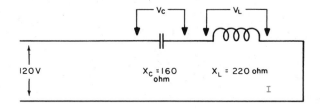

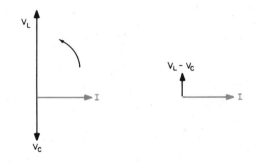

Fig. 15-12 Current and Voltage Relationship for Capacitor in Series with a Coil

combined effect on the total impedance. According to this, if X_L is made equal to X_c, nothing is left to limit the current. This is true except that real circuits have resistance.

With the help of figure 15-13, reconsider the previous circuit, made more realistic by including 25 ohms resistance, as a resistor in series with the ideal coil. The vector diagram shows the resistor voltage in phase with I, along with coil and capacitor voltages as before. The oppositely-directed V_L and V_c combine by subtraction to make the shorter $(V_L - V_c)$ vector shown with V_R in the next diagram. Finally, the triangle of the last diagram shows why (Total V)^2 = $(V_L - V_c)^2$ + V_R^2. In this expression, substituting amps x ohms for each V, it converts to $I^2 Z^2 = I^2$ $(X_L - X_c)^2 + I^2 R^2$. Dividing out the I^2 leaves a general formula for impedance in a series circuit: $Z^2 = (X_L - X_c)^2 + R^2$.

Using this formula, the impedance of the circuit of figure 15-13 can be found:

Z^2 = $(220 - 160)^2 + 25^2 = 60^2 + 25^2$
= 3 600 + 625 = 4 225.

$Z = \sqrt{4\ 225}$ = 65 ohms, the total opposition of all three parts in the circuit. Current $I = E/Z = \frac{120}{65}$ = 1.85 amps.

From the 1.85 amps, the voltage across each part can be found if needed: V = 1.85 x 160 = 296 volts across the capacitor. V across X_L = 1.85 x 220 = 407 volts. For the resistor, V = IR = 1.85 x 25 = 46 volts.

What happens if we make $X_L = X_c$? One way of doing this is by removing the coil of 220-ohm reactance and inserting one of 160-ohm reactance. This makes $X_L - X_c = 0$, but the 25-ohm resistor is still there to limit the current. The impedance of the circuit has become 25 ohms, and the current is $\frac{120}{25}$ = 4.8 amps. The voltage across the capacitor = $I X_c$ now becomes 4.8 x 160 = 768 volts along with 768 volts in the opposite direction across X_L.

There is another way to make $X_L = X_c$ in the above circuit without changing coil or capacitor. If there is a variable-frequency power source, make $X_L = X_c$ by changing the frequency of the alternating voltage supply. X = $2\pi fL$ (with L meaning henries inductance) and $X_c = \frac{1\ 000\ 000}{2\pi f\ C}$ (with C in microfarads).

If $X_L = X_c$, then $2\pi fL = \frac{1\ 000\ 000}{2\pi f\ C}$ and $4\pi^2 f^2 LC = 1\ 000\ 000$.

Taking the square root of each side of this equation, $2\pi f\sqrt{LC} = 1\ 000$.

$f = \frac{1\ 000}{2\ \pi\ \sqrt{LC}}$ with L in henries and C in microfarads. Using the previously given values, L = 0.58 h and C = 16.5 μf, LC = 9.57.

$f = \frac{1\ 000}{2\ \pi\ \sqrt{9.57}} = \frac{1\ 000}{6.28 \times 3.1}$ = 51 hertz

By changing the frequency from 60 to 51 hertz, the reactance of the capacitor has increased from 160. The inductive reactance has decreased from 220 until the two become equal, $X_L = X_c = 186$ ohms. If voltage is still 120 volts, the impedance equals the resistance in the circuit, which is 25 ohms. The current in the circuit is greater than at any other frequency, with higher voltages developed across the coil and the capacitor than would appear at any other frequency.

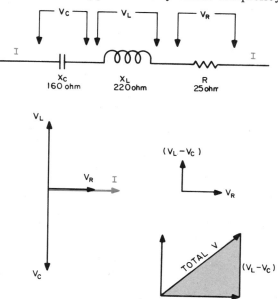

Fig. 15-13 Current and Voltage Relationship for Capacitor in Series with a Coil and a Resistor

A circuit in which $X_L = X_c$ is called a *resonant circuit.* *Resonance* in general means the production of maximum response to a vibration. *Electrical resonance* is the development of maximum power at a certain frequency.

The adjustment of a coil-and-capacitor circuit to make $X_L = X_c$ for a desired frequency is called *tuning.* Turning the tuning knob on a radio receiver moves the plates of a variable capacitor, changing its capacitance. The antenna coil indicated in figure 15-14 may appear as a large flat loop of about 30 turns of wire on the back of the receiver, or it may consist of many turns of wire wound on a ferrite rod about the size of a pencil. For now, think of the antenna coil as acting like the secondary coil of a transformer, in which each of a hundred broadcasting stations would like to induce a current. Adjustment of the variable capacitor makes the receiver antenna circuit responsible to only one. For example: Perhaps there is an antenna coil of inductance L = 180 microhenries, that is, 180 millionths of a henry, or 0.000180 h. The capacitor is set so its capacitance is 140 micro microfarad, which is 140 millionths of a microfarad, 0.001140 μf (also called 140 $\mu\mu$f). Resonant frequency of the circuit can be found, $f = \dfrac{1\ 000}{2\pi\sqrt{LC}}$, if L is in henries and C in microfarads.

$$f = \frac{1\ 000}{6.28\sqrt{0.00018 \times 0.00014}}$$

$$= \frac{1\ 000}{6.28\sqrt{0.0000000252}} = \frac{1\ 000}{6.28 \times 0.000159}$$

$$= \frac{1\ 000}{0.001} = 1\ 000\ 000.$$

Fortunately, that answer came out even. 1 000 000 hertz is also called 1 000 kilohertz, (kHz); 1 000 kHz = 1 megaHertz, (MHz). In the above circuit, a 1 000-kilohertz transmitter can stir up an ac of 1 million cycles per second because the $X_L = X_c$ of the circuit for that frequency. At other frequencies, one of the reactances is larger than the other,

introducing more ohms reactance-opposition to currents in other frequencies than 1 000 kilohertz.

POWER FACTOR CORRECTION

Capacitors are often added to commercial power circuits for power-factor correction. To learn the meaning of power-factor correction, review the behavior of an ac circuit containing inductors and resistors. The inductors might represent the inductive ballasts in a bank of fluorescent lamps (Chapter 14), or they might represent induction motors (Chapter 16).

For figure 15-15, an inductive load is connected to the 120-volt ac line through a wattmeter. The coils consist of 4 ohms of inductive reactance along with 3 ohms of resistance, so the impedance, Z, is 5 ohms. ($Z^2 = 4^2 + 3^2$). The current is 120/5 = 24 amps. The wattmeter in the circuit indicated 1 728-watts true power, because I^2R (= $24^2 \times 3 = 1\ 728$) is the energy-consuming part of

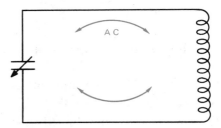

Fig. 15-14 Tuning Circuit

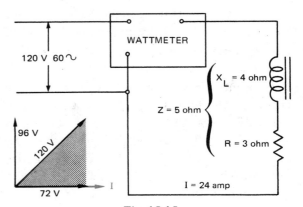

Fig. 15-15

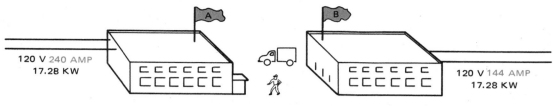

Fig. 15-16

the circuit. The 4-ohm purely inductive reactance consumes no energy. If 120 volts is multiplied by 24 amps, the result is 2 880. The answer could be called apparent watts, but a better unit for it is 2 880 volt-amperes. This reserves the word watts for cases where it means actual power. The *power factor* is the ratio $\dfrac{\text{true watts}}{\text{volts x amps}}$. In this example, the power factor is $\dfrac{1\,728}{2\,880} = 0.6$. Power factor is a number, which multiplied by volts x amps, makes the answer for watts correct for an ac circuit.

Because true power I^2R, and volt-amperes, or apparent power, $= I^2Z$, the power factor is also equal to $I^2R \div I^2Z = R/Z$.

Notice that in figure 15-15, the 120-volt line carries 24 amperes, yet delivers to the circuit only 1 728 watts. This is the same power that could be delivered by 120 volts and 14.4 amps to a purely resistive load, which has power factor of 1. Figure 15-16 represents a larger-scale comparison of two commercial energy users. Building A contains equipment with power factor = 0.6. It needs 240 amps at 120 volts, and uses energy at the rate of 17.28 kilowatts. 240 x 120 x 0.6 = 17 280 watts or 17.28 kilowatts. Building B's power factor is 1. It needs 144 amps at 120 volts and uses energy at the same rate as building A. (144 x 120 x 1 = 17 280 watts or 17.28 kilowatts.)

Power factor is of no concern to the ordinary home user of electrical energy, for this electric bill is based only on total kilowatt-hours of energy. Power factor is of concern to large commercial and industrial users of energy, because their energy costs are based not only

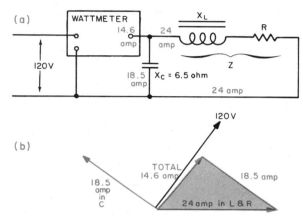

Fig. 15-17 Power Factor Correction

on total kilowatt-hours, but also on the maximum volt-amperes that the power company delivers to them. In figure 15-16, building A will be charged more money because it costs more to deliver 240 amps than it does to deliver 144 amps. The operator of building A buys some capacitors and installs them across the line in the building to reduce costs.

Compare the experimental circuit of figure 15-17 with figure 15-15. In 15-17 a capacitor (X_c = 6.5 ohm) has been added across the 120-volt line, putting it in parallel with the inductive load. The X_L and R still connect to the 120-volt supply and carry 24 amps. This current lags the 120 volts by the same vector angle as was shown in figure 15-15. The addition of the capacitor has introduced another current, $I - E/X_c = 120/6.5 = 18.5$ amps, which leads the voltage by 90 degrees, as is to be expected of capacitor current. The total of these two currents is shown in the drawn-to-scale vector diagram as 14.6 amps, which is nearly in phase with the voltage. The big accomplishment of introducing the capacitor is the reduction of the line current. Applying this to the problem of figure 15-16, building A is dealing with current ten times larger than those of figure 15-17. The manager of building A should buy capacitors large enough to carry 185 amps, connect them in parallel with the machines, reduce the line current, and thereby reduce the electric bill. This use of capacitors is common practice. In a few years the savings will exceed the cost of the capacitors. Installation of capacitors has resulted in what is called power factor correction. In the new situation (figure 15-17),

$$PF = \frac{1\ 728 \text{ watts, as before}}{120 \text{ volts x } 14.6 \text{ amps}} = \frac{1\ 728}{1\ 752} = 0.98^+$$

which is close enough to the ideal 1.0 to be satisfactory.

Power factor is of no concern in the household electric bill, so the homeowner has no need for P F correction. Fluorescent lighting ballasts, motors, and other magnetic devices are not affected when there are capacitors in parallel with them.

POINTS TO REMEMBER

- Two nearby conductors separated by dielectric form a capacitor.
- $C = \frac{Q}{V}$ defines capacitance.
  C = capacitance, measured in farads.

- Q = charge, measured in coulombs.
  V = Volts.
- A capacitor can be charged by a dc source, but it does not conduct dc.
- The capacitor stores energy in its electric field; it may remain charged when removed from a circuit.
- A capacitor conducts ac; its ohms opposition is called capacitive reactance.
  $$X_c = \frac{1}{2\pi f\ C}.$$
- Alternating current at a capacitor leads capacitor voltage by 90°. The net power used during one complete ac cycle is practically zero.
- Impedance, $Z = \frac{E}{I}$.
- Series-circuit impedance,
  $$Z = \sqrt{(X_L - X_c)^2 + R^2}$$
- Maximum current exists in a coil-and-capacitor circuit when X_L equals X_c. This condition is called resonance.
- Resonant frequency, $f = \frac{1}{2\pi L\ C}$
  (L in henries, C in farads).
- Power factor $= \frac{\text{watts of true power}}{\text{volts x amperes}}.$
- In a simple series circuit, power factor $= \frac{R}{Z}.$

REVIEW QUESTIONS

1. What is capacitance? In what unit is it measured?

2. What is capacitive reactance? In what unit is it measured?

3. A certain capacitor is connected to a 120-volt 60-hertz line. An ammeter in series with it records 0.2 amp.
   (a) Find reactance of the capacitor.
   (b) Find capacitance of the capacitor.

4. (a) Find reactance of a 1 μf capacitor for 159-hertz ac.
   (b) Find reactance of a 1 μf capacitor for 159 000-hertz ac.
   (c) Is a capacitor a better conductor for high frequencies or for low frequencies?

5. Draw a simple vector diagram showing the phase relationship of the 120 volts and the 0.2 amps referred to in question 3.

6. A 20 μf capacitor has been charged from a 200-volt dc source; how many coulombs of charge does it have? If the capacitor is discharged in one hundredth of a second, calculate the average discharge current.

7. A capacitor, C = 0.133 μf, is connected in series with a 500-ohm resistor in an ac circuit where the frequency is 1 000 hertz.
   (a) Find reactance of the capacitor. (Check the arithmetic, the answer should be close enough to 1 200 ohms to use that number for the next calculations.)
   (b) Find the impedance of the series combination of capacitor and resistor.
   (c) If the source voltage is 13 volts, find the current in the circuit.
   (d) Find voltage across capacitor and voltage across resistor.
   (e) Draw a vector diagram showing the phase relation of capacitor voltage and resistor voltage. Add to the diagram the vector showing the 13-volt total voltage. Also add a vector showing the current.
   (f) According to the diagram, the circuit current leads or lags the 13 volts by about how many degrees?

8. A capacitor and inductor are in series on a 120-volt 60-hertz line. The capacitor has X_c = 100 ohms. The inductor has X_L = 100 ohm and R = 8 ohms.
   (a) Find impedance of the series circuit.
   (b) Find current in the circuit.
   (c) Find voltage across the capacitor.
   (d) Estimate voltage across the inductor:
       (1) exactly same as that on capacitor
       (2) slightly more than on capacitor
       (3) slightly less than on capacitor
       (4) 120 volts
       (5) less than 120 volts

9. Find the power (true watts) used by:
   (a) the capacitor in question 7
   (b) the resistor in question 7

10. Find the power (true watts) used by:
    (a) the capacitor in question 8
    (b) the inductor in question 8

11. What is the purpose of power-factor correction? How is it done? Does it make any difference whether capacitors are added in series or added in parallel? Explain.

12. Name one of the most valuable uses for capacitors.

13. What is a resonant circuit?

14. Should a homeowner be concerned with power factor?

15. Is power factor important to a person operating a factory where electricity is used to run machines? Justify the answer.

RESEARCH AND DEVELOPMENT

Experiments and Projects on Capacitors and Their Uses

INTRODUCTION

In Chapter 15, the types, construction, and principle of operation of capacitors were discussed. Definitions and formulas for computations were given. You learned that a capacitor connected to a dc source will charge to the value of the source voltage and then the current will stop flowing. Also, when the same capacitor is connected to an ac source, current will continue to flow. Another important fact is that a capacitor will remain charged and it can be dangerous until it is discharged. The experiments which follow provide an opportunity to work with and study the characteristics of capacitors. The projects afford an opportunity to learn how they are constructed, used, and tested.

EXPERIMENTS

1. To study the action of a capacitor during the charge and discharge period.
2. To observe the dc and ac charge values in a capacitor.
3. To determine the relation between the voltage and current in an ac series circuit having a resistor and a capacitor.
4. To observe the current value of a circuit with a coil and capacitor connected in parallel.

PROJECTS

1. Take an old radio apart and examine the capacitors. Take several capacitors apart to find out how they are made and of what materials. Make a capacitor of aluminum foil and waxed paper and test it. Use it in a crystal radio circuit and see how it works.
2. Assemble a series circuit such as illustrated in figure 15-10, with components of a different capacity. Test and record data. Make calculations and draw a vector diagram.
3. Assemble a parallel circuit such as illustrated in figure 15-11, with components of a different capacity. Test and record data. Make calculations and draw a vector diagram.
4. Make a direct-current magnetizing device using large capacitors as energy source.

EXPERIMENTS

Experiment 1 CHARGE AND DISCHARGE OF A CAPACITOR

OBJECT

To study the action of a capacitor during the charge and discharge period.

APPARATUS

1 - Capacitor bank rated at about 50 microfarads
1 - Stopwatch
1 - DC voltmeter (0-250 or 0-300 V)
1 - DPST switch

PROCEDURE

1. Connect the voltmeter in series with the capacitor as shown in figure 15-18. Determine from the ohms per volt specifications, the resistance of the voltmeter. In this circuit, the voltmeter acts as a series resistor, as well as a voltmeter.

2. Close the line switch and record the voltmeter readings at 5-second intervals. Subtract the voltmeter readings from the voltage supply to obtain the actual voltages across the capacitor terminals while it is on charge.

3. When the voltmeter indicates zero, open the line switch.

4. Connect the voltmeter across the capacitor terminals as illustrated in figure 15-19, and read the instrument immediately.

5. Read and record the voltmeter readings every 5 seconds after the first reading while the capacitor is on discharge.

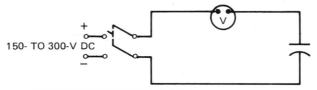

CONNECTIONS FOR CHARGING THE CAPACITOR

Fig. 15-18

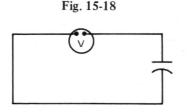

CONNECTIONS FOR DISCHARGING THE CAPACITOR

Fig. 15-19

OBSERVATIONS

Use a table similar to the one shown here to record your observations.

OBS. NO.	CHARGE		DISCHARGE	
	TIME IN SECONDS	E CAP	TIME IN SECONDS	E CAP
1				
2				
3				
4				
5				
6				
7				
8				
9				
10				
11				
12				
13				
14				
15				
16				
17				
18				

GRAPH

On graph paper, plot a curve showing (1) relationship between the capacitor terminal voltage and time in seconds during the charging period, (2) the relationship between the capacitor terminal voltage and time in seconds during the discharge period.

QUESTIONS

1. Why is it necessary to subtract the voltmeter readings from the supply voltage to determine the capacitor terminal voltage when the capacitor was on charge?

2. What effect does the value of internal resistance of the voltmeter have on the time for charge and discharge?

3. a. What is meant by the term time constant, as used in capacitor work?
   b. What time period in seconds would equal one time constant for the circuit used in this experiment?

4. Using the necessary computations, show that the voltage, obtained in the observation data for one time constant when the capacitor was charging, was an accurate reading.

5. a. When is a capacitor assumed to be fully charged?
   b. Was the capacitor used in this experiment fully charged at the end of the time period given in the (a) part of this question?

6. Show, with the necessary computations, that the voltage obtained in the observation data for one time constant was an accurate reading when the capacitor was discharging.

Experiment 2 EFFECT OF CAPACITANCE ON DIRECT CURRENT AND ALTERNATING CURRENT

OBJECT

To observe the dc and ac charge values in a capacitor.

APPARATUS

1 - Incandescent lamp, 25-watt
1 - Lamp receptacle
1 - Capacitor, 10-μf
1 - DC voltmeter (0-150 V)
1 - AC voltmeter (0-150 V)
1 - SPST switch
2 - B-batteries, 45-volt or equivalent dc source

PROCEDURE

1. Connect the dc circuit according to the diagram in figure 15-20.

2. Close the switch and observe the lamp.

3. Test the voltage drop across the lamp and across the capacitor.

4. Open the switch and check voltages again.

5. Connect the ac circuit according to figure 15-21.

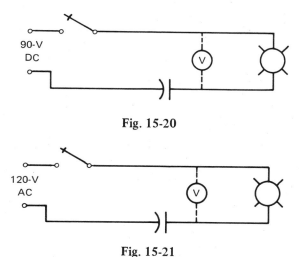

Fig. 15-20

Fig. 15-21

6. Close the switch and observe the lamp.

7. Test the voltage drop across the lamp and across the capacitor.

8. Open the switch and see what happens.

OBSERVATIONS

Use a table similar to the one shown here to record your observations.

OBS.	SWITCH	LAMP	VOLTAGE	CAPACITOR	VOLTAGE
DC	CLOSED				
DC	OPEN				
AC	CLOSED				
AC	OPEN				

QUESTIONS

1. When the switch was closed in the dc circuit, what happened to the lamp? Explain.

2. When the switch was opened in the dc circuit, what happened to the lamp?

3. What happened to the lamp in the ac circuit when the switch was closed?

4. What was the voltage across the lamp in the dc circuit? How can the reading on the meter be accounted for?

5. What was the voltage drop across the lamp in the ac circuit? Did it differ proportionally from that in the dc circuit? Explain the answer.

Experiment 3 AC CAPACITANCE CIRCUIT

OBJECT

To determine the relation between the voltage and current in an ac series circuit having a resistor and a capacitor.

APPARATUS

1 - AC voltmeter (0-150 V)
1 - AC ammeter (0-10 A)
1 - Lamp bank with a variety of lamps: 15-, 25-, 40-, and 60-watt
1 - Capacitor, not electrolytic, 10- to 20-μf or larger
1 - Wattmeter (0-150 or 0-250 W)
1 - DPST knife switch

PROCEDURE

1. Connect apparatus as shown in the circuit diagram, figure 15-22.

2. Adjust the resistance of the lamp bank so that the voltage drop across it is nearly equal to the voltage drop across the capacitor.

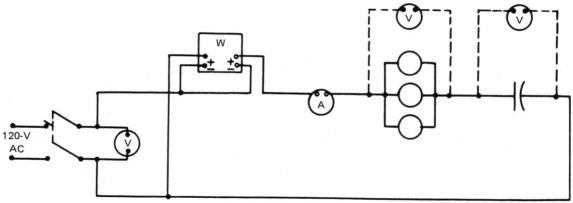

Fig. 15-22

3. Read and record the line voltage, voltage drop across the lamp bank, voltage drop across the capacitor, and the line current.

Note: Do not touch the terminals of the capacitor once the circuit has been energized. To discharge the capacitor, open the line switch and connect the capacitor terminals with a short insulated jumper wire.

OBSERVATIONS

Use a table similar to the one shown here to record your observations.

OBS. NO.	LINE VOLTS	LAMP BANK VOLTS	CAPACITOR VOLTS	WATTS	LINE CURRENT	CIRCUIT CONDITION
1						X_c Equal To R
2						R Greater Than X_c
3						One 50 μf Capacitor, No R

4. Increase the value of resistance and repeat Step 3.

5. Remove the resistance and, using capacitance only, repeat Step 3. (Use only one 50-microfarad capacitor for this observation.)

CALCULATIONS   (Using Observations No. 1 and No. 2)

Use a table similar to the one shown here to record your calculations.

OBS. NO.	LINE VOLTS	Z	R	X_c	C	PF	PF ANGLE
1							
2							

SAMPLE CALCULATIONS

$$E\text{ Line} = \sqrt{E_R^2 + E_C^2} \qquad X_c = \frac{E_C}{I} \qquad PF = \frac{W}{VA}$$

$$Z = \frac{E\text{ Line}}{I} \qquad\qquad\qquad PF = \cos \text{Angle } \Theta$$

$$R = \frac{E_R}{I} \qquad C = \frac{1}{2\pi F X_c} \qquad \text{Angle } \Theta =$$

VECTORS

Draw to scale vector diagrams for Observations No. 1 and No. 2.

QUESTIONS

1. Why did the line current lead the line voltage in the circuit used in this experiment?

2. Explain why the algebraic sum of the voltage drop in this series circuit does not equal the line voltage.

3. Explain the reasons for the extremely low wattmeter reading in Observation No. 3.

Experiment 4 COIL AND CAPACITOR IN PARALLEL

OBJECT

To observe the current in a circuit with a coil and capacitor connected in parallel, figure 15-23.

APPARATUS

1 - Coil of several hundred turns of #18-#24 AWG magnet wire with a laminated iron core surrounding coil. See sketch.
1 - AC capacitor, 10-μf
1 - AC capacitor, 20-μf
1 - AC capacitor, 40-μf
1 - SPST switch
1 - AC ammeters (0-10 A)
1 - AC variable power supply

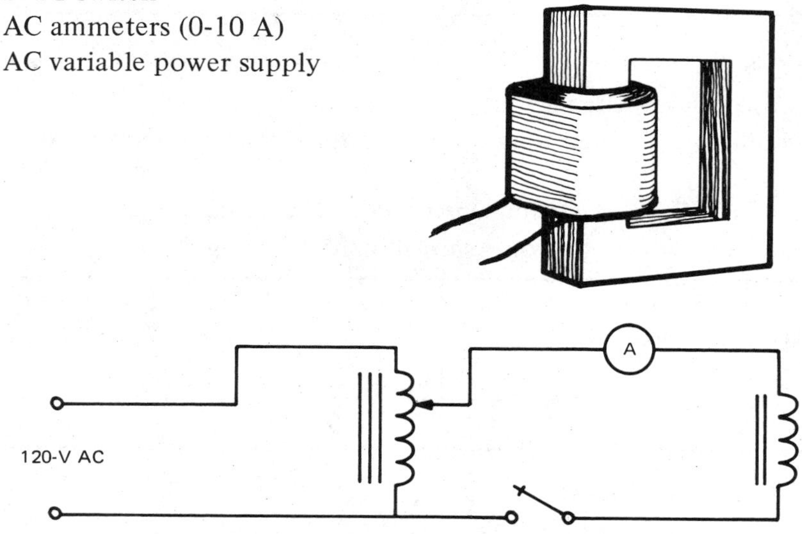

Fig. 15-23

PROCEDURE

1. Use the same ac voltage as was used in Experiment 1, Chapter 14, to get a reasonable current in the coil.

2. If the current in Step 1 was about 1 amp, connect a capacitor of about 20 microfarads in parallel with the coil, figure 15-24. (If current was 1/2 amp, use a 10-microfarads capacitor. If the current was 2 amps, use a 40-microfarads capacitor.)

3. Measure capacitor current and line current. Connect an ammeter in series with the capacitor, figure 15-25.

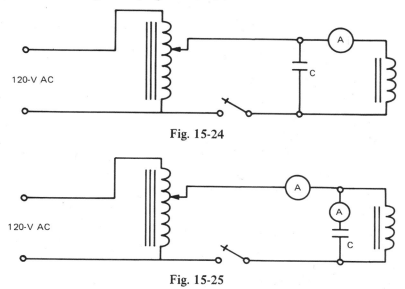

Fig. 15-24

Fig. 15-25

QUESTIONS

1. Should line current equal the sum of coil and capacitor? Why or why not?

2. What effect does the iron core have on the function of the coil?

PROJECTS

CAPACITOR DISCHARGE MAGNETIZING DEVICE

This device, figure 15-26, which uses the stored-up energy in capacitors to energize the magnetizing coil is powerful and rapid acting.

Most of the parts required can be salvaged from discarded equipment or obtained from surplus property sources.

MATERIALS

1 - Switch, SPDT knife switch (S_1)
1 - Neon lamp, Ne-2 (L_1)
1 - Resistor, 25-ohm, 10-watt, or substitute a 100-watt lamp (R_1)
1 - Resistor, 110 000-ohm, 1/2-watt (R_2)

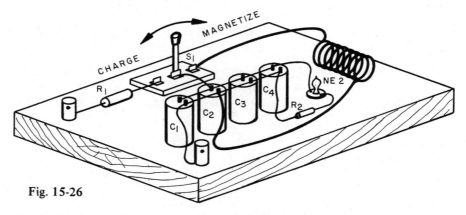

Fig. 15-26

1 to 4 - Capacitors, 1 000-1 500-μf each, 200-volt dc (C_1, C_2, C_3, C_4)

2 - Binding posts

1 - Coil, 10-turn #10 or #8 wire. Make coil large enough to take bar magnets and tools you wish to magnetize.

Radios, televisions, and other electronic equipment have large capacitors. These capacitors retain their charge after the power is disconnected. A charged capacitor can give a severe shock. To protect the worker and test equipment, they are discharged before work is begun. This is done by shorting the capacitor terminals to the chassis with a screwdriver that has an insulated handle.

PROCEDURE

Obtain the parts.

Plan the arrangement of the parts, figure 15-27, and make an appropriate base.

Assemble the parts and wire.

Note: Make all connections as short as possible and mechanically and electrically secure. Use #10 wire for connections from switch to coil, coil to capacitors, and capacitors to switch.

Test: Connect the magnetizer to the direct current supply and throw the knife switch to the left to charge the capacitors. When they are charged, the Ne-2 lamp will light. Next, place the piece to be magnetized inside the coil and throw the switch to the right, and the magnet will be magnetized.

A charged capacitor can give a severe shock. When the switch is thrown to the magnetizing position, the capacitors discharge and provide the energy in the coil to magnetize. Large capacitors in radio and television circuits, and in other electronic equipment, should be discharged before work is begun. This may be done by shorting the capacitor terminals to the chassis with a screwdriver that has an insulated handle.

Evaluate.

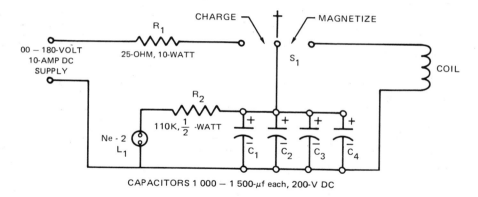

SCHEMATIC FOR CAPACITOR DISCHARGE MAGNETIZING DEVICE

Fig. 15-27

Chapter 16
AC motors

The discussion of motors in Chapter 13 described the universal motor. The *universal motor* is a common type of motor which works on either ac or dc. Since the armature and field coils are in series, the armature current reverses at the same instant that the field reverses. The armature continues turning in the same direction. Because of this factor, the motor can be used on either current. Universal motors are used to a large extent in small portable equipment.

INDUCTION MOTORS

Large motors and motors permanently mounted to run something at fairly constant speed are often *induction motors.* These motors drive washing machines, dryers, furnace fans, refrigerators, bench grinders, table saws, and other equipment. If you disassemble an induction motor, you will not find brushes to carry current to the rotating part of the motor. The current in the rotor (rotating part) is induced in the rotor itself by the changing magnetic field of the stationary coils in the iron frame. In some ways, the rotor is comparable to the secondary coils of a transformer, in which current is induced by the alternating magnetic field of the primary coils with no wire connection

between the two. The rotor is a short-circuited, low-resistance secondary winding with its conductors arranged across the magnetic field so that the current in them can cause rotation.

The rotor consists mainly of a cylindrical assembly of iron laminations, but this section will be more concerned with the conductors (copper or aluminum) embedded in the iron. The aluminum fins apparent in figures 16-1 and 16-2 act as fan blades to move cooling air through the motor. The bars and end rings are the essential electric conductors. This arrangement is often referred to as a *squirrel cage.*

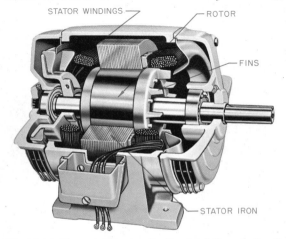

Fig. 16-1 Essential Parts of a Squirrel-Cage Induction Motor

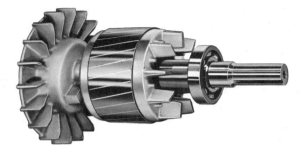

Fig. 16-2 Cage Rotor for an Induction Motor

Fig. 16-3 Stator of an Induction Motor

THREE-PHASE AC

Induction motors used in mills and factories are nearly all three-phase induction motors. Review chapter 10 on the description of commercial alternators that generate alternating current in stationary windings that surround a rotating magnet. All commercial power alternators have three sets of stationary coils embedded in the iron stator. Figure 16-5 shows, in simplified fashion, the arrangement of these three coils. Actually, they look like those in figure 16-3. The generation of voltage in each coil is the process described in Chapter 10, but with the added effect that the three coils are connected to produce three separate alternating currents which are 120° out of step with each other.

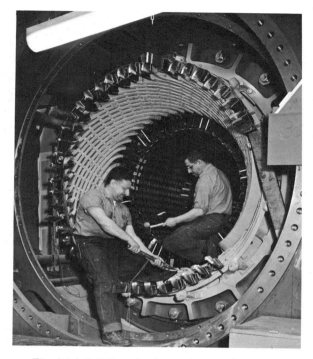

Fig. 16-4 Building the Stator for an Alternator

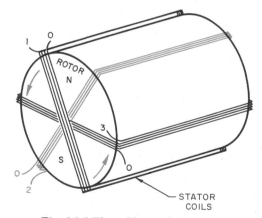

Fig. 16-5 Three-Phase Alternator

Looking at figure 16-5 again, as the counterclockwise-rotating north (N) pole of the magnet sweeps past the top of the coil marked 1, 0, a peak voltage is induced. The peak is labeled 1 in figure 16-6. The south (S) pole of the rotor passing by the bottom of coil 1 helps produce that peak, too. During this half-cycle of voltage for coil 1, above the axis in figure 16-6, the direction of voltage is such as to push electrons out of the coil-end number 1.

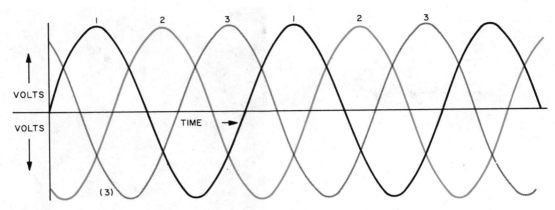

Fig. 16-6 Wave Pattern of Three-Phase Voltages

The north pole of the rotor next sweeps past the left side of coil 3, generating voltage in a direction to pull electrons into the coil-end numbered 3, the voltage maximum being shown below the horizontal axis at point (3) in the graph. When the north pole of the rotor passes the numbered side of coil 2, the generated voltage pulls electrons into the coil at 0 and pushes them out the wire-end marked 2. This is shown by the graph above the axis at peak 2. Peak 3 occurs when the north pole of the rotor passes the right side of coil 3, pushing electrons out of the wire-end numbered 3.

The symmetrical coil arrangement in the generator produces the symmetrical sine-wave pattern of three out-of-phase voltages of figure 16-6. Make use of them in a circuit like that of figure 16-7.

The three coils represent the coils of the generator, with the wire-ends numbered "0" in figure 16-5, brought together to a common connection to the three equal resistors. Generator coil 1 is in series with resistor 1, and sine-wave 1 of figure 16-8 represents the current in line-wire 1. Similarly, sine-wave 2 is the current in line-wire 2, and sine-wave 3 is the current in coil and resistor 3. The common connecting wire, called the *neutral,* may be connected to the earth, if desired. Assume the wave-tops in figure 16-8 represent a maximum current of 10 amps, and from that estimate the amount of total current in the neutral wire.

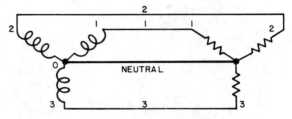

Fig. 16-7 Three-Phase Circuit

Assume that current to the right in a line wire is shown above the axis in figure 16-8, and values below the axis represent current to the left. At time-instant a, wire 1 carries its 10-amp maximum to the right, while waves 2 and 3 each carry 5 amps to the left. At that instant there is no current in the neutral wire. At time b, wire 2 is at zero amps. Wire 1 carries about 8 1/2 amps to the right and wire 3 carries the same 8 1/2 amps back to the left, so there is no current for the neutral. At c, wires 1 and 2 both have 5 amps to the right, that is a 10-amp total. However, the 10 amps is carried back by wire 3, and there are 0 amps in the neutral wire. At d, the current in wire 1 has dropped to zero. Wire 2 carries about 8 1/2 amps to the right. Wire 3 carries 8 1/2 amps back to the left. There is no current in the neutral. By now, how much current there is in the neutral wire at time e, or any other instant, can be figured out. As long as resistors 1, 2, and 3 are of equal value, carrying equal currents as read by a meter, their total is zero

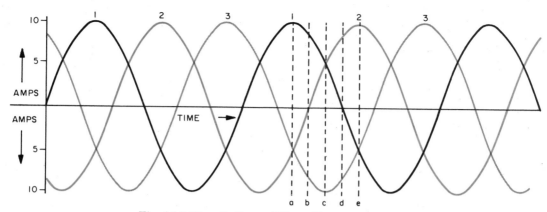

Fig. 16-8 Wave Pattern of Three-Phase Currents

and the neutral fourth wire is not needed to make the circuit work, thereby using only half as much copper connecting wires as three separate single-phase circuits. If the load resistors are unequal, the fourth wire is necessary for the total of the three currents, which is no longer zero.

THREE-PHASE INDUCTION MOTOR

A three-phase induction motor has three stationary windings connected like the resistors of figure 16-7. The three similar windings carry equal currents, so only three wires into the motor are required. Figure 16-5 can also serve as a simplified diagram of the three coils in the stator winding of the three-phase motor. As in the generator, these coils are spread out and imbedded in slots in the iron.

The three coils in the stator are shown at the left in figure 16-9(A). The current directions are those of time-instant a in figure 16-8. They use above-axis values from 16-8 to indicate electrons entering the numbered end of the coil, and below-axis values to show electrons leaving the numbered end of a coil. The unmarked coil ends in figure 16-9(A) tie together inside the motor. Wires at the far end of the stator are not shown, to keep the picture simple. Current direction in the wires lying in the slots is important. It will be used to find the direction of magnetic field inside the stator. The gray arrows on the outside surface of the stator indicate current direction in wires in the nearby slots. Notice that all three slots on the left have current away from the viewer. The three slots on the right have current toward the viewer. Remember the left-hand coil rule from Chapter 6: "with the fingers pointing in the direction of the current, the extended thumb points in the direction of magnetic field inside the coil." Now think of the stator assembly not as three coils, but as one unit. Imagine a left hand big enough, figure 16-9(A), to grasp the entire assembly just as it might grasp a coil that consisted of three turns of wire. Observe that the fingers reach around the stator in a direction parallel to the instantaneous current direction in stator wiring. As a result, the thumb direction indicates the nearly vertical upward direction of magnetic field inside the stator for this one instant of time.

Figures 16-9(B) through 16-9(E) show current directions at later instants corresponding to instants b through e on figure 16-8. In 16-9(B), coil 2 shows no current direction because at that instant its current is zero. At time D the current in coil 1 has dropped to zero. Apply the left hand to figures 16-9(B), (C), (D), and (E). You should arrive at directions of magnetic field as shown in

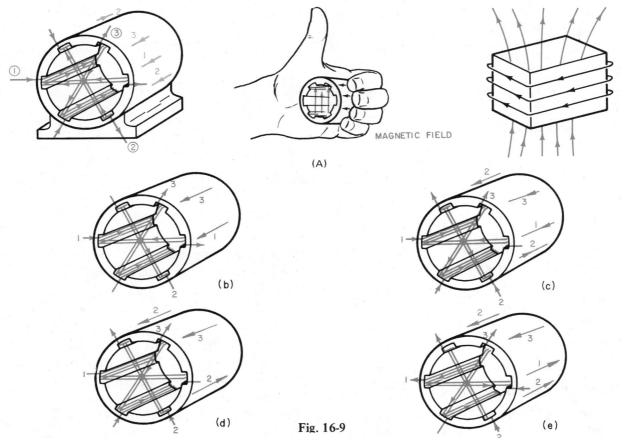

MAGNETIC FIELD

(A)

(b)

(c)

(d)

Fig. 16-9

(e)

figure 16-10(A) through (E). The important thing to notice is that the magnetic field lines rotate in one direction. This magnetic field is of constant strength. If details of current direction and field direction are carried out for one entire cycle of the ac sine wave, we find that the magnetic field makes one complete rotation inside the stator during one ac cycle. If this simple motor is supplied with 60-hertz ac, the field rotates at 3 600 rpm.

Figure 16-11 shows the conductors of the squirrel-cage rotor placed inside the stator. Stator windings are not shown. Only the rotating magnetic field they produce is shown. When the stator is energized, the magnetic field pattern (shown in color) sweeps across the rotor conductor as the field rotates. Motion of the magnetic lines induces voltages in the horizontal conductors of the rotor. The directions of the voltages are shown in figure 16-11.

Figure 16-12 shows only the rotor of figure 16-11, with current directions. This induced current magnetizes the rotor iron, producing poles N and S on the cylindrical surfaces. Comparing the position of these poles with the poles on the interior of the stator, the attraction and repulsion of poles causes the rotor to turn counterclockwise, the same direction that the magnetic field lines rotate. Or, the rotation can be accounted for with the help of 16-12(B). The black arrow represents one conductor in the top of the rotor. Two magnetic lines of force (color) sweep across it to the left. The induced current in the wire opposes this motion (Lenz's Law) by its own magnetic field (the little circle). Current in the wire in this way pushes against the approaching magnetic field, and the approaching magnetic field pushes equally hard against the

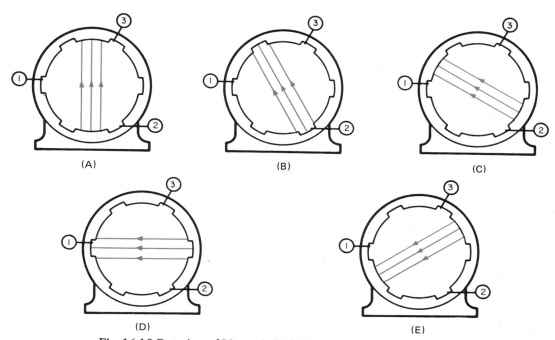

Fig. 16-10 Rotation of Magnetic Field in a Three-Phase Motor Stator

current-carrying wire. This latter push moves the current-carrying wire to the left, contributing to counterclockwise rotation. Although the field of this motor rotates at 3 600 rpm and tries to drag the rotor around with it, the rotor speed does not get up to 3 600. If it did, there would be no relative motion of field and rotor conductors. The field would not be cutting across the rotor conductors. Hence, no current would be induced in the rotor. Rotor speed of 3 400 to 3 500 rpm can be expected from the 3 600 rpm field.

Reversal of any pair of line connections to the three-phase motor reverses the direction of rotation of the magnetic field, and therefore, reverses the rotation of the rotor.

SYNCHRONOUS MOTOR

If a motor is to run at exactly 3 600 rpm, the same stator can be used. The stator should be used, however, with a rotor that has its magnet poles developed by an external dc source, rather than by induction. Such a motor is called a *synchronous motor*. Some mention has been made of the similarity of

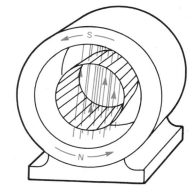

Fig. 16-11 A Stator with Squirrel-Cage Rotor in Position

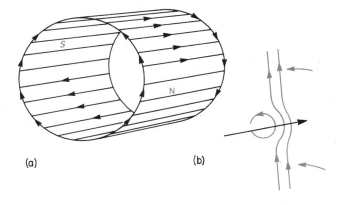

Fig. 16-12

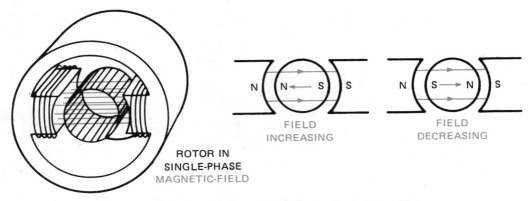

Fig. 16-13 Rotor in the Magnetic Field of a Single-Phase Motor

construction of dc motors and generators. Here is another example of similar construction. An alternator can be run as a synchronous motor because they are alike in structure.

Three-phase synchronous motors are seldom found in small shops. They are usually found in steel mills, paper mills, or cement mills. Sizes run from 20 hp (15 kw) up to thousands of hp (1 Mw). They are used on machinery that runs at constant speed. Driving dc generators is a common use. The power factor of a synchronous motor depends on the strength of the dc field of the rotor. The dc must be supplied from an external rectifier or small dc generator. By increasing the direct current fed to the rotor, the power factor can be brought up to 1.00. By further increase, this motor will make its current lead the line voltage, as occurs in a capacitor circuit. Induction motors are low power factor machines. Power factor correction is obtained in some mills by running synchronous motors along with induction motors.

Motor Speeds

A speed of 3 500 rpm is rather high for a large motor because the natural tendency of all moving objects is to travel in a straight line. If rotating parts of the motor suddenly start traveling in straight lines due to insufficient holding-together forces, the motor or other things nearby may be damaged. By arranging

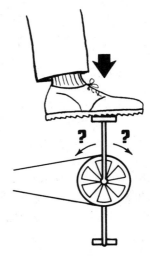

Fig. 16-14 Dead Center

coils on the stator so that several pairs of poles are formed, slower speeds can be obtained. Figures 16-11 and 16-9 show a two-pole stator winding for 3 600 rpm; with 4 poles the speed is 1 800 rpm; with 6 poles, 1 200 rpm, and so on.

Summarizing the behavior of three-phase squirrel-cage motors, the three sets of coils produce a magnetic field of constant strength that rotates at constant speed, producing good torque by inducing current in the rotor. They are mechanically simple, having no sliding contacts to produce maintenance problems.

Instead of a squirrel cage, some three-phase induction motors use windings on the rotor iron which connect to slip rings on the rotor shaft. The amount of induced current

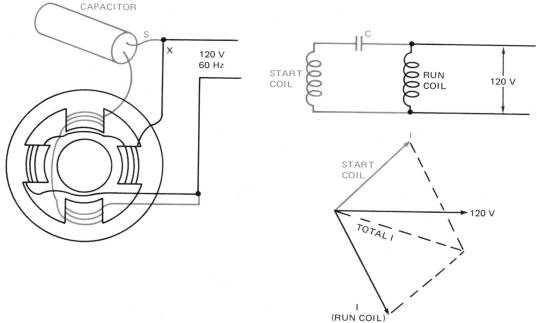

Fig. 16-15 Single-Phase Stator Circuit, Capacitor Starting

in the rotor windings is then controlled by external variable resistors which connect to brushes that rest on the slip rings. By this means, higher starting torque and some speed adjustment can be achieved. Wound-rotor motors are built in sizes from 1/2 hp (0.4 kw) to hundreds of hp (100 kw).

SINGLE-PHASE MOTORS

This discussion started with mention of induction motors used at home. Homes are not supplied with three-phase current. If three wires come into a house, it is three-wire, single-phase current (see Chapter 5).

If a squirrel-cage rotor is placed inside a stator that is supplied with single-phase ac, figure 16-13, plenty of current will be induced in the rotor conductors by transformer action, the rotor acting as a low-resistance secondary circuit. But that current will not of itself produce rotation. When the alternating field is increasing, the induced current in the rotor tries to oppose the increase of field, and builds up magnetic poles repelling the field poles. When the alternating field is decreasing, the

induced current opposes the decrease of field, producing poles that attract the field poles. Because the rotor poles are directly in line with the field poles, no turning force is developed. The rotor is merely squeezed and stretched many times per second. Compare this to the effect of a straight downward push on a bicycle pedal in the position shown in figure 16-14. If the foot is pushing exactly vertically toward the center, no torque is developed, but only a little nudge in one direction or the other is needed to start the rotation.

To give the rotor of the induction motor the little push to start it rotating, something must make the magnetic field itself turn, somewhat as it does in the three-phase motor. A common way of doing this is shown in figure 16-15. Another set of windings (in color) in series with a capacitor is added to the stator. The *run coil*, which is connected directly to the 120-volt line, has both resistance and inductive reactance. Its current lags the applied voltage (see vector diagram). The circuit shown in color has resistance, inductive

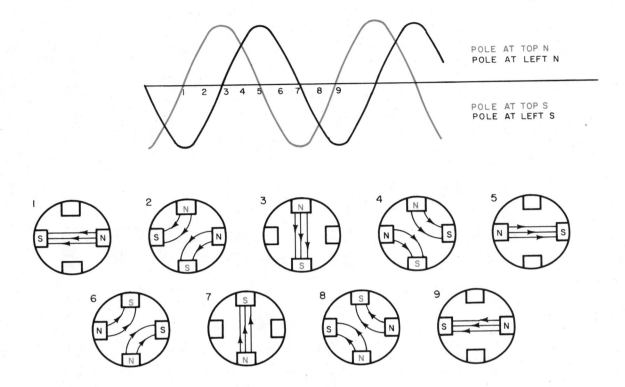

Fig. 16-16 Sine Wave of Single-Phase Motor Operation

reactance, and enough capacitive reactance to make the current in this coil-and-capacitor portion *lead* the applied voltage.

The two currents are shown in sine-wave form in figure 16-16. The contribution of each current to magnetic field inside the stator is shown for nine time-instants. Looking at the sequence of nine magnetic-field diagrams, notice the counterclockwise turning of the direction of the field lines due to the out-of-phase alternations of the magnetic poles. The rotating magnetic field drags the rotor around with it by the same action described for the three-phase induction motor.

CAPACITOR-INDUCTION MOTOR

The single-phase induction motor diagrammed in figure 16-15 is called a capacitor motor. This title implies that the capacitor circuit is used the entire time that the motor is running. Induction motors commonly used for household and small shop tasks are capacitor-start motors, in which the capacitor and starting coil are energized only during the few seconds while the motor is picking up speed. At about 3/4 of rated speed a switch (at point S in figure 16-15) opens and stays open while the motor is running. One type of switch mechanism is shown in figure 16-17.

The conducting portion of the switch is mounted on the stationary end-bell of the motor. Its own springiness tends to keep the contacts open. But when the rotor is stopped or running slowly, a pair of coil springs mounted on the rotor hold a sliding plastic collar against the switch, thus holding the contacts together. As rpm increases, a pair of small weights moves outward and slides the collar away from the switch, permitting the contacts to fly open.

It is usually easy to fix an electric motor when it does not work. When a motor is plugged in it sometimes hums but does not start turning. Pull the plug, so it will not

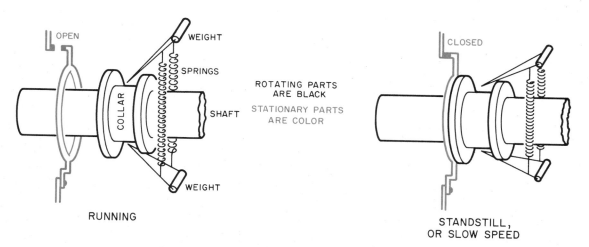

Fig. 16-17 Single-Phase Motor Starting Switch

overheat and smell. The hum is caused by vibration in the iron as its magnetism alternates, indicating that there is a current in the motor. Lack of rotation may have a mechanical cause. Turn the motor pulley by hand and see if it is free to turn. Perhaps it has not been oiled recently, so that it has square ball bearings or none at all. If the rotor turns freely by hand, plug it in again and spin the pulley by hand. The rotor may then gradually pick up speed. But its own failure to start indicates trouble in the starting-coil circuit, and the most likely spot for trouble is the starting switch. It is not difficult to remove the end bell of the motor and examine the starting-switch mechanism. Contacts may be burned or dirty and fail to close effectively. Dirt may interfere with free operation of the sliding collar.

Overload Protection

If a motor stops after being overworked for a while, do not assume it is broken. Many motors are provided with a heat-operated device that opens the motor circuit when the motor is overheated. After cooling, a push button can reset the thermal protection and close the circuit. Unwisely, a few motors may have a device that recloses the circuit automatically. Because of this, the power should always be turned off whenever a motor stops running while in use.

Running Torque

Earlier it was pointed out that an alternating magnetic field will not produce torque on a stationary rotor to start it moving. Once started, however, the alternating field does produce torque on the rotor. Unlike the smooth and steady torque on a rotor in a three-phase field, the torque on the single-phase operated rotor is pulsating. The lengthy details of how that torque arises are not discussed in this book. One approach to the explanation involves thinking about three items: (1) transformer emf generated in the rotor because the field is alternating; (2) speed emf generated in the rotor because it is moving through the field; and (3) inductance of the rotor causes rotor current to lag emf that produces it. These varying quantities, considered together, can account for the continued turning.

SPLIT-PHASE INDUCTION MOTOR

The type of motor called *split-phase* looks like the motor of figure 16-15 without the capacitor. As in the capacitor-start motor, the starting winding is of small diameter wire placed at the inner surface of the stator. Its ratio of resistance to reactance is higher than that of the running winding. The running winding is of larger diameter wire, and it has more inductive reactance because it is buried in deeper slots in the stator iron. Being more inductive and less resistive, the run-coil circuit has a current that lags the voltage by nearly 90°, while the starting coil current does not lag as much. As shown in the vector diagram of figure 16-18, start-coil current leads run-coil current, but the angular difference is less than in the capacitor-start motor. These two currents together produce a succession of twists on the magnetic field, similar to the sequence shown in figure 16-16. Once this jerkily rotating field has accelerated the rotor, the internal starting switch opens and the motor continues to operate on only the running winding. Comparing two motors of the same horsepower rating, the split phase motor is cheaper, requires a higher starting current, and develops less starting torque than the capacitor-start motor.

Shaded-Pole Motor

The shaded-pole motor uses a still cheaper way of obtaining a little field-rotation to start the rotor turning. This little motor is appro-

priate to run toys or small fans, or anything that starts easily and takes little power to run. A copper loop (shown in color, figure 16-19) is placed in a slot cut in the face of each field pole so that the loop surrounds half of the pole. During a part of the ac cycle when field-coil current is increasing, the increasing magnetic field induces a current in the single-turn *shading loop* which opposes the increase of field, that is, a current opposite in direction to the current in the field coil. So, while field current is increasing, relatively few magnetic lines form in the shaded-pole area. When field-coil current starts to decrease, this change induces a current in the shading loop that *opposes the decrease*. Shading loop current in the *same* direction as field current opposes decrease of magnetic field, producing field to add to the failing magnetic field of the main coil. The net result is that magnetic field sweeps across the pole face as shown by figure 16-19(A) and (B) during one half-cycle, producing a counterclockwise twist of field. During the next half-cycle, current and fields reverse Again a counterclockwise shift of field lines occurs as the field forms first through the unshaded portion of the pole face, then moves to the shaded-pole area.

Repulsion-Induction Motor

Some form of *repulsion motor* may be seen in use driving a machine that must start under heavy load, such as a high-pressure pump, elevator, or printing press. The basic repulsion motor uses an armature similar in appearance to a dc armature (Chapter 15) placed in a single-phase stator, figure 16-13. The alternating field induces alternating voltages in each armature coil, but it induces no current unless something is connected to the commutator. Connect a pair of brushes to the ends of a copper wire and place them against the commutator, figure 16-20. The amount and directions of ac permitted in the

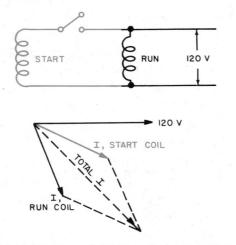

Fig. 16-18 Starting a Split-Phase Induction Motor

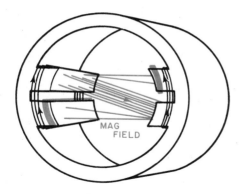 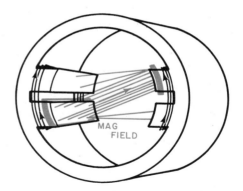

Fig. 16-19 Magnetic Fields in a Shaded-Pole Motor

armature will be determined by the position of the brushes. In the simple repulsion motor, the pair of brushes is fastened to a brush-holder assembly. The entire assembly is adjustable in that it can be rotated to any desired position, controlling torque and direction of rotation of the armature. In some respects this motor behaves like a dc series motor. It has high starting torque, and a two-pole repulsion motor at no load will run faster than 3 600 rpm on 60 hertz. Its speed changes when the load changes. Speed and torque at a constant load can be controlled by adjusting brush position. The *repulsion-induction* motor includes a squirrel cage on its rotor, along with the commutator-type winding of the repulsion motor. It has the desirable high torque along with more constant speed. In the *repulsion-start, induction run* motor, as the motor approaches rated speed, the brushes lift off the commutator. At the same time, a short-circuiting ring connects all the commutator bars together, so it runs as a squirrel-cage induction motor.

To keep the diagrams simple on the preceding pages, two-pole motors with a 3 600 rpm magnetic field have been assumed. Induction motors built for lower speeds have more field pole windings: 4 poles for speeds near 1 800 rpm; 6 poles for 1 200; 8 for 900 rpm,

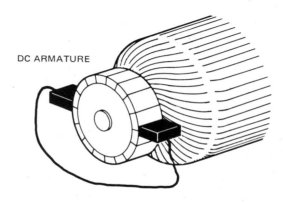

Fig. 16-20 DC-Type Armature Used in a Repulsion-Induction Motor

and so on. The count of poles means only *running* coils.

For example, 8 coils are readily visible in the end view of the stator of a 1 725 rpm capacitor-start motor, but it is properly called a four-pole stator, figure 16-21.

POINTS TO REMEMBER

- Three-phase ac lines provide three voltages 120° apart. Three-phase ac produces a rotating magnetic field in the stator windings of 3-ϕ motors.

- Induction motors have rotor current induced by transformer action and motion of rotor conductors through the magnetic field.

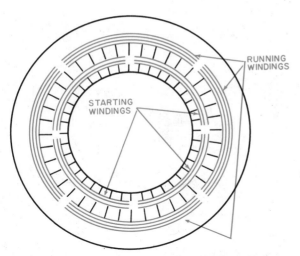

Fig. 16-21 End View of Stator, Capacitor-Start Motor

- Three-phase squirrel-cage induction motors run at fairly steady speed, have good starting torque, and high starting current.

- Three-phase wound-rotor induction motors can provide speed variation and high starting torque with lower starting current.

- Large synchronous motors have a dc field in a three-phase stator. Rpm is constant, determined by line frequency and number of stator poles.

- The universal motor is a small series motor that works on ac or dc.

- The simple split-phase motor has low starting torque and high starting current.

- The capacitor-start motor has higher starting torque and lower starting current.

- Repulsion-type motors have still higher starting torque.

- Shaded-pole motors are useful only for low power jobs.

SINGLE-PHASE MOTORS

Type	Usual hp range (watts)	Starting current	Starting torque	
Universal (ac series)	0.01 to 1 (7W to 746 W)	high	high	High speed, small size for motor power, speed varies with load.
Capacitor-start	1/8 to 20 (100 W to 15 kW)	medium	high	Nearly constant speed with varying load. Good general-purpose motor.
Split-phase	1/20 to 1/3 (40 W to 250 W)	high	low	Cheap, simple. Nearly constant speed.
Repulsion-induction	1/6 to 10 (100 W to 7.5 kW)	med. low	very high	Larger size for motor power. More maintenance. Used when high torque is required.

Fig. 16-22

REVIEW QUESTIONS

1. How does an induction motor differ from a series motor in construction and in operational characteristics?

2. Explain how the addition of a starting coil and a capacitor function in starting a single-phase induction motor with a squirrel-cage rotor.

3. What is the difference between a split phase and an induction motor with a capacitor attached? Give some advantages and disadvantages of each.

4. Describe the stator and the rotor in a shaded-pole motor. What is the purpose of the shaded pole? Give several uses for these motors.

5. How is the direction of rotation changed in a repulsion-induction motor?

6. What are the advantages of a repulsion-induction motor?

7. What is the fundamental difference between a repulsion-induction motor and a capacitor-start induction motor?

8. Draw a wiring diagram of a shunt motor and of a series motor.

9. Can a split-phase motor be started if the starting windings are burned out? What would happen if the power was turned on?

10. What is the purpose of the centrifugal switch in a split-phase motor?

11. What procedure is used to calculate the horsepower of a motor?

12. How is the speed of an induction motor related to its construction?

13. What type of motor would you suggest for each of these jobs? Give your reason for each choice.
    (1) Washing machine
    (2) Bench grinder
    (3) Pump for farm water system
    (4) Driving blower on warm air heating system

14. Three-phase ac could be obtained from an automobile alternator by connecting directly to the alternator windings, before rectification. What advantages and disadvantages do you suggest for using three-phase ac motors to operate ventilating fans and windshield wipers in the car?

15. In your own home or neighborhood, where would you look to find each of the types of motors described in this chapter?

16. For each of the devices listed, state in order what you would check if the motor would not run when plugged in with the switch turned on.
    1. Hand electric drill
    2. Vacuum cleaner
    3. Dryer

17. How would you go about changing wires in a capacitor-start motor so it would run in the reverse direction?

RESEARCH AND DEVELOPMENT

Experiments and Projects on Alternating-Current Motors

INTRODUCTION

Wherever alternating current is available throughout the world, the various kinds of ac motors provide convenient, reliable power. In Chapter 16 you have had an opportunity to learn important facts about the design, construction, operating principles, and use of the different types of ac motors. After studying these facts, you should be aware of the reasons why certain motors will function satisfactorily in a number of situations, while the use of other motors is limited to particular situations.

When you do the suggested experiments, you will discover the reasons why some motors operate effectively under one type of load and why they would not be satisfactory for another. Also, you will see why certain motors operate satisfactorily under certain starting and running loads. The suggested projects afford an opportunity to learn more about the several kinds of single-phase motors and how to connect and operate them. Also, you may want to design and build an experimental ac motor, or repair one that has been damaged.

EXPERIMENTS

1. To test the current consumption of a small series-wound motor running free and under load.

2. To test the current consumption of a fractional-hp (less than 0.5-kw), 115-230-volt ac induction motor running free and under load. Make the test with the motor operating on 115 volts and on 230 volts.

3. To test the starting torque of a 115-volt ac, 1/4-hp (approximately 0.2-kw) split-phase motor. Make the same test on a 1/4-hp (approximately 0.2-kw) capacitor motor.

4. To test the starting torque and rpm under load of a repulsion-induction motor.

5. To test the current consumption and operating characteristics of a synchronous motor.

PROJECTS

1. Disassemble a fractional-hp (less than 0.5-kw) ac induction motor and examine the parts. Clean and reassemble.

2. Disassemble an automotive starting motor and examine the parts. Clean and reassemble.

3. Connect and operate a fractional-hp (less than 0.5-kw) ac motor. Change the direction of rotation of the motor.

4. Connect and operate a shaded-pole motor. Change the direction of motor rotation.

5. Design and build a synchronous motor.

6. Locate and repair damaged starting coil on a split-phase induction motor.

7. Connect a universal motor with a rheostat in the circuit to control the speed.

EXPERIMENTS

Experiment 1 CURRENT CONSUMPTION, SERIES MOTOR

OBJECT

To find the current and power consumption of a small series-wound motor running free and under load.

APPARATUS

1 - Fractional-hp (less than 0.5-kw) series-wound motor such as an electric hand drill

1 - AC ammeter (0-10 A)

1 - Convenience outlet, wire, and plug

PROCEDURE

1. Connect ammeter in current supply as shown in figure 16-23.

2. Run drill free for a short interval and record the current consumption.

3. Drill a hole with a small drill and record ampere reading under load.

4. Drill with a large drill and record ampere reading under load.

OBSERVATIONS

Use a table similar to the one shown here to record your observations.

OPERATING CONDITION	AMPS
MOTOR RUNNING FREE	
MOTOR UNDER LIGHT LOAD	
MOTOR UNDER HEAVY LOAD	

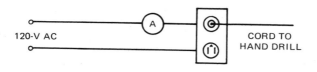

Fig. 16-23

QUESTIONS

1. Did the motor consume more current under a light load than when running free?

2. Was there a change in current when the load was increased?

3. How do you account for the current consumption under each operating condition?

Experiment 2 CURRENT CONSUMPTION, AC INDUCTION MOTOR

OBJECT

To find the current consumption of a single-phase fractional-hp (less than 0.5-kw), 115-230-volt ac induction motor running free and under load, operating at each voltage.

APPARATUS

1 - DC automobile generator
1 - Single-phase, capacitor-start, induction-run motor, 1/2-hp (375-watt), 115-230-volt
1 - Mounting board and fastening devices
1 - DPST switch
1 - AC ammeter (0-10 A)
1 - AC voltmeter (0-300 V)
1 - Wattmeter
2 - 4″ (100-mm) pulleys
1 - Belt

PROCEDURE

1. Mount motor and generator on mounting board, attach belt, and hand test operation.

2. Select 115-volt leads on motor. See information on dual voltage motors, pages 407-408. Mount meters in circuit as shown in figure 16-24.

3. Make connections and have your instructor check your work.

4. Run motor under load. Record amps when motor starts and after it has been running at full speed. Record voltage under same conditions.

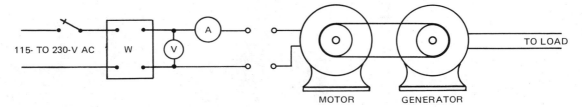

Fig. 16-24

5. Remove belt, start motor, and take some readings as suggested above.

6. Detach the wires from 115-volt line, select 230-volt leads on motor, and make new connections according to diagram. Have your work approved and connect motor to 230-volt source.

7. Run the motor and record the amps and volts at start and at full speed.

8. Stop motor, replace belt, and hand test.

9. Start motor and record data as in Step 7.

OBSERVATIONS

Use a table similar to the one shown here to record your observations.

115 VOLTS	AMPS—VOLTS	230 VOLTS	AMPS—VOLTS
START FREE		START FREE	
RUN FREE		RUN FREE	
START UNDER LOAD		START UNDER LOAD	
RUN UNDER LOAD		RUN UNDER LOAD	

QUESTIONS

1. Did your observations agree with the data on the nameplate?

2. Which method of operating the motor is the most economical: the 115-volt or the 230-volt system?

3. Calculate the cost of operating the motor for 10 hours on 115 volts and on 230 volts.

Experiment 3 MOTOR POWER AND TORQUE

OBJECT

To find actual horsepower of a motor operating under load, and the starting and running torque.

APPARATUS

1 - 1/4-hp (approximately 0.2-kw) split-phase motor
1 - V-belt pulley, 2" to 4" (50 mm to 100 mm)
1 - Pc. appropriate sash cord
2 - Spring scales (0-24 lb.)
1 - SPST switch

PROCEDURE

1. Connect apparatus as shown in figure 16-25.

2. Start motor and then change the tension on spring scales while you watch the ammeter. Notice the current depends on spring-scale forces. Finally, adjust the tension so that the ammeter reads about the same as the rated current stated on the motor nameplate.

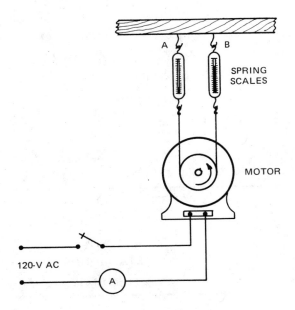

Fig. 16-25

3. Record reading on ammeter when motor is under load.

4. Stop the motor. Start it again and record ampere reading as it starts under load.

5. Find the horsepower from data recorded while motor was operating under load.

    Suggested Procedure for Finding Horsepower

    Use difference in readings on spring scale. For example, Scale A reads 10 lb., and Scale B reads 2 lb. Therefore, 10 – 2 = 8-lb. drag on pulley. Work done per minute equals 8-lb. force x distance traveled by pulley slot pulling on belt:

    Distance per minute   = rpm x π x diameter of pulley
    Example: 3″ diameter pulley = 1/4 ft.
    Distance per minute   = 1 800 rpm x 3.14 x 1/4
                                    = 1 413 ft.

    Work done = 1 413 ft. x 8 lb. = 11 300 ft.-lb. per minute
    1 hp = 33 000 ft.-lb. per minute 11 300 ÷ 33 000 = about 1/3 hp

6. Compare horsepower rating of motor and actual horsepower under load.

7. Compare ampere rating on motor and ampere reading on ammeter when motor was starting under load.

8. Make the same ampere comparison with motor running under load.

9. Calculate the starting torque and running torque of the motor. Refer to Chapter 13.

OBSERVATIONS

Use a table similar to the one shown here to record your observations.

LB. DRAG	SCALE	MOTOR RATING	STARTING	UNDER LOAD
	A			
	B			
AMPERE READING				

QUESTIONS

1. How did the horsepower with motor under load compare to the motor rating?
2. Did the amperes recorded on the ammeter when motor was under load agree with the motor rating? If it varied, what was the variation?
3. Did the ampere reading when the motor was running free differ from the ampere rating on the motor nameplate? If so, how do you account for the variation?
4. What was the starting torque of the motor? The running torque?

PROJECTS

CONNECT AND OPERATE A FRACTIONAL-HORSEPOWER (Less Than 0.5-kw) AC MOTOR

Connecting a fractional-horsepower (less than 0.5-kw) ac motor to the proper current supply is an interesting and comparatively easy task if one knows how to proceed. The following diagrams of split-phase and capacitor-type motors will aid you in understanding the arrangement of the basic connections and also how to change the direction of rotation.

> When wiring a circuit or connecting an appliance, use appropriate wire, large enough to carry the current.

Split-Phase Induction Motor

The running winding of a split-phase motor is of a relatively heavy insulated copper wire and is placed at the bottom of the stator slots. The starting winding is of a relatively small-size wire and is placed near the top of the slots above the running winding coils.

Both windings are connected in parallel to the single-phase line when the motor is started. After the motor has accelerated to approximately 2/3 to 3/4 rated speed, the starting winding is automatically disconnected from the line by means of a centrifugal switch. Split-phase motors require a large starting current and have a low starting torque.

The direction of rotation may be changed by reversing the connections of either the starting winding or the running winding. *Note:* Some types of single-phase motors are nonreversible because the connections to their windings are inaccessible.

Figure 16-26 shows the coil windings and switch on a split-phase induction motor, sometimes called a "resistance-start," induction-run motor.

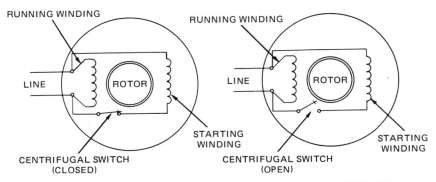

CONNECTIONS OF THE CENTRIFUGAL SWITCH AT START AND RUN

Fig. 16-26

The Capacitor-Start, Induction-Run Motor

The construction of the capacitor-start, induction-run motor is practically the same as for a resistance-start, induction-run motor, except that a capacitor is connected in series with the starting windings. This capacitor is usually mounted in a metal casing on top of the motor, although it may be mounted in any convenient external position on the motor frame. In some cases, the capacitor is mounted inside the motor housing. The capacitor provides higher starting torque than is possible with the resistance-start, induction-run motor. Furthermore, the capacitor helps to limit the starting surge of current to a lower value than is found with the straight split-phase motor.

To reverse the direction of rotation of a capacitor-start, induction-run motor, the leads of the starting-winding circuit are reversed. As a result, the magnetic field developed by the stator windings rotates around the stator core in the opposite direction, and the rotor's rotation is reversed. The direction of rotation may also be reversed by interchanging the two running winding leads, figure 16-27.

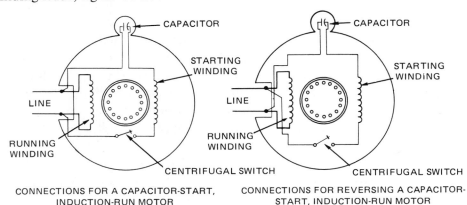

CONNECTIONS FOR A CAPACITOR-START, INDUCTION-RUN MOTOR   CONNECTIONS FOR REVERSING A CAPACITOR-START, INDUCTION-RUN MOTOR

Fig. 16-27

The Capacitor-Start, Capacitor-Run Motor

The capacitor-start, capacitor-run motor is similar to the capacitor-start, induction-run motor, except that the starting winding and capacitor are connected in the circuit at all times. This motor has very good starting torque. In addition, since a capacitor is used in the motor at all times, the power factor at rated load is practically 100 percent, or unity.

Several different designs for this type of motor are all basically the same in their operation. One kind of capacitor-start, capacitor-run motor has two stator windings which are placed 90 electrical degrees apart. The main or running winding is connected directly across rated line voltage, figure 16-28. A capacitor is connected in series with the starting winding. This second winding in series with the capacitor is also connected across rated line voltage. There is no centrifugal switch, since the starting winding is energized at all times when the motor is in operation.

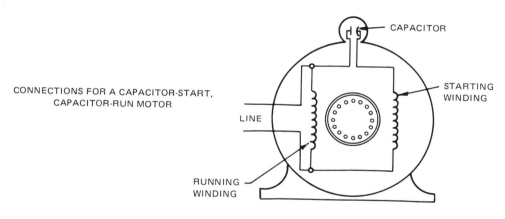

CONNECTIONS FOR A CAPACITOR-START, CAPACITOR-RUN MOTOR

CAPACITOR

STARTING WINDING

LINE

RUNNING WINDING

Fig. 16-28

Dual-Voltage Motors

Single-phase motors often have dual-voltage ratings of 115 volts and 230 volts. The running winding consists of two sections, each rated at 115 volts. One section of the running winding will be marked T_1 and T_2, while the other section will be marked T_3 and T_4. If the motor is to be operated on 230 volts, the two 115-volt windings are connected in series across the 230-volt line. If the motor is to be operated on 115 volts, the two 115-volt windings are connected in parallel across the 115-volt line. The starting winding, however, will consist of only one 115-volt winding. The leads of the starting winding are marked T_5 and T_6. If the motor is to be operated on 115 volts, both sections of the running winding are connected in parallel with the starting winding.

Figure 16-29(A) illustrates the circuit connections for a dual-voltage motor connected for 115 volts. For a 230-volt operation, the connection jumpers are changed in the terminal box so that the two 115-volt running

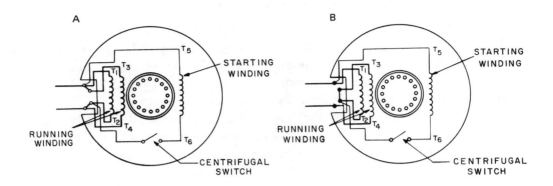

Fig. 16-29

windings are connected in series, figure 16-29 (B). The two sections of the running winding are connected in series, across the 230-volt line. It will be noted that the 115-volt starting winding is connected in parallel with one section of the running winding. If the voltage drop across this section of the running winding is 115 volts, the voltage across the starting winding will also be 115 volts.

MATERIALS

1 - Split-phase motor, 1/4-hp (approximately 0.2-kw), 115-volt ac

or

1 - Capacitor-start, capacitor-run motor, 1/4- or 1/3-hp (0.2-kw to 0.25-kw), 115-volt ac

1 - Starting switch

or

1 - Capacitor-start, induction-run motor, 1/4- or 1/3-hp (0.2-kw to 0.25-kw), 115-volt ac

1 - Dual-voltage motor, 1/4- to 1-hp (0.2-kw to 0.8-kw), 115- and 230-volt ac Wire and other accessories

PROCEDURE

Select a motor, appropriate switch, wire, and accessories.

Read the information on the nameplate of the motor.

Remove the cover plate of the terminal box and examine the terminal block. Note the number of terminals and how each is marked. Compare the information with that on the nameplate. Next, study the schematic diagram for the type of motor selected.

Make the connections according to the information on the motor, or from a schematic diagram.

Have your instructor check the connections. After they have been approved, attach the motor to the current supply and run it.

After the motor has run for awhile, open the switch and, as it coasts to a stop, notice the direction of rotation. Record the direction of rotation in terms of clockwise or counterclockwise.

Disconnect the motor from the current supply.

Make a schematic diagram for the connections to reverse the rotation. Make the connections according to your diagram and have them approved by your instructor. Run the motor again and check the rotation.

Evaluate.

DESIGN AND BUILD A SYNCHRONOUS MOTOR

This simple synchronous motor, figure 16-30, is comparatively easy to build.

It is designed to run on alternating current from a low-voltage transformer.

It is not self-starting and has very little torque.

Its principal use is to illustrate the constant speed characteristic of a synchronous motor.

The rpm of the motor is determined by the number of poles on the rotor.

MATERIALS

1 - Machine screw, #8-32 x 2″ (M4 x 50 mm L)
2 - Machine screw nuts, #8-32 (M4)
1 - Machine bolt, 3/8″ x 2 1/2″ (M10 x 65 mm L)
1 - Machine bolt nut, 3/8″ (M10)
2 - Fiber discs, 1/16″ x 1″ (1.5 mm x 25 mm)
2 - Pieces soft steel, 1/8″ x 1/2″ x 2 1/2″ (3 mm x 12 mm x 64 mm)
1 - Piece soft steel, 1/8″ x 1″ x 2 1/4″ (3 mm x 25 mm x 57 mm)
2 - Fahnestock clips
3 - Pieces sheet iron, #18, #20, or #22 gauge, 3/8″ x 3″ (9 mm x 76 mm), or 1 piece 3″ x 3″ (76 mm x 76 mm)
1 - Piece wood, 3/4″ x 2″ x 5 1/4″ (20 mm x 50 mm x 133 mm)
2 - Wood screws, #6 x 1″ RH
2 - Wood screws, #5 x 3/4″ RH
2 - Wood screws, #4 x 1/2″ RH
   Magnet wire, #24 gauge, AWG Formvar

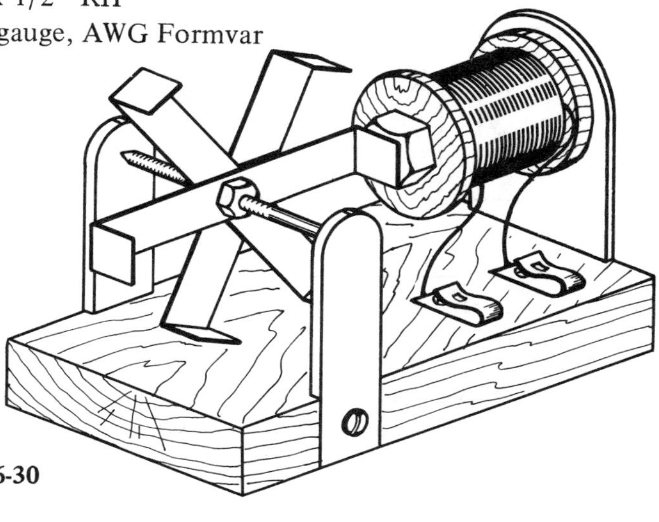

Fig. 16-30

PROCEDURE

Study the relationship of the various parts in figure 16-31 and make each according to the dimensions given in the materials list.

Plan the coil to be 2" (50 mm) long, with 1/2" (12 mm) of the bolt extending beyond the disc. For suggestions on planning and winding a coil, refer to Chapter 6.

Notice that the ends of the rotor shaft are turned or ground to a 60° point, and that the bearings are center punch marks in the steel brackets.

The rotor may be made of three pieces of sheet metal or laid out and cut from one piece.

The shaft may be made from a long #8-32 (M4) machine screw, or from a piece of round stock. The shaft must be the correct length so it will rotate freely.

Assemble the base and upright and attach the coil.

To accurately position the rotor bearings, check the length of the finished rotor blades and lay out the distance from center of the bearings to the head of the bolt according to your calculations. The clearance between the head of the bolt and the rotor should be 1/32" (1 mm).

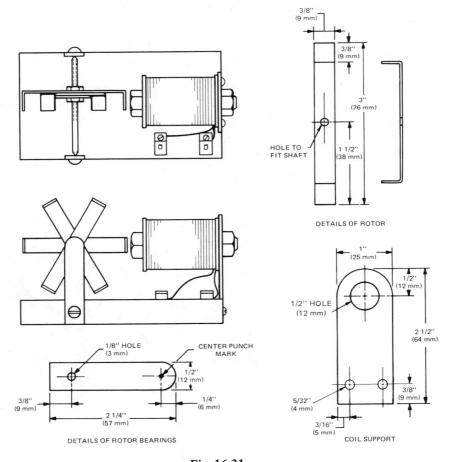

Fig. 16-31

Remember, the coil can be moved toward the rotor poles, but not away from them.

To make an adjustment, the rotor can be raised or lowered.

Assemble the parts, make necessary adjustments, and test.

The rotor poles must be evenly spaced, and the ends should line up with the center of the coil.

Connect the motor to an ac transformer of approximately 6 volts.

Spin the rotor to start the motor running. Several trials may be required as it must rotate at the correct speed before it will continue to run.

You can determine the correct speed by using the following formula:

$$rpm = \frac{Frequency \times 120}{Numbers\ of\ poles}$$

Note: The speed of a synchronous motor may be changed by changing the number of poles.

Evaluate.

ALTERNATE DESIGN FOR SYNCHRONOUS MOTOR

The design of this motor, figure 16-32, will permit the use of the solenoid coil described in Chapter 6.

Since the metal parts are easily fabricated, innovations of the basic motor can be readily made.

MATERIALS

1 - Pc. wood, 3/4" x 3 1/2" x 6" (20 mm x 90 mm x 150 mm)
1 - Pc. round metal, 3/16" x 2 3/4" (5 mm x 70 mm)

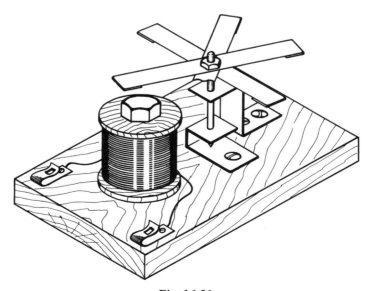

Fig. 16-32

2 - Machine nuts, #10 x 24 (M5)

3 - Pcs. sheet iron, #18, #20 or #22 gauge, 3/8" x 4" (9 mm x 100 mm)

1 - Pc. sheet iron, #18, #20 or #22 gauge, 5/8" x 3 3/4" (16 mm x 95 mm)

1 - Pc. sheet iron, #18, #20 or #22 gauge, 5/8" x 2 1/4" (16 mm x 57 mm)

1 - Machine bolt, 3/8" x 3" (M10 x 75 mm)

1 - Machine bolt nut, 3/8" (M10)

2 - Fahnestock clips

5 - Wood screws, #4 x 1/2" RH

1 - Solenoid coil

PROCEDURE

Study the drawing in figure 16-33 and make the parts according to the specifications on them and in the list of materials.

Notice that the rotor shaft has a point on one end and that it is threaded about an inch on the other end.

After completing all the parts, mount the coil on the base. Place the bearings so the 3/8" (9 mm) fold on the ends of the rotor poles are centered over the bolt head. Adjust the height of the rotor so the poles swing about 1/32" (1 mm) above the bolt head.

1. Check to see that they are all evenly spaced.

2. Attach the coil to a low-voltage ac transformer of about 6 volts.

3. Spin the rotor to start the motor running.

Several trials may be required as it must rotate at the correct speed before it will continue to run.

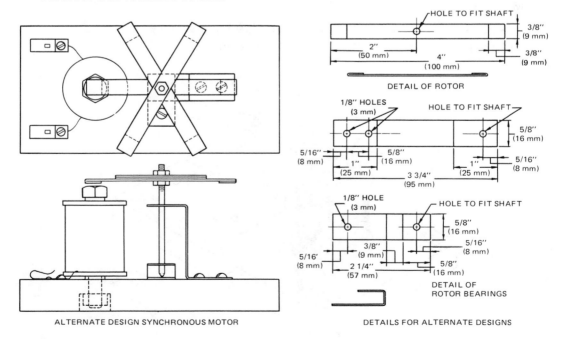

ALTERNATE DESIGN SYNCHRONOUS MOTOR   DETAILS FOR ALTERNATE DESIGNS

Fig. 16-33

Determine the speed by using the following formula:

$$\text{rpm} = \frac{\text{Frequency x 120}}{\text{Number of poles}}$$

Note: The speed of a synchronous motor may be changed by changing the number of poles.

Evaluate.

Chapter 17
Rectifiers

SEMICONDUCTORS

Chapter 1 mentioned materials called semiconductors. These materials are halfway between good insulators and good conductors. The elements silicon and germanium, and some metallic compounds, are classified this way. Silicon and germanium are elements whose atoms contain 4 electrons in the outermost electron shell. At very low temperatures these electrons are tightly held in place, and the pure element is an insulator. If the temperature is raised, more and more electrons become energetic enough to become free-moving, like the electrons in a metal. So, unlike metals, a rise in temperature decreases the resistance of a semiconductor.

N-Type Semiconductors

The most interesting and useful semiconductors are those with very small amounts of added impurities. Suppose a small amount (1 part in 100 million) of an element whose atoms contain 5 electrons in the outermost shell, such as arsenic (As), is added to silicon (Si) or germanium (Ge). The impurity atoms fit into the regular crystal pattern of the germanium atoms, figure 17-1, with four of the five outer

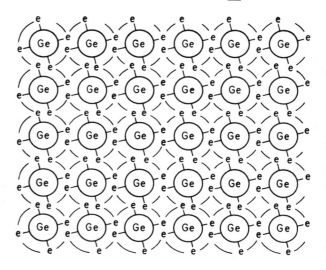

(Ge) includes the atom's nucleus (+32) and 28 of its electrons, which are not of interest to us. The four outermost electrons of each atom, sometimes called valence-bond electrons, are of interest. Notice above that each Ge is surrounded by a shell of 8 electrons, four of its own plus one electron from each of four neighboring atoms. These interlocking shells of 8 form the stable, low-energy electron pattern which we use in describing the behavior of rectifiers, transistors, and many semiconductor applications.

Fig. 17-1 Pattern of Electron Arrangement in Pure Germanium at 0° K

electrons of the impure atom fitting into the tightly held electron pattern set by the germanium atoms. The fifth electron does not fit into the pattern set by the 4 electrons of each germanium atom. This fifth electron is free to wander around loosely, like an electron in a metal. Such a mixture of a semiconductor with a small amount of 5-electron impurity is called an *N-type* semiconductor, figure 17-2. The N refers to the negative free electrons. The mixture itself is electrically neutral.

P-Type Semiconductors

If we add to a pure semiconductor a small amount (1 part in 100 million, again) of an element such as boron or gallium or indium, whose atoms have three electrons in their outermost ring, the result is an electron pattern in the solid similar to figure 17-3. The 3 electrons of boron (B) fit into the 4-electron pattern set by germanium or silicon, but the pattern is incomplete. This means there is a hole in the electron pattern because the boron atom did not have enough electrons to fill the 4-electron pattern. Although this type of semiconductor mix, called P-type, has no free electrons like a metal or an N-type semiconductor, it still is a better conductor than pure germanium

or silicon. Figure 17-4 is a simplification of figure 17-3, showing only electrons and holes in the P-type semiconductor. The "e's" represent electrons and the dotted circles are holes. Suppose a small potential difference is applied to this material, tending to drive electrons to the right. Although these tightly held electrons cannot wander around freely like electrons in a metal, they can slip over to occupy a hole in the electron pattern of a nearby atom.

Figure 17-4(B) shows the result of such electron movement. Compare this with figure 17-4(A). In the second row, an electron has slid over to occupy a hole, leaving a hole in the space it formerly occupied.

In the fourth row, an electron from the energy source has occupied the hole at the left. Two other electrons have moved a space to the right, leaving holes behind, and the electron at the right end of row 4 has left the crystal, being attracted to the + of the energy source.

Junction of P-and-N Semiconductors

Observe what happens when a piece of N-type material is put in contact with a piece of P-type material, figure 17-5. Some of the energetic free-moving electrons of the N-type scatter over into the P-type, but as

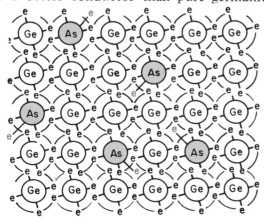

Fig. 17-2 Pattern of Electron Arrangement in N-Type Semiconductor

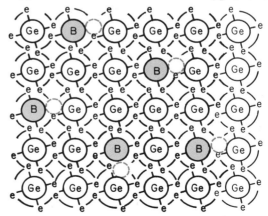

Fig. 17-3 Pattern of Electron Arrangement in P-Type Semiconductor

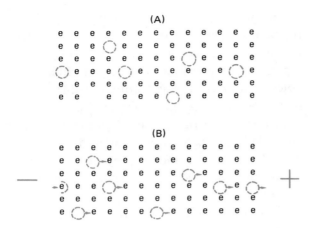

(A)

(B)

Fig. 17-4 Simplified Pattern of Electron Arrangement in P-Type Semiconductors

soon as they do, they must take a seat, that is, occupy one of the holes in the electron pattern in the P-type material. When an electron occupies a hole, it loses energy and becomes like any other tightly held electron in the semiconductor.

One result of this electron movement is that the N-material becomes positively charged by electron loss at the junction and the P-material becomes negatively charged.

Another important result is the development of a high-resistance layer at the junction. The removal of free electrons from the N-material is the removal of its conducting ability. The filling of vacant spaces in the P-type material destroys its conducting ability too, because electrons in the P-material must have vacant spaces available to move into, if they are going to move. Far from the junction, the P-material still has holes, and the N-material still has free electrons.

P-N JUNCTION AS RECTIFIER

If this P-N combination is connected to an energy source, with polarity as shown in figure 17-6, there is no current. The attempt to push more electrons into the already negative-charged P-type material may merely fill a few more holes in the P-type material. The attempt

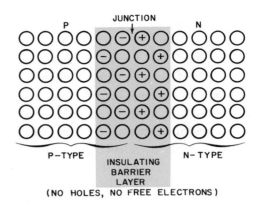

○ = NEUTRAL ATOMS

⊖ = NEGATIVE–CHARGED P-TYPE IMPURITY ATOMS, THE –ELECTRON NOW FILLING A HOLE

⊕ = POSITIVE–CHARGED N-TYPE IMPURITY ATOMS, HAVING LOST A FREE ELECTRON

Fig. 17-5   Pattern of Electron Arrangement in a P-N Junction

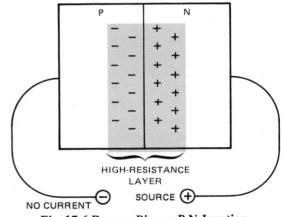

Fig. 17-6 Reverse Bias on P-N Junction

to remove more electrons from the already positive-charged N-material would be a removal of conducting ability; so the applied potential difference widens and strengthens the insulating barrier.

Looking at figure 17-6, the minus signs on the P-side of the junction represent electrons trapped in holes in the P-material. Holes exist in the blank area at the left. The plus signs on the N-side of the junction represent positive-charged atoms that have lost their free-moving electrons. In the blank area at the right, there are free electrons.   The high-resistance layer has an electron structure much like that of figure 17-1, with no holes and no free electrons.

The electrons in the P-region do not return to the N-region where they came from when the junction was formed because conduction electrons in the N-region possess a relatively high level of energy in order to be free of their parent atom. Electrons in the P-type material have relatively low energy levels. They do not have energy enough to float freely through the crystal. Therefore, it is easy for electrons in the N-region to lose energy and slide into the P-region. However, electrons in the P-region must gain energy from the outside in order to move into the N-region. Scientists call this energy difference a *potential hill*. Cars can coast down a hill, but it takes energy to go back up. If enough energy is supplied (by high voltage, for example), the junction will conduct electrons from P-material to N-material just as someone strong enough can push a vehicle uphill. Except in the case of zener diodes, which will be discussed later, this direction of conduction is undesirable.

If the P-N junction is connected to a source with polarity as shown in figure 17-7, the attempt to push electrons into the N-material is aided by the positive charge at the N-side of the junction. The attempt to remove electrons from the P-material is aided by the negative charge of the P-material. Remember that electrons spilled over from N-type to P-type of their own accord when the junction was made. The applied potential now helps them go in the direction that they tend to go of themselves. However, in order for electrons to cross the junction, they must tear themselves away from the positively charged N-region and overcome the repulsion of the negatively charged P-region. For germanium, this requires about 0.2 volts; for silicon, 0.6 volts. These two threshold voltages are very useful in distinguishing germanium from silicon semiconductors. The technician simply con-

nects the diode in a forward-conducting circuit and measures the voltage drop across the junction as in figure 17-7. The series resistor protects the diode from excessive current.

In summary, the P-N junction conducts readily from N to P. However, an attempt to make electrons move from P to N develops a barrier to stop such motion. The connection of a P-N junction to a voltage source with polarity as in figure 17-7, is called *forward bias*. The nonconducting situation of figure 17-6 is called *reverse bias*. Placing diodes in series increases their ability to withstand reverse voltages.

This one-way conduction ability of some combinations of materials had been discovered, but not understood, many years ago. The oldest widely used dry-disc rectifier is the copper-oxide type. Electrons flow readily from electron-rich copper metal into a cuprous-oxide (P-type) layer formed on the copper surface. Contact with the oxide is made by a soft lead disc pressed against the oxidized surface. Electrons flow from the oxide into the copper only with great difficulty. Selenium rectifiers (a series of rectifying cells) were used for many years in 120-volt ac radios. However,

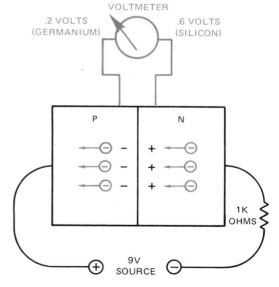

Fig. 17-7 Forward Bias on P-N Junction With "Threshold" Voltage Indicated

such rectifiers are being replaced with silicon ones as they malfunction. The selenium rectifier is much bulkier, produces more unwanted heat, and drops a much higher forward voltage when conducting than does its modern silicon replacement.

The word *diode*, meaning two-electrode, is used as a general name for any of the great variety of rectifying devices, whether they are germanium P-N junctions, selenium junctions, vacuum tubes, etc.

Like germanium, silicon is an element whose atoms contain four electrons in the outermost shell. Pure silicon forms an electron pattern like that of germanium, and can be "doped" by addition of impurity atoms to form N-type or P-type silicon. Silicon junction diodes are widely used for high current (up to 500-amp) rectifiers, especially in locations where other diodes would be damaged by high temperature, figure 17-8. Silicon diodes have very low resistance forward-biased, so the power loss ($I^2 R$) in the diode is reasonably low.

RECTIFIER CIRCUITS

A great variety of needs exists for dc of various voltages. Radio receivers and transmitters, television and other communication equipment, and amplifiers require a dc supply for operation of the tubes or transistors they contain. If they are to be operated from the ac power lines, rectifiers are required. Because it is easier to change the voltage of ac than it is to change dc voltage, the first step in the circuit may be a transformer if the required dc is far from 120 volts.

In figure 17-9, the transformer might be a step-down from 120 to 15 volts for a battery charger, for example. In this circuit, the dc is not as smoothly continuous and steady as the current a battery can produce. Because the rectifier conducts only on alternate half-cycles, the current is pulsing. A graph of current against time for this rectifier is shown in

figure 17-10. For battery charging, this pulse current is satisfactory. For some other uses, such as energizing a transistor amplifier, it cannot be used unless the pulses are smoothed out into a steady current. A capacitor of sufficient size across the line that supplied the load is the first step, figure 17-11. But some 60-hertz pulsing will still appear at the load, because the capacitor discharges and recharges 60 times per second. The full-wave rectifier of figure 17-12 recharges the capacitors 120 times per second. The use of two capacitors and a resistor or inductor in the filter section improves the smoothing out of voltage so well that this arrangement is widely used.

THE VACUUM DIODE

One hundred years ago, electricity was a great mystery. Between 1850 and 1900, a

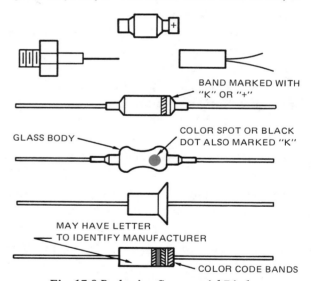

BAND MARKED WITH "K" OR "+"

GLASS BODY

COLOR SPOT OR BLACK DOT ALSO MARKED "K"

MAY HAVE LETTER TO IDENTIFY MANUFACTURER

COLOR CODE BANDS

Fig. 17-8 Packaging Commercial Diodes

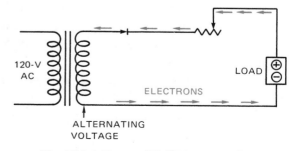

120-V AC

LOAD

ELECTRONS

ALTERNATING VOLTAGE

Fig. 17-9 A Form of Half-Wave Rectifier

great many experimenters investigated the conduction of electricity through the space inside glass tubes from which most of the air had been pumped. They saw the effects of streams of electrons, and measured those effects. They called those streams *cathode rays*, not yet knowing that they consisted of fast-moving individual particles or electrons. By 1897, J.J. Thompson, an English physicist, had built a tube basically similar to a TV picture tube and measured the weight per charge ratio of the particles now called electrons. Thompson's results caused him to be credited with the discovery of electrons.

By 1900, J.A. Fleming, who was also British, had developed a device called the *Fleming valve*, figure 17-13. He developed it by following up on Edison's unpursued

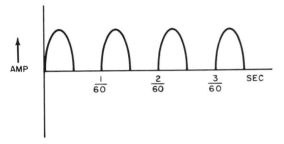

Fig. 17-10 Graph of Current of Half-Wave Rectification

discovery in 1883 of a mysterious current from the filament of the lamp to a third wire sealed into the lamp. Fleming's valve is called a *vacuum diode*. A diode is a two-electrode tube. Any white-hot object — a match flame, a hot wire, or the sun — has electrons that are so energetic that many fly entirely away from the atoms that once held them. This process is called *thermionic emission*, meaning the giving off of charged particles due to heating. In an ordinary electric lamp, the hot filament is surrounded by an invisible swarm of electrons. If a positive-charged metal plate is placed inside the lamp bulb, these electrons are attracted to it continuously as more electrons come boiling out of the hot wire. We now have a diode. The filament is one electrode and the plate is the other. Vacuum diodes will work for the circuits of figures 17-11 and 17-12. The filament is usually heated by ac from another secondary winding on the transformer. Many older TV receivers were energized by a full-wave rectifier tube using two plates and a single cathode filament heated by a separate 5-volt winding in a circuit similar to figure 17-12. A half-wave vacuum tube rectifier with separately heated cathode is shown in detail in figure 17-14.

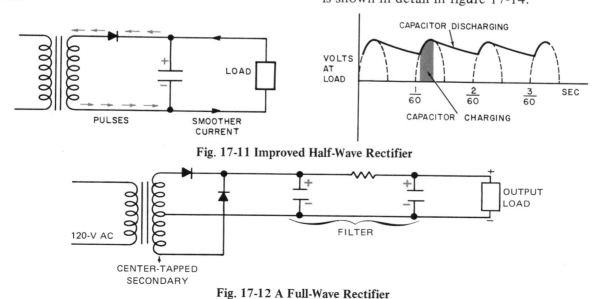

Fig. 17-11 Improved Half-Wave Rectifier

Fig. 17-12 A Full-Wave Rectifier

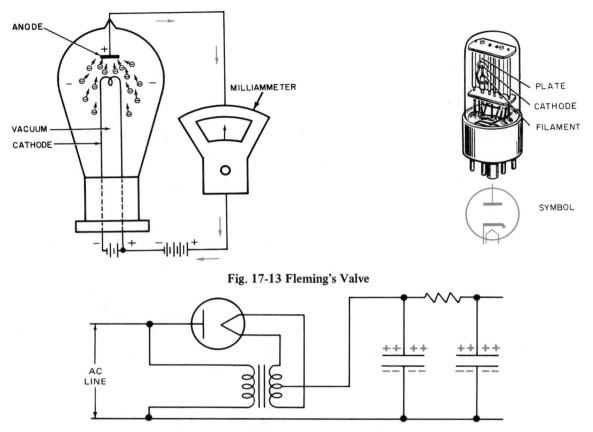

Fig. 17-13 Fleming's Valve

Fig. 17-14 Vacuum Tube Rectifier With Separately Heated Cathode

Reverse Voltage Applied to a Diode

Vacuum diodes withstand several hundred volts without any reverse conduction from the anode plate to the cathode; some of them are built for several thousand volts. During the nonconducting half-cycle of a rectifier circuit, the diode is subjected to the reverse voltage of the line plus the voltage on the filter capacitor. Therefore, the diode must be able to block a reverse voltage of about twice the peak of the forward voltage. Vacuum tube diodes are used in a great deal of equipment where high-voltage rectification is needed. For example, 20 000 volts are applied to a TV picture tube through a vacuum tube diode.

Many semiconductor diodes have lower reverse voltage ratings than tubes. The reverse voltage rating of the diode must be observed to

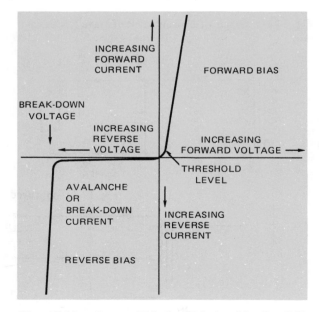

Fig. 17-15 Current-Voltage Relationship for P-N Junctions

avoid ruining it by overheating. Figure 17-15 is a graph of current against voltage for a semiconductor diode. It is not scaled numerically because amounts vary widely for different diodes, although the curves are all of this form. The voltage in the graph is the voltage drop across the diode itself, not the entire circuit voltage. The upper right portion shows that current increases rapidly as forward voltage is increased above the threshold level. The lower left portion shows that with reverse voltage there is a tiny leakage current. This current is due to a fact not mentioned before. Heat energy in a semiconductor at normal temperatures permits a very small number of electrons to jump out of their place in the low-energy pattern. This results in P-material having a few free electrons and N-material having a few holes. This effect is of such slight relative importance that it has not been mentioned before. For this slight minority of electrons loose in the P-type side, reverse bias on the diode starts them moving. Their motion is the leakage current. If enough voltage is applied, they begin moving fast enough so that their collisions with low-energy electrons in the crystal knock some electrons loose. The newly loosened free electrons accelerate and jar others loose in a chain reaction called an *avalanche of electrons*. This is similar to the process of sudden ionization in a gas in Chapter 8. This avalanche current ruins most diodes, because the heat developed from volts x amps upsets the uniform atomic arrangement in the crystal.

ZENER DIODES

One type of silicon diode, however, is built to take the avalanche and serve a purpose as it does so. These zener diodes are available with break-down voltage ratings of from about 2 volts to 300 volts. For example, suppose there is some spot in an electronic circuit where a 5-volt potential difference is wanted. Nothing over 5 volts is wanted, but some changes in circuit behavior might make it rise higher. A zener diode with a 5-volt break-down rating is connected across the points where not over 5 volts is to be maintained. If the voltage rises above 5, the diode suddenly conducts enough current so that the 5-volt drop across the diode is maintained.

VOLTAGE REGULATOR

The zener diode not only protects a device from sudden high voltages, but it can also keep the voltage from falling when a power-using device draws more current. Figure 17-16 is an example of a zener diode voltage regulator which provides a constant 9 volts to a transistor radio even if the voltage of the automobile system fluctuates between 10 and 15 volts, or a different brand of radio draws 175 milliamperes instead of 150 milliamperes. Analyze how this happens.

Assume that the diode can carry 300 milliamperes without producing excessive heat. Since for normal operation 12 volts are supplied and 9 volts are dropped across the load with the zener in parallel, 3 volts

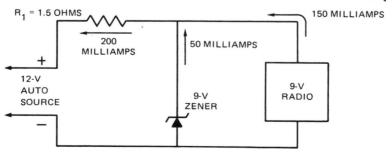

Fig. 17-16 Zener Diode Voltage Regulator

must be left to appear across R_1. The current in R_1 must be 200 milliamperes (3 volts ÷ 15 ohms). If the radio draws 150 milliamperes, there must be 50 milliamperes in the diode.

Notice that the 9-volt zener diode is operated with reverse-biased break-down voltage. Figure 17-15 shows that the voltage drop across a diode in this condition changes very little for wide changes in diode reverse current. The ideal characteristic current-voltage line of a zener diode is vertical under these conditions. If the auto system voltage climbs to 15 volts and the zener maintains its controlled 9-volt drop, then there must be a 6-volt drop on R_1 (15 – 9 = 6) and 400 milliamperes through it. With 150 milliamperes in the load, the diode current must now be 250 milliamperes because the current in R_1 must equal the current in the diode plus the current in the load. Since the diode can handle 300 milliamperes without destruction, it has protected the radio from a rising voltage source.

If auto voltage falls to 10 volts, there is only a 1-volt drop on R_1, no current in the diode, and still 150 milliamperes at 9 volts on the radio.

If a different radio is connected to the system, which requires a different current, the diode conducts either more or less to maintain the constant 9-volt output over a wide load range.

PHOTOSENSITIVE SEMICONDUCTOR APPLICATIONS

When light strikes a semiconductor, some of the energy of the light may be given to low-energy electrons in the solid, jarring them loose from the electron pattern so that they are temporarily free moving. This increase in number of free electrons, and incidental availability of holes they leave behind, greatly improves the conductivity of the semiconductor. Cadmium sulfide, germanium, silicon, lead sulfide, and copper oxide may be used in this

way, making photoconductive cells. The cells are used with a dc source and a relay so that the presence or absence of light may actuate the performance of some desired task. The resistance of a cadmium sulfide cell, for example, may be over 1 megohm in darkness, and about 100 ohms when brightly lit.

The *solar cell* is an improved type of photovoltaic cell consisting of an N-type wafer of silicon doped with arsenic, on the surface of which boron has been added to form a transparently thin (0.001-inch or 0.0254-mm) layer of P-type silicon.

When light strikes atoms in the junction layer, energy of the light sends some electrons out of holes, making the electrons free moving. Some of these electrons uselessly lose their energy by collisions and fall back into a hole. Some of the energetic electrons usefully wander into the N-type silicon before losing their energy. Once into the body of the N-type wafer, they can do work flowing through an external circuit, rather than getting back home directly through the high-resistance barrier at the junction.

The open-circuit voltage of a solar cell is about half a volt. Voltage and efficiency (to 12 percent) are fairly independent of the amount of illumination; more light produces

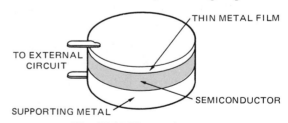

Fig. 17-17 Photovoltaic Cell

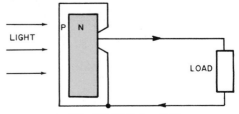

Fig. 17-18 Solar Cell

more current. Solar batteries are used to energize electronic equipment in artificial satellites. The communication satellite, Telstar, is powered by 3 600 solar cells. Solar cells, figure 17-18, have been successfully used in experiments in which they maintained a charge on storage batteries used for rural telephone systems. Emergency phones on some highways are radiophones energized by solar cells.

GAS DIODES AS RECTIFIERS

In Chapter 8 the behavior of electrons in a conducting low-pressure gas was described. By proper choice of electrode material and arrangement, gas tubes can be made that will start conduction in one direction readily, but in the reverse direction only with great difficulty. Such tubes, called *gas diodes,* can be used in rectifier circuits. The gas may be argon, neon, or mercury vapor. Old-type battery chargers using a mercury-vapor diode can still be seen. Compared with vacuum tubes of the same physical size, gas tubes

carry larger currents, and the voltage drop across the tube is low. When an alternating voltage is applied between cathode and plate, the gas diode starts conducting as soon as the voltage is high enough (a few volts) to ionize the gas. Conduction continues until the end of the half-cycle, when the plate becomes negative. At that time, electrons and positive ions recombine, and ionization has to be started again in the next half-cycle.

Some gas diodes are built to maintain a fairly constant voltage drop across the tube even when the tube current changes considerably. Such tubes are *voltage regulators.* Their function has been largely replaced by zener diodes.

Boxing in the cathode by a *grid*, which is really a cylinder with a hole in it, makes a gas triode that enables the timing of the start of conduction to be controlled. Holding the grid sufficiently negative prevents the ionizing flow of electrons from starting. It is easy to arrange an ac phase control circuit where this

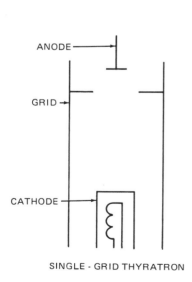

SINGLE - GRID THYRATRON

Fig. 17-19 Hydrogen Gas Thyratron Tube

tube blocks current until some chosen time during a half-cycle and then reduces the negative charge on the grid enough to permit electrons to start flowing from cathode to anode. Conduction continues through a load until the end of the half-cycle, when the polarity of the line reverses and turns the tube off. This gas triode is called a *thyratron,* figure 17-19, and was used widely for accurate motor and welding controls.   The phase control functions of thyratrons has been largely replaced by solid state devices which will be discussed in detail in Chapter 20.   The thyratron is used to initiate conduction in huge mercury diodes called *ignitrons. Huge* implies power capabilities up to one million watts for a single diode.

POINTS TO REMEMBER

- The atoms of semiconductor elements have 4 electrons in the outermost electron-shell of the atom.

- Addition of a 5-outer-electron impurity provides loose electrons for conduction; the mix is called N-type.

- Addition of a 3-outer-electron impurity increases conduction by providing holes in the electron structure that electrons can move into; the mix is P-type.

- P-type and N-type materials in contact form a rectifying diode; electrons move readily from N toward P inside the diode.

- Electrons emitted from the cathode of a vacuum diode travel to the plate when the plate is more positive than the cathode, but do not travel in the reverse direction when the cathode is more positive than the plate.

- Half-wave rectifiers produce this form of voltage:

- Full-wave rectifiers produce this form of voltage:

- Vacuum diodes are used in high-voltage low-current rectifiers, gas diodes are used in high-current rectifiers.

- Zener diodes are used to regulate load voltages.

- Zener diodes operate in a reverse breakdown condition.

REVIEW QUESTIONS

1. Name several semiconductors.

2. Which elements contain only four electrons in the outermost shell of the atoms?

3. What happens to the electrons in these semiconductors when they are cold? When they are heated?

4. What is meant by N-type semiconductors? What does the N refer to in the mixture?

5. Explain what happens to the atom structure in semiconductors such as silicon or germanium when a small amount of impurities whose atoms contain five electrons in their outermost shell are added.

6. Tell what happens to these semiconductors when an impurity whose atoms contain three electrons in the outermost shell is added.

7. Define the semiconductor mixture called P-type. How does this P-type differ from the N-type?

8. What happens when a piece of N-type material is put in contact with a piece of P-type material?

9. A P-N junction conducts readily from N to P, but an attempt to make electrons move from P to N develops a barrier to stop such motion. Why?

10. Explain what is meant by the terms *forward bias* and *reverse bias.*

11. Tell how a diode rectifier works and explain the difference between half-wave and full-wave rectifying of current.

12. What is the reason for using a zener diode in certain circuits?

RESEARCH AND DEVELOPMENT

Experiments and Projects on Semiconductors and Rectifiers

INTRODUCTION

Semiconductors are materials in a halfway zone between good insulators and good conductors. In them, electrons are held firmly in a fixed state. They remain in this fixed state until an input signal of sufficient energy is applied to free them. In Chapter 17 you learned about the electron arrangement in semiconductors, how junctions are made between different materials, and the result. You also learned how various components are used in rectifier circuits, the action of electrons in vacuum tubes, as well as other semiconductor applications, and the function of gas diodes as rectifiers.

The experiments which follow provide an opportunity to examine the function of semiconductors and to study the waveforms of rectifiers. In assembling the projects you will use various components to produce a circuit which serves a purpose such as a battery charger or dc power supply.

EXPERIMENTS

1. Examine the function of the P-N junction as a rectifier.

2. Study the waveforms of half-wave rectification.

3. Study the waveforms of full-wave rectification.

4. Study filter capacitors in a power supply circuit.

5. Observe the zener diode in a regulating circuit.

PROJECTS

1. Prepare a chart to illustrate the position of electrons and their movement in N- and P-material.

2. Assemble a silicon rectifier as would be used in a battery charger for a 120-volt circuit and test the output voltage.

3. Assemble and test a half-wave rectifier.

4. Assemble and test a full-wave rectifier.

5. Assemble a circuit in which a thyratron tube is used to control a mechanism.

6. Assemble a dc power supply.

EXPERIMENTS

Experiment 1 FUNCTION OF THE P-N JUNCTION AS A RECTIFIER

OBJECT

To understand the function of a P-N junction in a rectifier circuit.

APPARATUS

1 - Pilot lamp, 6-volt
1 - Germanium diode, IN34 or equivalent
1 - Resistor, 68-ohm, 1/2-watt (56 or 82)
1 - Battery, 6-volt
1 - Lamp socket, miniature

PROCEDURE

1. Assemble the apparatus according to the circuit diagrams, figures 17-20(A and B).

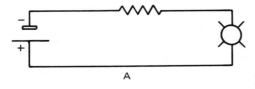
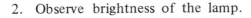

A   **Fig. 17-20**   B

2. Observe brightness of the lamp.

3. Reverse the connections on the battery and observe the light.

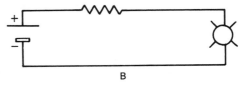

Fig. 17-21

4. Connect the circuit as in figure 17-21.

5. Observe light from lamp.

6. Connect circuit as in figure 17-22.

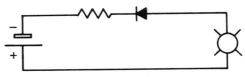

7. Observe the quality of light from lamp.

Fig. 17-22

OBSERVATIONS

Use a table similar to the one shown here to record your observations.

	DEGREE OF LIGHT		
FIG.	**BRIGHT**	**DIM**	**OUT**
17 - 1A			
17 - 1B			
17 - 2			
17 - 3			

QUESTIONS

1. Was the brightness of the light the same when the terminal connections from the battery were reversed? What does this condition prove?
2. What happened when the diode was placed in the circuit in reverse-bias position? Explain your answer.
3. How can you determine if a diode is connected in reverse bias?
4. Explain the change in the illumination from the lamp when the diode was connected in forward bias.

Experiment 2 HALF-WAVE RECTIFICATION

OBJECT

To study the waveforms produced in a half-wave rectifier circuit.

APPARATUS

1 - Oscilloscope
1 - Transformer, 115-volt ac, 12-volt center tapped, 25- to 50-watt
1 - Germanium diode, IN34 or equivalent
1 - Resistor, 10 000-ohm, 1/2-watt

PROCEDURE

1. Connect the apparatus according to the diagram, figure 17-23.
2. Turn on the current.
3. Adjust the oscilloscope and observe the waveform.

Fig. 17-23

OBSERVATIONS

Draw the wave pattern on graph paper.

QUESTIONS

1. What is the purpose of a half-wave rectifier?

2. Is there a disadvantage in the use of a half-wave rectifier? Explain your answer.

Experiment 3 FULL-WAVE RECTIFICATION

OBJECT

To study the waveforms produced in a full-wave rectifier circuit.

APPARATUS

1 - Oscilloscope
1 - Transformer, 115-volt ac, 12-volt center tapped, 25- to 50-watt
2 - Germanium diodes, IN34 or equivalent
1 - Resistor, 10 000-ohm, 1/2-watt

PROCEDURE

1. Connect the apparatus according to the diagram, figure 17-24.

2. Turn on the current.

3. Adjust the oscilloscope and observe the waveform.

OBSERVATIONS

Draw the wave pattern on graph paper.

QUESTIONS

1. How does the waveform of a full-wave rectifier differ from that of a half-wave rectifier?

2. What is the important advantage of a full-wave rectifier?

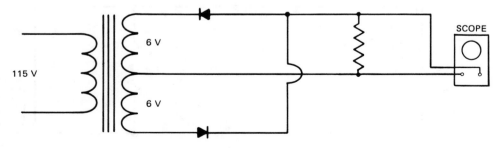

Fig. 17-24

Experiment 4 FILTER CAPACITORS IN POWER SUPPLY CIRCUIT

OBJECT

To study the effect of a filter capacitor in a power supply circuit.

APPARATUS

1 - Oscilloscope
1 - Transformer, 115-volt ac, 12-volt center tapped, 25- to 50-watt
2 - Germanium diodes, IN34 or equivalent
1 - Resistor, 10 000-ohm
1 - Capacitor, 20- to 50-μf

PROCEDURE

1. Assemble each of the circuits, figures 17-25 and 17-26, and observe the quality of the waveforms on the oscilloscope.

OBSERVATIONS

Draw and label each waveform on graph paper.

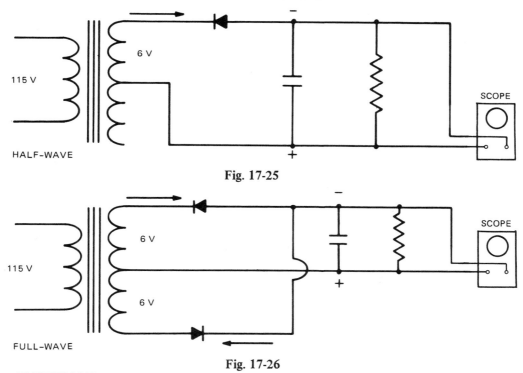

Fig. 17-25

Fig. 17-26

QUESTIONS

1. What effect did the inclusion of the capacitor have on the waveform in the half-wave rectifier? In the full-wave rectifier?

2. Did you notice any significant difference in the quality of the waveforms? In other words, did the capacitor help one more than the other?

Experiment 5 ZENER DIODE REGULATOR

OBJECT

To observe the regulation characteristics of a zener diode.
To plot the curve of zener voltage versus load current.

APPARATUS

1 - Zener diode of 6- to 8-volt, 50-milliwatt rating
1 - Resistor, 6 800-ohm, 1/2-watt
1 - Potentiometer, 25 000-ohm
2 - DC milliammeters (0-1 mA)
1 - DC voltmeter (0-10 V)
1 - DC power source, 12-volt

PROCEDURE

1. Connect the circuit as shown in figure 17-27.

2. Energize the circuit, adjust R_2 back and forth to insure proper connection of the zener, and observe the voltmeter. The voltmeter should remain near the rated value of the zener as this is done.

3. Remove R_2 (open the circuit) and record the readings of the zener current and voltage.

4. Reconnect R_2 and adjust its value until ammeter A_2 reads 0.1 mA.*
Record the value of zener voltage and current.

    *Current range should conform to manufacturer's specifications for the diode used.

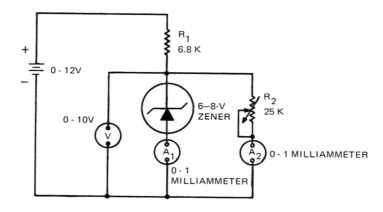

Fig. 17-27

5. In like manner, increase the ammeter A_2 current in steps until a maximum of 1 mA* is read, recording zener voltage and current for each step.

6. Plot the curve of zener voltage versus load current on graph paper.

OBSERVATIONS

Use a table similar to the one shown here to record your observations.

1 * 1	1 * 2	V 1
	OPEN 0 ma	

QUESTIONS

1. Explain the two extremes of the curve. Why aren't they in a straight line with the center portion?

2. How would you connect a circuit to increase the regulation voltage?

3. What difficulties would there be in increasing the current rating by putting the zeners in parallel?

4. Could R_1 ever be eliminated? Explain.

PROJECTS

ASSEMBLE A SILICON RECTIFIER

This silicon rectifier, figure 17-28, is a dc power supply with a low-voltage output that may be used in a wide range of experimental activities and for charging 12-volt storage batteries.

The design of the rectifier may be modified to include an ammeter for indicating the charging rate. The circuit can be changed to provide for charging 6-volt or 12-volt batteries.

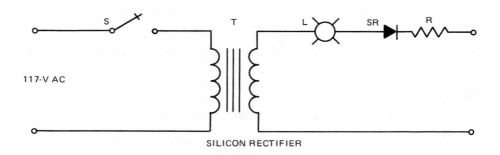

Fig. 17-28

MATERIALS

1 - Vector board base, or metal chassis and case
1 - Transformer, 117-volt primary, 36-volt secondary, 5-amp
1 - Silicon diode, Radio Shack #276-1142
1 - Lamp, G.E. #1130 or equivalent
1 - Lamp base
2 - Binding posts
1 - Line cord and plug
1 - SPST switch
1 - Resistor, 5-ohm, 15-watt

PROCEDURE

Examine the schematic diagram in figure 17-28, read the bill of materials, and obtain the electronic components.

Plan a layout for the parts. This should be about the same as the schematic. Ask your instructor to check your plan.

If it is a take-home project, design and fabricate a chassis and case.

Mount the silicon rectifier on a heat sink.

Assemble the parts on the vector board or chassis.

Wire according to figure 17-28.

Check all connections and the circuit. Have your work approved.

Test and evaluate.

FULL-WAVE RECTIFIER FOR VACUUM TUBE CIRCUITS

The full-wave rectifier shown in figure 17-29 filters the current and may be used as a power supply for radio transmitters. The model performs satisfactorily and is comparatively easy to assemble. It produces high voltage necessary for transmitting tubes.

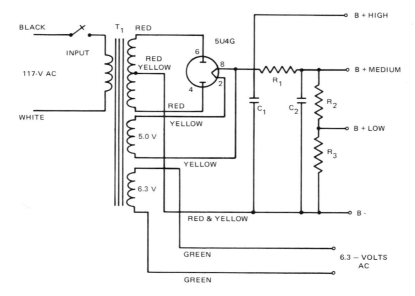

FULL-WAVE POWER SUPPLY

Fig. 17-29

MATERIALS

1 - Power transformer, 70-ma, 350-volts each side of center tap (T_1)

1 - Resistor, 2 500-ohm, 2-watt (R_1)

1 - Resistor, 15 000-ohm, 25-watt (R_2)

1 - Resistor, 15 000-ohm, 25-watt (R_3)

2 - Electrolytic capacitors, 8-μf, 450-volt (or a dual section capacitor) (C_1, C_2)

1 - Rectifier tube, 5U4G

1 - Octal tube socket

1 - SPST switch

1 - Terminal strip or binding posts

1 - Tie point

1 - Line cord and plug

PROCEDURE

Plan the arrangement of the parts.

Select standard chassis of appropriate size or fabricate one from sheet metal.

Lay out and drill holes in chassis. When finished, remove all burrs.

Mount the parts and wire according to the schematic diagram, figure 17-29.

Check all connections to make certain that they are mechanically secure and the wires attached according to figure 17-29.

Mark the terminals and have your work approved.

Test and evaluate.

PLAN AND ASSEMBLE A CIRCUIT IN WHICH A THYRATRON TUBE IS USED TO CONTROL A MECHANISM

A thyratron tube is used in circuits designed to control motors, lamps, and other electrical devices.

When using a circuit such as the one in figure 17-30, it is essential that the relay contacts have sufficient capacity to withstand the load.

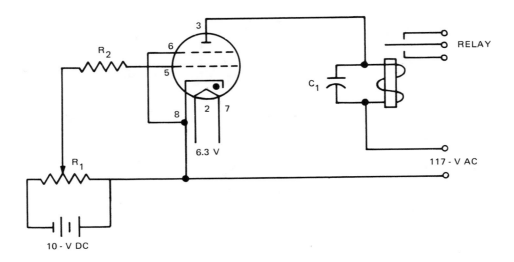

Fig. 17-30 Thyratron Tube Control Circuit

MATERIALS

1 - Potentiometer, 10 000-ohm (R_1)
1 - Resistor, 5 000-ohm (R_2)
1 - Capacitor, 1-μf (C_1)
1 - Relay, 2 500-ohm
1 - Octal tube base
1 - Thyratron tube, #2050
1 - Voltage divider
1 - Filament transformer
1 - Low-voltage dc supply
1 - Vector board or chassis
1 - SPST switch

PROCEDURE

Plan a layout, according to figure 17-30, to be assembled on a vector board or in a metal chassis.

Assemble the parts and wire. Check all connections to determine whether they are mechanically secure and the wires attached according to the schematic diagram.

Connect a fractional-horsepower motor or other device to the relay. *Note:* The tube should be allowed to heat for 20 seconds before connecting the load.

Have your work approved.

Test and evaluate.

PLAN AND BUILD A DC POWER SUPPLY

This dc power supply is a comparatively heavy duty unit that has enough capacity to satisfy the requirements for most home workshop enthusiasts. It may be used for many purposes in the school shop as well as at home. It is not intended for use as a power supply for radio work. However, the unit could be altered to serve for radio work by the addition of a resistor and a switch on the line between C_1 and C_2.

See note at end of procedure.

MATERIALS

1 - Metal chassis
2 - SPST switches, 10-amp (S_1, S_2)
2 - Capacitors, 1 000- to 1 200-μf, 250-volt (C_1, C_2)
1 - DC ammeter (0-10 A)
1 - DC voltmeter (0-200 V)
1 - Fuse holder
1 - Fuse, 3AB — 10-amp (F)
1 - Resistor, 10-ohm, 100-watt, wire-wound (R_1)
1 - Resistor, 15 000-ohm, 25-watt (R_2)
1 - Autotransformer, 0- to 140-volt output, 10-amp, Superior #1168 or equivalent (T)
4 - Silicon rectifiers, 15-amp, 200 PIV (Peak Inverse Voltage)
1 - Attachment plug, parallel-ground
2 - Insulated terminals, one red, one black Allied 47C6329 & 47C6332

PROCEDURE

Secure the necessary parts, study the schematic diagram, figure 17-31, page 436, and make a layout drawing for placing the parts in a chassis.

Mount the components in the chassis and wire.

Note: It is suggested that the diodes be mounted on a heat sink with cases insulated from it as explained in the manufacturer's instructions.

When you have finished wiring, check all connections to see if they are mechanically secure. After you are satisfied that all the connections are tight, check the circuit against the schematic diagram.

Have your work approved, then test and evaluate.

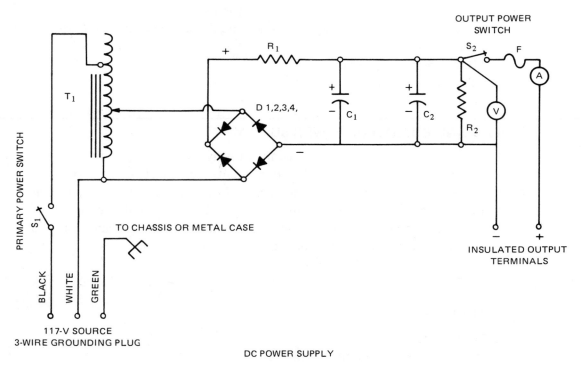

Fig. 17-31

Note: Direct-current filtering can be improved with the addition of a 100-ohm, 15-watt resistor and an SPST switch as indicated in figure 17-32.

With the switch open, the power supply would deliver about 0-130 volts at 0.5 amps and about 0-180 volts at 0.01 amps.

With the switch closed, the unit would deliver about 0-130 volts at 5 amps and about 0-170 volts at 1 amp.

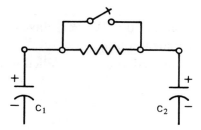

Fig. 17-32

Chapter 18

Amplifiers

AMPLIFICATION

Amplifiers are increasers. In the broadest sense, to amplify something means to make it larger or stronger or clearer or more intense. The word amplifier is most often used to name one of the essential sections of a radio or record player or TV receiver, in which voltage, current, or both (power) may require enlargement. Although the human amplifier, diagrammed in figure 18-1, has been used for thousands of years, electronic amplifiers have great advantages. The energy source can be better controlled by the input signal, giving a faithful reproduction at a desired amount of power output. Electrically, *signal* means a small voltage, usually alternating, that tells the amplifier circuit what to do. The signal voltage controls in detail the fluctuations of current from a more powerful energy source in the tube or transistor circuit.

There are three major types of solid state amplifier devices commonly available on the consumer market. They are the junction transistor, the field effect transistor, and the operational amplifier. Each of these devices and the basic principles of the vacuum tube amplifier will be presented in this chapter.

SIGNAL INPUT ENERGY SOURCE OUTPUT

Fig. 18-1 Human Amplifier

THE JUNCTION TRANSISTOR

The part of Chapter 17 dealing with electron behavior at P-N junctions should be reviewed because it applies to transistors.

Suppose a germanium crystal is formed with a thin layer of P-type element in the middle and N-type material well supplied with free electrons at the ends. Figure 18-2 shows this structure, called an N-P-N transistor, with a connection to a dc source. Recalling the behavior of P-N junctions as rectifiers, let the small circles represent electrons that spill over from the N-material into the P-material. The positive signs represent atom cores, now positively charged, in the N-material that supplied electrons. The nonconductive layer at the right-side P-N junction is intensified by the reverse bias applied by the 24-volt battery. Figure 18-2 shows only a very slight conducting condition (back leakage of a diode).

In figure 18-3, the N-P junction at the left is given a 2-volt forward bias to cause conduction. When that connection is made, electron movement starts every place in the circuit at once. To avoid confusion, start from the negative terminal of the 2-volt battery and see what happens to electrons as their movement is traced through the circuit. Electrons from the negative wire move through

the resistor into the conductive N-type material, supplying electrons to neutralize the positive charge at the first N-P barrier, and into the P-material, just as would be expected of a rectifying junction when it is conducting forward-biased. But, once into the very thin P-layer, those electrons may either: (1) drop into holes in the P-layer and slide out to wire B, completing their circuit to the positive 2-volt side of the left-hand battery; or (2) attracted by the positive 24-volt charge on the N-material at the right, head toward the second P-N junction and into the N-material, completing their circuit through the resistor to the 24-volt battery. At least 95 percent of them take this second choice, as could be guessed by looking at voltages. To make this 24-volt route more favored, the thin P-layer has a very small fraction of impurity atoms. This means that there are few holes in the pattern for electrons to fall into. The electrons going toward the 24-volt side go through fast and free moving. They do not have time to look for a place to settle into. Sometimes the name *electron diffusion* is applied to this movement of electrons through the P-layer.

Figure 18-2 showed electrons ^o_o sitting in the P-layer. They did not move to the right and around to the positive side of the

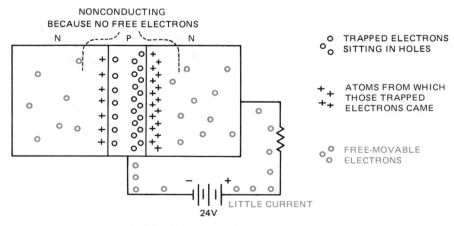

Fig. 18-2 N-P-N Transistor

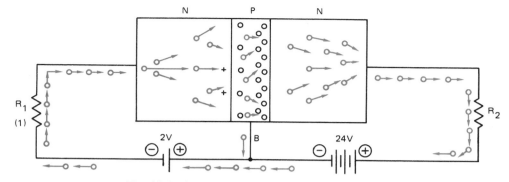

Fig. 18-3 Electron Movement in N-P-N Transistor

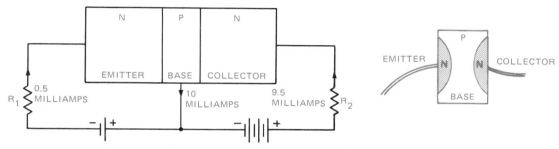

Fig. 18-4 Currents in the Three Layers of the N-P-N Transistor

24-volt battery because they were sitting with low energy, trapped with the rest of the low-energy electrons in the electron pattern and could not enter the high-energy levels available for free electrons in the adjacent N-material whose electron structure is already complete. In figure 18-3, the group of electrons entering from the N-material at the left are energetic free-moving conductive electrons. If this source of energetic electrons is removed, the current into the 24-volt positive source is reduced to leakage levels again.

Figure 18-4 shows names that have been given to the three layers of the germanium transistor, along with numbers showing relative amounts of electron current.

The name *base* was used for the center section by the first transistor-makers for the crystal of metal into which alloying impurities were introduced. Mechanically, they started with the base and built other elements on it.

Perhaps you have already noticed the possibility of using this little piece of N-P-N germanium as an amplifier. In this transistor,

the introduction of current from the 2-volt battery permitted a larger energy source, the 24-volt battery, to produce current. Large energy is controlled by small energy; that is amplification.

Instead of using the N-P-N sandwich, many transistor amplifiers use one that is P-N-P. The collector circuit, as before, is the power output circuit. The input signal is fed into the emitter-base junction. Relate the names emitter, base, and collector to the flow of control, rather than to flow of electrons.

Figure 18-5(A) shows the charged, non-conductive barriers developed at the base-collector junction of a P-N-P transistor. The base N-region is thin and made from crystals with very few impurity atoms. Most of the free electrons from these impurity atoms are lost to fill holes in the adjacent P-region. This leaves a number of atoms with a positive charge in the base region even though there are no holes in the crystal structure. In order to be free in such a situation, an electron would have to possess a relatively large amount of energy, more than that possessed by trapped electrons in the adjacent P-crystal. At the same time, the trapped electrons in the P-region (collector) built up a negative charge which repels further entry into the crystal of electrons from the 24-volt negative source.

Connect a small forward-biased source to the base-emitter junction, figure 18-5(B). High energy conduction electrons from the 2-volt negative source enter the N-crystal, diffuse through it and are attracted to the positively charged P-emitter region. These electrons fill holes in the P-region, lowering the energy needed by electrons already in the emitter to leave the crystal. Electrons leave the P-type emitter returning to the 2-volt positive source. The flow of electrons into the N-type base neutralizes some of its positive charge and lowers the level of energy needed for free-conduction electrons to exist. In fact, it lowers this required energy level below the level of trapped electrons in the P-collector region. These electrons leave the P-region and come flooding across the N-base region to slide into low energy holes in the emitter region.

One electron injected into the base will neutralize it sufficiently to let many electrons travel from the collector to the emitter.

Thus, a small constant current from base to emitter will allow a much larger current from collector to emitter. This is *current amplification.*

Alpha and Beta

Junction transistors, even those of the same type and from the same manufacturer differ quite widely in their ability to amplify current. Two related properties of a transistor have been defined to help analyze and design circuits and to communicate about the amplifying ability of the transistor.

Alpha (α) is defined as the ratio of collector current to emitter current:

$$\alpha = \frac{I_c}{I_E}$$

In figure 18-5(B), 95 percent of the electrons leaving the emitter entered the collector. The rest of the emitter current (0.5 milliamperes) entered the base. Therefore, the following relations exist:

$$I_E = I_C + I_B, \text{ and } I_C = \alpha \times I_E$$

For this transistor, $\alpha = 0.95$. This is a relatively poor transistor by modern standards. Alphas above 0.99 are not uncommon.

Beta (β) is defined as the ratio of collector current to base current: In the circuit of figure 18-5(B), the base current of

(A)

(B)

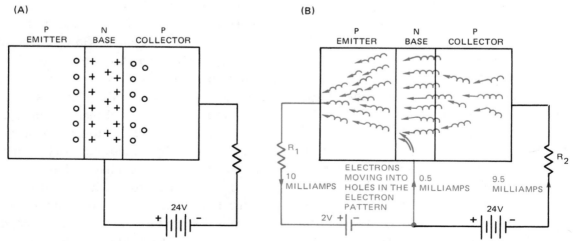

Fig. 18-5 Electron Movement in the P-N-P Transistor

0.5 milliampere allowed a collector current of 9.5 milliamperes. Beta for this transistor is $\frac{9.5}{.5}$ = 19.

$$\beta = \frac{I_C}{I_B} \quad \text{and} \quad I_C = \beta \times I_B$$

Beta is sometimes called the *forward current transfer ratio* in transistor specification manuals and catalogs.

Alpha and Beta for the same transistor are related by the formulas:

$$\beta = \frac{\alpha}{(1 - \alpha)} \text{ and } \alpha = \frac{\beta}{(\beta + 1)}$$

Alpha and Beta are not really fixed constants associated with a given transistor. They vary slightly with temperature, circuit voltages, and currents. Therefore a transistor specifications manual or catalog will list an *average* Beta figure for a given line of transistors at a given temperature, collector current, and emitter-to-collector voltage. The maximum and minimum Beta figures for a given code numbered transistor are often published.

SIMPLE TRANSISTOR AMPLIFIER CIRCUIT

Symbols for transistors, as used in circuit diagrams, are shown in figure 18-6. These symbols may be rotated to other positions in the circuit schematic. They differ only in the emitter symbol; the arrow points away from the base for N-P-N and toward the base for P-N-P.

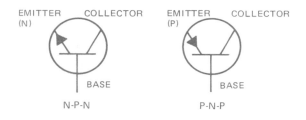

Fig. 18-6 Symbols for Junction Transistors

Assume that one wished to operate a relay so that when light struck a solar cell it turned on an external ac circuit. The photocell delivers only 0.4 volts when in the light. One has available an N-P-N transistor with a Beta of 75. The relay has 300 ohms of internal resistance. It closes at 6 volts and remains closed unless the voltage falls to 3 volts.

If the solar cell is placed in the emitter-base circuit of the transistor and the relay is placed in series with the collector and a 9-volt battery, the task is easily accomplished. Figure 18-7 is an example of such a dc amplifier. When the photocell is in the dark, the base-emitter circuit has no current and therefore there is only leakage current in the emitter-collector circuit or in the relay coil.

When light strikes the solar cell it produces a voltage of 0.4 volts while delivering a small current. In order to close the relay, collector current must be at least 20 milliamperes. Since collector current is β times the base current, the base-emitter current must be at least 20 ÷ 75 = 0.266 milliamperes = 266 microamperes. Remembering that the forward-biased threshold voltage of a germanium diode is 0.2 volts, this much of the 0.4 volts from the solar cell appears between the emitter and base. That leaves 0.2 volts for R_1. R_1 must not be greater than $R = 0.2 \div 266 \times 10^{-6}$ or 750 ohms. Making R_1 a 1 000-ohm rheostat will allow for sensitivity adjustment under a wide range of light and temperature conditions.

In this circuit, small voltage changes in the input cause much larger voltage changes in the output. The relay contacts can operate any other circuit in complete isolation from the amplifier itself.

AC Amplification

In amplifying ac signals, one must set up a dc level of operation for the transistor

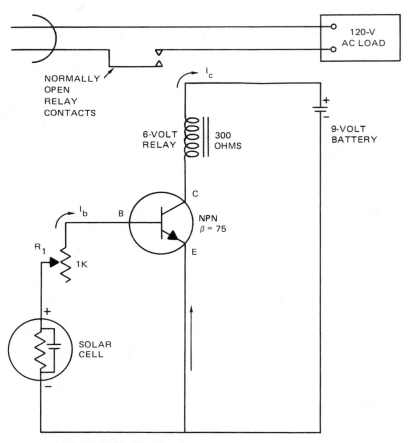

Fig. 18-7 Basic DC Junction Transistor Amplifier

and then introduce the ac signals to systematically increase and decrease the dc levels of operation. The dc emitter-to-collector voltage must be greater than the ac peak signal after amplification to prevent distortion.

A basic ac amplifier is shown in figure 18-8. The coupling capacitors C_1 and C_2 allow ac signals into and out of the amplifier without shorting out dc currents or voltages. The circuit makes it possible to use a simple 4-ohm speaker as a high-output microphone. Connecting the output of this amplifier to the input of any hi-fi amplifier will result in an efficient public address system.

The transistor is a germanium N-P-N with a Beta of 50, for example, a no. 2N1302. Resistors R_L and R_e are usually chosen to operate the transistor at about one-half its

current capacity and with about one-half the applied supply voltage across emitter and collector. In this case the transistor is biased at a collector current of 3 milliamperes and 4.2 volts emitter to collector. This provides plenty of range for changing voltages and currents when the signal source is connected. The transistor has a Beta of 50. Therefore, base current is: 3 milliamperes ÷ 50 = 60 microamperes (μA). The voltage at the base is 0.5 volts, which includes 0.2 volts for the threshold level of a germanium diode and 0.3 volts for the emitter resistor drop. This means that the dc base-to-ground resistance as viewed from the base input is 0.5 volts divided by 60 microamperes or 8 333 ohms.

A voltage divider is formed by R_1 and R_2 to provide forward bias to the base circuit. R_2 also provides a low resistance path to the

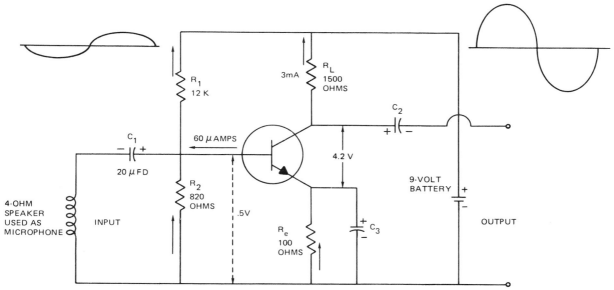

Fig. 18-8 Basic AC Junction Transistor Amplifier

ground for any back leakage from the collector-base junction. If R_2 is approximately one-tenth the dc resistance value of the path through the base to emitter and on to ground, most of any leakage current will choose R_2 rather than reenter the transistor. R_2 is chosen to be 820 ohms because this is a standard carbon resistor value near one-tenth of 8 333. The choice of 820 is arbitrary; 1 000 ohms or 750 ohms would also work. The lower the value chosen, the more temperature stable the amplifier will be. However, more of the input will be shunted to the ground and battery life will be reduced.

With R_2, chosen to be 820 ohms, connected in parallel with the base which has a potential of 0.5 volts, the current through it must be:

0.5 divided by 820 or 0.61 milliamperes
= 610 μamp

The current in R_1 must equal the current in R_2 plus the base current:

610 microamperes + 60 microamperes =
670 microamperes in R_1

Since R_1 and R_2 are across the 9-volts source, R_1 must have a voltage drop of 8.5 volts; and R_1 = 8.5 volts ÷ 670 μamp = 12 700 ohms.

A close standard carbon resistor is 12K. This will work well, causing other assumed quantities to be only slightly in error when the circuit is tested.

The functions of R_e and C_3 have not been explained. Remember that the dc voltages in an amplifier must be greater than the peaks of the ac signal voltages or distortion will occur. The voltage between base and ground must always be in a forward direction or the transistor will be nonconducting. Connecting the emitter directly to ground (without R_e) will result in a base-to-ground bias of 0.2 volts. A small input signal could easily reverse bias the transistor input for part of the cycle. Adding R_e and its voltage drop between the emitter and ground increases the dc voltage base to ground so that the amplifier can handle larger input voltage signals without distortion. C_3 provides a low impedance path for the input signal around R_e.

When the 4-ohm speaker is momentarily producing a voltage to drive electrons down its coils and along the common lead up toward C_3, it adds to the dc emitter-to-base electron flow, adds electrons to the transistor side of C_1, and draws electrons away from the negative side of C_1. This increase in base current allows a much greater increase in collector current. The voltage drop on R_L increases because of the increased collector current. The transistor appears to be a lower resistance device than it did without the signal, and its share of the 9-volt source decreases. This decrease causes the output voltage to be less positive. On the next half-cycle when the 4-ohm speaker produces an electron surge toward the negative side of capacitor C_1, electrons leave the positive side of C_1 and follow the base lead into the transistor. The surge is carried by capacitor C_3 and electrons complete the circuit through the ground to the speaker again. This surge of current subtracts from the existing base current, allowing fewer electrons from the emitter to the collector. The transistor appears to have raised its resistance emitter to collector. Therefore, its voltage is increased. This results in a positive output voltage. This kind of amplifier has an output signal 180° out of phase with respect to the applied input signal. That is, when the input signal is on positive half-cycle the output will be on a negative half-cycle.

Since the transistor is a current-operated device, the greater the current surges in the input base-to-emitter circuit, the greater the output current change will be. Current changes in R_L develop the voltage changes which are reflected in the output. An increase in R_L will therefore increase the amount of amplification. However, if R_L is made too large, the dc bias of the transistor (voltage emitter to collector) cannot be great enough to allow ac signal changes to occur without distortion. In other words, increasing R_L raises the amount of amplification but decreases the size of the input signal that can be handled by the amplifier without overdriving or distortion.

THE FIELD EFFECT TRANSISTOR

Imagine an oblong piece of N-type semiconductor completely surrounded by a ring of P-type semiconductor as in figure 18-9. This component is a junction field effect transistor and has three leads; the source, the drain, and the gate. The two ends of the N-type material (source and drain) are connected in series with a voltage source and resistor load.

With the gate control signal at "0" volts, an electron current is set up from the main power source, through the source lead, through the base channel, out the drain lead, through R_L, and back to the positive side of the source of power. Current in the circuit will be limited by the crystalline resistance

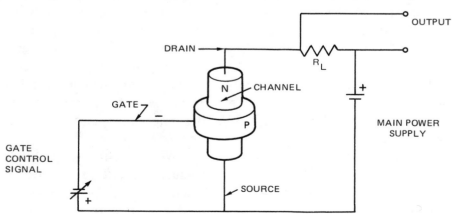

Fig. 18-9 N-Channel Field Effect Transistor Amplifier

between source and drain and the resistance of the load resistor, R_L. The voltage of the main source will be divided between R_L and the N-channel. By adjusting the control signal voltage to make the P-gate region negative with respect to the N-channel, it is possible to restrict the area in which free electrons can move through the N-channel because of the negative electric field associated with the filling of holes in the P-region. Electron flow between the P- and N-regions does not occur because they represent a reverse-biased diode.

Restricting the flow of electrons causes the resistance of the channel to rise, the total current to fall, and the voltage across R_L to fall. In fact, it is entirely possible to restrict the flow enough to stop it completely. The voltage necessary to create this condition is appropriately called *pinch-off voltage.*

The gate to source input of a field effect transistor represents the high resistance of a reverse-biased diode. This device has largely replaced smaller vacuum tubes in applications such as measuring circuits where only a very tiny current can be drawn from the signal source. With a given gate voltage, only a limited number of free electrons can take part in a source-to-drain current. Increasing or decreasing the main battery voltage has little effect on this number. Therefore, source-to-drain current is relatively independent of

supply voltage over a wide range, but highly dependent on gate voltages. The gate must have a strong enough negative dc bias to not approach "0" or be driven positive during any part of the ac input signal cycle.

A typical ac FET amplifier is shown in figure 18-10. Without an ac signal, R_L and R_S are chosen so that when connected in series with the known source-to-drain resistance, the dc main source voltage divides as shown. With no signal, there is no current in R_G, it has no voltage drop, and the gate is at the same potential as the lower end of R_S, which is 3 volts negative with respect to the source terminal. When the ac signal is connected, it adds one volt to the gate potential, and subtracts one volt from it on alternate half-cycles. The potential of the gate then varies from a negative 2 volts to a negative 4 volts with respect to the source. This change in potential increases and decreases the load current enough to cause a 10-volt peak-to-peak change in the output. The capacitor C_3 allows the sudden surges in output current to flow through it without appreciably upsetting the fixed 3-volt drop on R_S. C_1 and C_2 allow ac signals to enter and leave the amplifier without shorting out dc voltages.

Understand that the values of components and levels of currents and voltages in the preceding circuits are only examples of

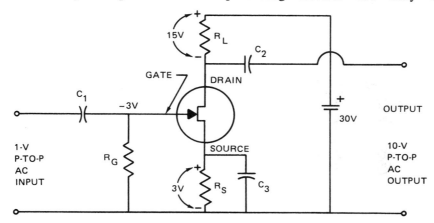

Fig. 18-10 Typical FET AC Amplifier

transistor amplifiers. There are many different kinds and sizes of transistors in many different package forms, and many ways to connect them as amplifiers. Detailed discussion of circuit principles and applications can be found in many texts devoted to this topic.

INTEGRATED CIRCUITS

Miniaturization of printed circuits through photographic processes and the analysis of crystalline connections into a complex circuit arrangement containing many individual transistors and resistors has led to the development of integrated circuits. Such devices may have numerous external connections and require special input, output, and power supply considerations. It is now possible to obtain a single crystal of complex structure which performs all the functions needed to operate a pocket calculator.

The integrated circuit has made possible significant achievements in space exploration, biomedical aids, military science, and the business world. It is rugged, reliable, and easily replaced in case of malfunction. Modern devices for logic circuits, counting, amplifying, oscillating, and photocontrol are commonly available on the consumer market with suggested circuit applications. A better feeling for how an integrated circuit is designed should be obtained by study of the structure of silicon controlled rectifiers (SCR) in the chapter on ac power control. Remember

that the available devices on the market can be rapidly destroyed through wrong connections if the manufacturer's directions and diagrams are not followed strictly.

THE TRIODE VACUUM TUBE AS AMPLIFIER

Electronic amplification was made possible by adding a third part, the *grid*, to the diode vacuum tube. Chapter 17 described the flow of electrons, given off by the heated cathode, from the cathode toward the positive-charged plate. The purpose of the grid is to control that electron flow. The next series of diagrams shows the effect of various grid-cathode voltages on the current to the plate.

In (1) of figure 18-12, although the heated cathode is maintaining a generous supply of electrons, the 24-volt negative charge on the grid repels them so much that no electrons get through to the positive plate. The current is cut off.

In (2), the grid is less strongly negative-charged. Occasional high-speed electrons approach the grid despite its repulsion and pass between the grid wires. Once through, they accelerate toward the positive plate, making a small current.

In (3), the negative charge on the grid is quite weak, and a greater number of electrons slip past the grid, attracted by the relatively strong positive charge on the plate. Note that in both examples (2) and (3) the

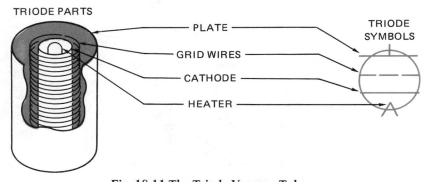

Fig. 18-11 The Triode Vacuum Tube

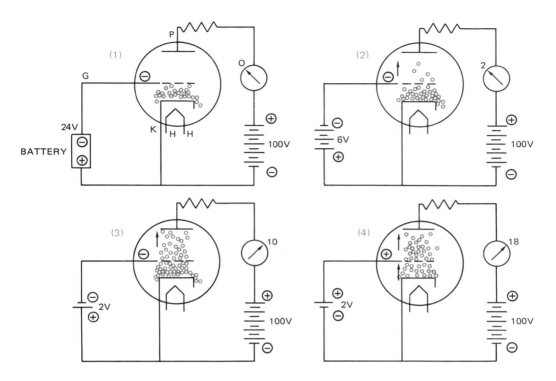

Fig. 18-12 Effect of Various Grid-Cathode Voltages on the Current to the Plate

6-volt or 2-volt battery maintains a voltage, but there is no current in this side of the circuit.

In (4), the grid is made 2 volts more positive than the cathode. The positive-charged grid now attracts many electrons from the cathode. Some of those electrons hit the grid wires and go through the 2-volt battery, forming a current in the grid circuit. The great majority of electrons near the grid sense the strongly (100-volt) positive-charged plate and are attracted to the plate, preferring it to the weakly positive-charged grid. The current in the plate circuit is larger than in the previous three examples.

The triode tube behaves very much like the field effect transistor which has taken over most of its functions. It is still used for high-voltage applications such as commercial radio transmitters. The circuit of figure 18-13 is outdated, but useful in explaining how a typical triode amplifier functions.

Some equipment should be added to the tube circuit in order to make use of it, perhaps as a record player. As a record rotates under the needle of the pickup, the needle is made to vibrate at various frequencies between 20 and 20 000. These vibrations of the needle, transferred to a crystal (piezoelectric, Chapter 12) cause the crystal to generate voltages that conform to the frequencies impressed on the record.

Suppose the crystal is vibrating 500 times per second, producing an alternating voltage of 1-volt maximum. This 1 volt alternately adds and subtracts to or from the 3 volts chosen for the grid battery, causing the grid itself to change from –2 volts to –4 volts, as compared to the cathode, 500 times per second. The plate current through the tube, controlled by this grid voltage, increases and decreases 500 times per second. This varying plate current might be allowed to pass through a speaker. In the speaker, the varying current

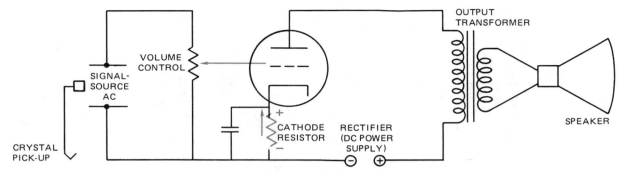

Fig. 18-13 One-Tube Phono Amplifier

in a coil of many turns of wire can magnetically vibrate a paper cone, producing a sound
vibration in the air of 500 times per second
(Chapter 6).

Actually this suggestion is not good, because the triode plate current is small and the
vibrating coil of the speaker is built for operation with 20 or 30 times that current. A set
of headphones, which is designed for a tiny
current, would work, but the power output is
small. The addition of 20 or 30 times as many
turns of wire on the speaker would make the
vibrating coil too heavy. The solution is to
use a step-down transformer. The low-current
high-voltage pulses in the primary induce
higher-current lower-voltage ac in the secondary, which powers the speaker.

The volume control is a high-ohm
(500 000, for example) rheostat connected as
a potential divider across the input. If the
slider is at the top of the resistor all of the
input signal is amplified. In the position
shown, however, about two-thirds of the
input voltage is applied to the grid.

The purpose of the cathode resistor is to
make the grid more negative than the cathode.
The grid is connected through the volume control to the lower end of the cathode resistor.
Since there is no dc in the volume control, the
grid is at the same potential as the lower end
of the cathode resistor. For example, if the
current through the tube is 40 milliamperes
and the cathode resistor is 75 ohm, a potential
difference of 0.040 x 75 = 3.0 volts is devel-

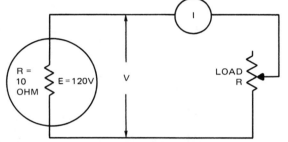

Fig. 18-14 AC Generator Connected to a Variable Resistor

oped across this resistor. Electrons flow
through a resistor from the more negative end
toward the more positive. That is, the lower
end of the resistor to which the grid connects
is 3 volts more negative than the cathode.

IMPEDANCE MATCHING

The use of the output transformer is an
example of the *impedance matching principle;*
maximum power (output) is delivered to the
load when the load impedance equals the
impedance of the source. This would produce
more volume in the speakers.

Consider for a moment an ac generator
with 10-ohms internal resistance, figure 18-14,
running at constant speed and developing
120-volts emf, which is connected to a load
with a resistance that can be varied. Calculations of power output at various loads are
listed in figure 18-15.

Items (3) and (4) provide higher power
output, but at low efficiency ($I^2 R$, watts lost
in the generator, is high) which is undesirable

Load Resistance	Total Resistance in Circuit	Current in Circuit	Volts at Load R	Watts Output
(1) 2 ohm	12 ohm	10 amp	20 V	200
(2) 5 ohm	15	8	40 V	320
(3) 10 ohm	20	6	60 V	360 max.
(4) 14 ohm	24	5	70 V	350
(5) 110 ohm	120	1	110 V	110

Fig. 18-15

in a big ac generator. But in an electronic amplifier where total power involved is small, a 5-watt output at 50 percent efficiency is preferable to 1 watt at 95 percent efficiency.

AMPLIFIER GAINS

The *gain* of an amplifier tells how many times greater the output signal is when compared with the same measurement of the input signal. Amplifiers may have voltage, current, or power gains. Power gain will equal voltage times current gain following the formula W = EI. For example, if the input signal to an amplifier were determined to be 5 millivolts peak-to-peak at 10 microamperes of current peak-to-peak when its output signal was 50 millivolts (peak-to-peak) at 750 microamperes (peak-to-peak), the voltage gain would be:

$$\frac{50 \text{ millivolts}}{5 \text{ millivolts}} = 10.$$

Current gain would be $\frac{750 \text{ microamperes}}{10 \text{ microamperes}} = 75$, and power gain would be 75 x 10 = 750. Gain is always calculated by the formula

$$gain = \frac{output}{input}$$

Amplifier gains are dependent upon the particular device, the values of circuit resistance, impedances, and the way the circuit is connected.

Efficiency

Efficiency is the ratio of the amount of useful power out of an amplifier compared

with the amount of total power used to operate the amplifier. An amplifier with a power gain of several hundred will not mean that power can be created from nothing. That amplifier and all others are operated from a given dc source which provides current and voltage.

$$\text{Efficiency (always less than 1)} = \frac{\text{useful power output}}{\text{total power used}}$$

POINTS TO REMEMBER

- In transistor amplifiers, reverse bias is applied to the collector-base junction, and small forward bias is applied to the emitter-base junction.

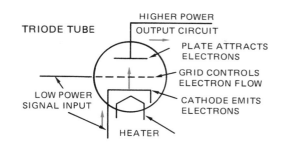

Fig. 18-16

- In any amplifier, a low-power signal controls higher power given out by an energy source.

- For maximum power output, the load impedance should match the impedance of the source of energy.

- Voltage gain of an amplifier circuit is the number of times that the input signal voltage is multiplied.

- Power gain is the number of times that the power input is multiplied by the amplifier.

REVIEW QUESTIONS

1. What is the difference between a diode and a triode vacuum tube? Draw the symbol for each tube and label the parts.

2. What is the purpose of a grid in a triode tube?

3. What is grid bias? Why is it important in a circuit?

4. Identify the parts of a field effect transistor that functions most like the cathode, grid, and plate of a vacuum tube.

5. What is meant by impedance matching and what is its purpose? Give an example.

6. The collector in a transistor serves a similar purpose as which element in a triode vacuum tube?

7. Why does an integrated circuit take up less space than a conventional circuit performing the same function?

8. What is the difference between N-P-N and P-N-P transistors in construction and in polarity of applied voltage?

9. Draw diagrams for N-P-N and P-N-P transistors and label the parts.

10. Make a diagram of a transistor amplifier circuit. Draw the same circuit using a triode vacuum tube.

11. Which is the power output circuit of a transistor?

12. Where does the input signal enter a transistor?

13. Name some of the advantages of transistors as compared to vacuum tubes.

RESEARCH AND DEVELOPMENT

Experiments and Projects on Amplifiers

INTRODUCTION

Many forms of amplifications, such as the megaphone, the bull horn, and the more sophisticated electronic public address systems, as well as those used in radios, phonographs, and televisions have been used through the years. In Chapter 18 you read the statements "amplifiers are increasers" and "in the broadest sense, to amplify something means to make it larger or stronger or clearer or more intense." Throughout the chapter you have had an opportunity to learn the basic principles of electronic amplification, the function of vacuum tubes and transistors, and how they are connected with other components to form an amplifier circuit.

In performing the suggested experiments, you can find out how amplification is achieved and learn more about the function of the parts required to make an amplifier.

The projects can be of the take-home variety or the assembly-disassembly type. Since the basic circuits are published in manuals, books, and magazines, schematic diagrams and lists of materials are not included for the projects in this unit.

EXPERIMENTS

1. Study the effect of the grid on the electron flow in a triode vacuum tube.
2. Study the effect of alternating current in an amplifying circuit.
3. To study dc transistor amplification.
4. To study integrated circuit amplification.

PROJECTS

1. Assemble a one-stage transistor signal amplifier.
2. Assemble a transistor oscillator.
3. Assemble a transistor resistance-coupled amplifier.
4. Assemble a transistor transformer-coupled amplifier.
5. Assemble a direct-coupled amplifier using an N-P-N and a P-N-P transistor.
6. Assemble a push-pull power audio amplifier with two transistors.

EXPERIMENTS

Experiment 1 THE TRIODE VACUUM TUBE

OBJECT

To study the effect of the grid on the electron flow in a triode tube.

APPARATUS

1 - Octal socket
1 - Vacuum tube - 6C5
1 - Filament transformer, 6.3-volt
1 - Tapped resistor, 30 000-ohm, 10-watt (R_1)
1 - Potentiometer, 5 000-ohm, 5-watt (R_2)

1 - DC milliammeter (0-20 mA) (M_1)
1 - DC voltmeter (0-20 V) (M_2)
1 - DC voltmeter (0-300 V) (M_3)
1 - DC power supply, 12- to 24-volt
1 - DC power supply, 350-volt
1 - Variable transformer, 0- to 140-volt

PROCEDURE

1. Review section on the vacuum diode.

2. Study figure 18-17 and make a plan for the placement of components.

3. Connect parts according to the diagram, figure 18-17.

4. Set all instruments for direct current and to the proper range.

5. Set the grid voltage for 12 volts.

6. Check the reading on the milliammeter. It should be zero or the grid voltage must be reversed. A zero reading on the milliammeter insures a negative grid with respect to the plate.

7. Reduce the grid voltage on one-volt steps. The reading on the milliammeter should appear at approximately four volts.

8. Record the milliamperes and plate voltage for each setting.

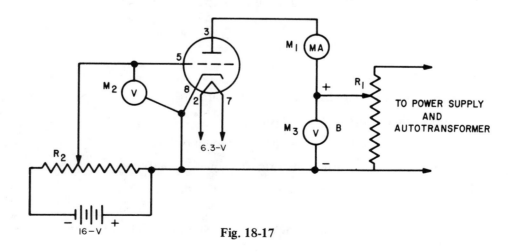

Fig. 18-17

OBSERVATIONS

Use a table similar to the one shown here to record your observations.

| | VOLTAGE | | MA |
	GRID	PLATE	PLATE
1			
2			
3			
4			
5			
6			
7			
8			
9			
10			
11			
12			

QUESTIONS

1. How can the flow of electrons from the grid to the plate be controlled?

2. What kind of current is used on the plate, ac or dc?

3. What causes the plate current to change?

Experiment 2 AMPLIFICATION FACTOR

OBJECT

To study an amplifying circuit.

APPARATUS

1 - Transistor, P-N-P (2N320 or 2N1415)
1 - Resistor, 90 000-ohm
1 - Resistor, 10 000-ohm
1 - Resistor, 900-ohm
1 - Resistor, 200-ohm
1 - Battery, 9-volt
4 - Binding posts
1 - Vector board

PROCEDURE

1. Arrange parts on the vector board.

2. Connect circuit in the schematic diagram, figure 18-18.

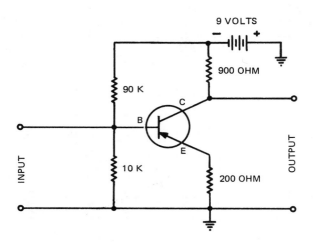

Fig. 18-18

3. Connect microphone or signal generator to the input and a headset to the output. (Keep signal generator output low).

4. Connect oscilloscope across the input and observe pattern.

5. Connect oscilloscope across the output and observe the larger amplified pattern.

6. Connect 20-microfarad electrolyte capacitor across the 200-ohm resistor (positive to ground) and observe the effect on amplifier output.

Note: This arrangement is called a common-emitter circuit. The ground symbol here does not require a connection to the earth. It merely indicates that the + of the battery is to be joined in a common connection with input and output terminals and two resistors. The 900- (or 1 000-) ohm resistor is called the load resistor. The 200 ohm is the emitter resistor. Changes in current in these resistors produce voltage changes at the output. The resistors shown at the left, 10 000 and 90 000, form a voltage divider across the 9-volt battery so that proper voltages are maintained between B and E, and B and C.

OBSERVATIONS

Using graph paper:

1. Draw the oscilloscope pattern for the input pattern.

2. Draw the output pattern.

QUESTIONS

1. Explain the term *amplification factor.*

2. What elements control the electron flow in a P-N-P transistor?

3. Describe the structure of a P-N-P transistor.

Experiment 3 DC AMPLIFICATION

OBJECT

To study dc amplification.

APPARATUS

1 - Transistor, germanium N-P-N, Radio Shack #276-2001
1 - Potentiometer, 1 000-ohm
1 - Transistor battery, 9-volt
1 - Relay (6-volt or lower, normally open contacts)
1 - Selenium solar cell, Radio Shack #276-115

PROCEDURE

1. Connect the circuit as shown in figure 18-7.

2. Using a flashlight and a voltmeter, determine the light and dark voltages produced by the solar cell.

3. Measure and record the voltage across the relay under both light and dark conditions of the solar cell. What is the effect of adjusting R_1?

4. Measure and record both base current and collector current under light-and-dark conditions. Notice that this transistor should have a beta figure higher than 75. Check to see if this is true by comparing I_B and I_C. Notice further that beta changes slightly with changes in collector load current.

5. Control a small motor by the relay contacts and a flashlight.

QUESTIONS

1. Identify practical situations where a light-controlled circuit would be desirable.

2. Design a circuit to turn a device off when the light strikes the solar cell.

3. From the ampere-hour rating of the battery, how long would one expect to operate this circuit in the on state before replacing the battery?

4. Design a power supply to replace the battery.

Experiment 4 INTEGRATED CIRCUIT AMPLIFICATION

OBJECT

To study an integrated circuit amplifier and effect of impedance mismatch.

APPARATUS

1 - Integrated circuit (Motorola MFC 8040)

2 - Capacitors, 1-mfd (C_1 and C_5)

1 - Capacitor, 100-mfd, 25-volt (C_2)

1 - Capacitor, 0.05-mfd (C_3)

1 - Capacitor, 0.1-mfd (C_4)

1 - Resistor, 72 000-ohm (R_1)

1 - Resistor, 270 000-ohm (R_2)

1 - Resistor, 110 000-ohm (R_4)

1 - Resistor, 100 000-ohm (R_3)

1 - Power supply

1 - Signal source, dynamic and crystal microphones

1 - Output transformer, 100-ohms to 4-ohms

1 - Speaker, 4-ohms

PROCEDURE

1. Connect the circuit as shown in figure 18-19.

2. Record the input voltage, output voltage, and voltage gain for each combination of input and output components.

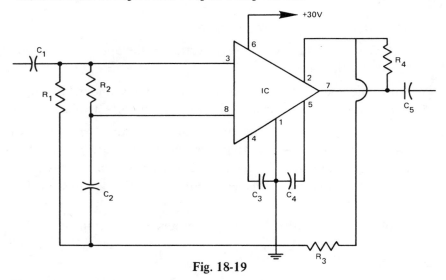

Fig. 18-19

OBSERVATIONS

Use a table similar to the one shown here to record your observations.

	OUTPUT COMPONENTS					
	Speaker			**Transformer and Speaker**		
Input components	Volts in	Volts out	gain	Volts in	Volts out	gain
Dynamic microphones						
Crystal microphones						

QUESTIONS

1. Knowing that dynamic microphones and speakers have low impedance while a crystal microphone has high impedance, what can you conclude about the input and output impedances of the amplifier?

2. What is the effect of lowering supply voltage to the amplifier?

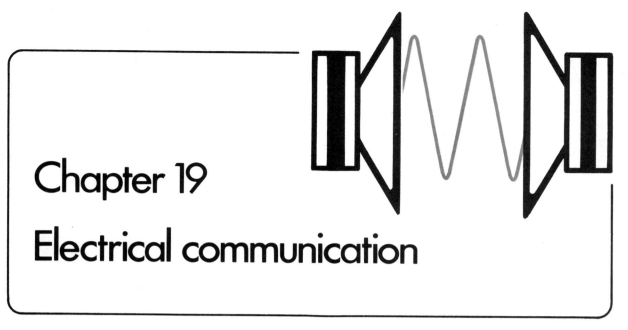

Chapter 19
Electrical communication

THE TELEGRAPH

In the 1750's an idea for an electrostatic-charge telegraph was proposed. It was to consist of twenty-six wires each labeled with a letter of the alphabet. At the sending end, the operator was to charge appropriate wires in succession. At the receiving end, a pith ball attracted to the wires would spell out the message. In the early 1800's, various schemes for using electrolytic effects in liquids were tried. Only two or three years before Joseph Henry showed Samuel Morse how to build a magnetic telegraph, Harrison Dyar operated a telegraph over an 8-mile (13-kilometer) line on Long Island, New York. The earth was used as one conductor and signals were delivered in coded form on litmus paper. The circuit for Joseph Henry's first attempts at magnetic signaling is illustrated in figure 19-2. For early long-line commercial systems, Henry developed the relay (figure 6-20 and figure 19-3) so that the small current in the long main line could control higher currents in the local receiver circuits. The upper portion of the keys shown in figure 19-3 is a spring-loaded switch which remains open unless it is pressed closed. All the keys are in series. The lower switch-level shown on the key is not spring loaded. When an operator

finished sending a message, the switch was closed so that another operator could start sending a message. Figure 19-4 illustrates how drastically radio receivers have changed and improved over the years.

THE TELEPHONE

The first practical telephone was patented by Alexander Graham Bell in 1876 after several years of work. The instrument

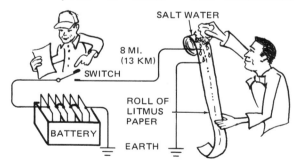

Fig. 19-1 The Electrochemical Telegraph, 1828

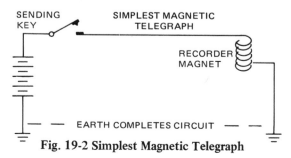

Fig. 19-2 Simplest Magnetic Telegraph

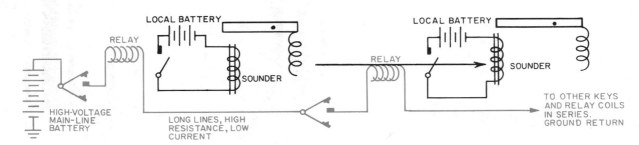

Fig. 19-3 Telegraph System

patented by Bell, figure 19-5, was an improvement on his 1875 model. The 1875 model lacked the L-shaped iron strip and was therefore magnetically ineffective. In using this phone, the user shouts into the horn at the left, causing the flexible leather diaphragm to vibrate in accord with the sound waves that are produced. Fastened to the center of the diaphragm is an iron strip. When the diaphragm and iron move closer to the magnet coil, the magnetic field of the coil increases because the air gap in the magnetic circuit is reduced. (See Chapter 6.) When the iron strip moves away from the core, the magnetic field is decreased. Expansion and collapse of magnetic field, in accord with vibrations of the diaphragm, induce an alternating voltage in the coil. This voltage alternately adds to and subtracts from the battery voltage, producing pulsations in the dc leading to the other instrument.

At the other instrument, the listener's ear is in the horn. The pulsating dc in the magnet coil produces a pulsating magnet. This pull on the pivoted steel strip fastened to the diaphragm causes the diaphragm to vibrate in the same manner as the vibrations originally impressed on the first instrument. In principle, this receiving action is the same as that of the more modern headphones described in Chapter 6. The basic disadvantage of this device lays in its action as a transmitter. All the energy delivered to the receiver had

Fig. 19-4 Radio Receivers, 1923 and 1967

to arise from voltages generated in the coil, which in turn came only from the energy of the sound of the speaker's voice. The battery warmed the coils and maintained the magnetic field, but it contributed nothing to the vibratory impulses applied to the receiver.

Many experimenters, including Edison, Berliner, Hunnings, and White worked on materials which would cause the transmitter

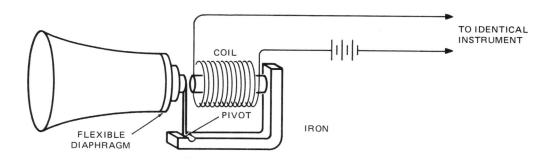

Fig. 19-5 Bell's Telephone Transmitter and Receiver, 1876

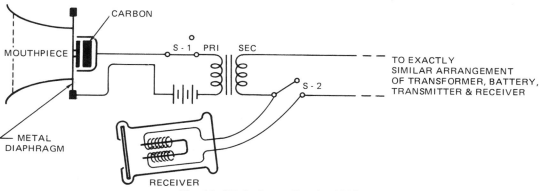

Fig. 19-6 Telephone Circuit, 1890

current to pulsate by causing the resistance of the circuit to change in accordance with sound vibrations, rather than by using induced emf. A compressed mass of granular carbon, called a *carbon button,* was the most successful material. It is still used, although it is different in size and shape. When the diaphragm moves to the right, figure 19-6, and compresses the carbon, the improved contact between carbon particles reduces resistance and permits the battery to produce more current. When the diaphragm bounces back to the left, the carbon has more resistance and current decreases. The amount of change of current is much greater than could be obtained with the previous device. Because the voice sound acts only as a controller, rather than as an energy source, and because it controls the greater energy of the battery, the carbon button and battery circuit are actually an amplifier. Emile Berliner introduced the step-up transformer into the the circuit to permit transmission over longer

distances with less loss of power, which is less $I^2 R$ loss. In figure 19-6 two switches are also shown in their positions when the telephone is in use. When the receiver is replaced on its hook, the movable hook opens S-1 and closes S-2. Though not shown in the diagram, an essential part of the 1890 phone was a hand-cranked ac generator. When operated, the generator connected to the line wires and energized ringers on all telephones on the circuit. It also lighted a signal lamp at a central switchboard. Although this circuit bears an 1890 date, it is essentially practical, and many such circuits are still in use.

The first commercial uses of vacuum tube amplifiers were in radio and telephone communication. Amplifiers made true long-distance telephoning possible. In 1915, conversation between New York and San Francisco was established.

The ears hear vibrations in the range from 20 per second up to about 17 000 per second. Telephone circuits of 60 years ago transmitted a rather narrow range of frequencies. Speech was understandable, but many of the frequency variations that make individual voices recognizable were missing. Most voice energy is in the range of 100 to 3 000 vibrations per second. To improve the frequency range, changes in design (but not principle of operation) have been made in carbon microphones, transformers, transmission lines and receivers throughout the entire network. Presently, receivers use a ring-shaped magnet instead of the horseshoe-type magnet. Direct current to operate the transmitter is supplied from a central station, so the phone user need not be supplied with dry cells in each phone. Transformer coils have been rearranged so that the speaker does not hear his own voice in his own receiver.

A great deal of effort has gone toward automation of the interconnection of telephones. When a dial-system handset is lifted from its holder, a spring-loaded switch disconnects the ringer from the line and connects the line to the carbon-button primary circuit through the dial mechanism. This circuit is energized with dc from the central office. (Series capacitors keep dc out of the receiver, but permit the alternating currents of dial tone or speech to get through the receiver circuit.) Turning the dial winds a spring. If we dial "5", as the dial unwinds the dc is stopped and started five times, which causes one of a set of central relays to step to contact number 5. This closes just one of a whole series of contacts that lead to the receiving telephone.

ELECTROMAGNETIC WAVES

In 1865 a young Scotsman, James C. Maxwell, influenced by Faraday's thoughts about the transmission of electric and magnetic force through space, devised mathematical formulas to interconnect and explain all that was then known about electricity and magne-

tism. Out of his formulas came the mathematical description of the transfer of energy by a wavelike disturbance in space. His formulas even gave the velocity of the wave. Maxwell wrote: "This velocity is so nearly that of light, that it seems we have strong reason to conclude that light itself (including radiant heat, and other radiations if any) is an electromagnetic disturbance in the form of waves propagated . . . according to electromagnetic laws."

About twenty years later, one of Maxwell's suggested "other radiations" was discovered and investigated by Heinrich Hertz, in Germany. Hertz, using high voltages from a spark coil on circuits containing coils, capacitors, and spark gaps, found he could produce high-frequency alternating currents in the circuits which sent out electric waves. These waves were similar to those now used in TV transmission, in that their frequency was a few hundred megahertz. Hertz's transmitting and receiving antennas might remind one of the simplest TV receiving antenna, just two short rods end-to-end, which is called a *dipole,* figure 19-7. The name dipole, or two-pole, is often applied to something that is positive on one end when the other end is negative.

With crude equipment, Hertz was able to show that these rays reflect from metal and can be reflected and focused by concave mirrors. This had been done with light and is now done with radar and microwaves. He found that they were refracted by passing through a triangular prism (like light) and that they are *polarized.* A polarized electric wave tends to kick electrons back and forth in one

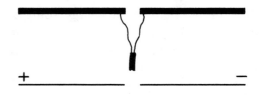

Fig. 19-7 Simple Dipoles

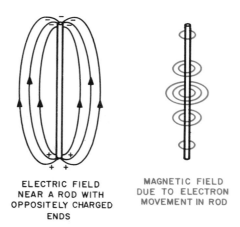

Fig. 19-8 Electric and Magnetic Fields

preferred direction. For example, TV broadcast stations send out waves that are horizontally polarized. For that reason, rooftop TV antennas have an arrangement of horizontal conductors, so the wave can shake electrons back and forth horizontally in the antenna rods. The combination of electric and magnetic waves is called *electromagnetic waves.*

Radiation of Electromagnetic Waves

To better understand what electromagnetic waves are, think about electrons vibrating up and down in a rod 50 or 60 times per second. When electrons are moving, the rod is surrounded by a magnetic field, figure 19-8, as was described in Chapter 6. As the current increases, circular magnetic lines of force expand outward from the wire and collapse back as the current decreases. When a cluster of electrons has been pushed to one end of the rod, leaving a deficiency at the other, an electric field exists near the wire, figure 19-8. When electrons run back to the positive end of the rod, the electric field collapses back into the wire, to reappear later in the opposite direction when electrons accumulate at the other end.

At low frequencies of vibrations, (below 20 kilohertz) energy stored in electric and magnetic fields seems to be returned to the

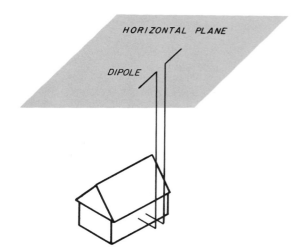

Fig. 19-9 Horizontal Dipole

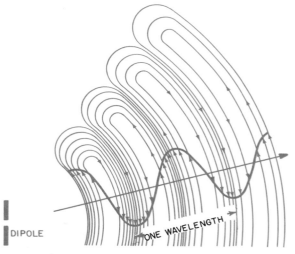

Fig. 19-10 Portion of Electric Field of Radiating Dipole Antenna

conductor when the field collapses. However, at higher frequencies (100 kilohertz and above) the energy in the fields appears to escape its source and speed off into space. The higher the frequency, the greater the tendency for energy to radiate rather than return. The radiated wave of energy travels away from its source at the speed of light, 186 000 miles per second.

Wave Patterns

The electromagnetic wave sent out from a dipole is a three-dimensional pattern of

Fig. 19-11 Dipole Antenna-Vertical Plane

electric and magnetic lines of force, moving together. Observe the lines of force that expand outward in the horizontal plane of the dipole, figures 19-9 and 19-10. In the horizontal plane, the pattern of radiating electric field lines, if visible, would look something like figure 19-10. The same pattern is present in any plane that includes the dipole, such as the vertical plane in figure 19-11. The *magnetic field* lines show up in planes perpendicular to the dipole, figure 19-6, 19-12. Maximum energy is radiated in directions perpendicular to the dipole at its center. Theoretically, no energy is radiated off in the directions pointed out by the ends of the dipole. The electric lines of figure 19-10 and the magnetic lines of figure 19-12 must be combined to form a picture of the electromagnetic wave radiating away from the transmitting dipole. Compare this combination to a succession of spherical balloonlike surfaces blowing up as they expand from the center of the dipole. Figure 19-13 is an attempt to picture just small portions of these spherical wave-fronts.

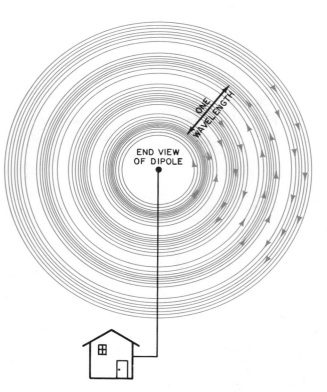

Fig. 19-12 Magnetic Field Around Radiating Dipole

Hold a similar receiving dipole parallel to the transmitting dipole. The receiving dipole is hit by alternating electromagnetic field patterns in succession, figure 19-14. Horizontal electric fields try to kick electrons back and forth horizontally, and vertical magnetic fields cut across the horizontal wire, also trying to move electrons back and forth in the wires of the horizontal dipole. This is the picture of energy transfer through space by electromag-

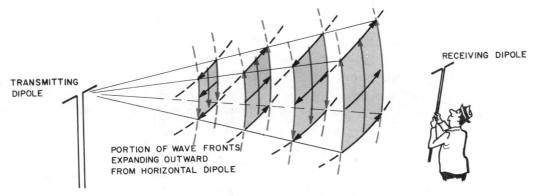

Fig. 19-13 Radiation Through Space

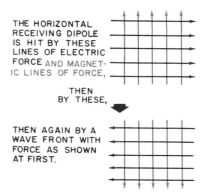

THE HORIZONTAL
RECEIVING DIPOLE
IS HIT BY THESE
LINES OF ELECTRIC
FORCE AND MAGNET-
IC LINES OF FORCE,

THEN
BY THESE,

THEN AGAIN BY A
WAVE FRONT WITH
FORCE AS SHOWN
AT FIRST.

Fig. 19-14

netic waves. Electromagnetic waves are alternations of magnetic and electric lines of force.

Selection of Wave Frequencies

In the preceding discussion, the Hertzian dipole has been used because it is a simple and frequently seen antenna. There are a hundred other ways of arranging wires for transmitting and receiving antennas. When a radio antenna is put up into the air, waves from many transmitters are sweeping across it, but the listener gets only one station because in any one region different transmitters radiate energy at different frequencies. The selection of one frequency is determined by a portion of the receiver circuit to which the antenna is attached, called the tuner. The tuning, which is the selection of a station operating at a particular frequency, is done by coils and capacitors.

In Chapter 15, properties of coils and capacitors such as, ohms reactance of a coil,

$X_L = 2\pi f L$; ohms reactance of a capacitor, $X_C = \dfrac{1}{2\pi f C}$; and, impedance of a coil and capacitor in series, Z, can be found from $Z^2 = (Z_L - X_C)^2 + R^2$. If X_L and X_C differ by a large amount, then $X_L - X_C$ is numerically large, and their series impedance is very high.

X_L and X_C depend not only on L (henries) and C (farads), but also on frequency. For example, consider what might happen if the effect of a small alternating voltage at different frequencies is tried on the circuit shown in figure 19-17. The table shows the results of calculating X_L and X_C ohms for several different frequencies. The impedance of the circuit, assuming resistance is small enough to forget, is given by the difference of X_L and X_C. Notice that at 700 kilohertz the impedance is small in comparison to that for other frequencies. From the trends of the columns of numbers, it may appear that X_L could equal X_C for some value of frequency between 700 and 800. A formula for finding that particular frequency, starting with the requirement that X_L equals X_C, is

$$2\pi f L = \dfrac{1}{2\pi f C} .$$

Multiply both sides by $2\pi f C$:

$2\pi f L \times 2\pi f C = 1$, or $4\pi^2 f^2 LC = 1$.

The square roots of each side of the equation are equal, so $2\pi f \sqrt{LC} = 1$, which can be rearranged to $f = \dfrac{1}{2\pi \sqrt{LC}}$

with L in henries and C in farads. If 150

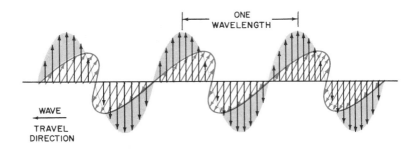

ONE
WAVELENGTH

WAVE
TRAVEL
DIRECTION

Fig. 19-15 Graph of Electromagnetic Wave

frequency in kilohertz (kH)	X_L ($2\pi fL$)	X_c ($\frac{1}{2\pi fC}$)	difference, $X_L - X_c$ or $X_c - X_L$
400 kc	377 ohms	1 250 ohms	873 ohms
500	472	1 000	538
600	566	833	267
700	660	715	55
800	850	545	305
1 000	942	500	442
1 000 kH	1 036 ohms	454 ohms	582 ohms

319μμf   150μh

Fig. 19-16

millionths of a henry is put into this formula for L and if 319 micro-microfarads are put into this formula for C, f works out to about 727 000 hertz, or 727 kilohertz. At that frequency, $X_L = X_C$, so the current in the part of the circuit containing the coil and capacitor will be limited only by the small resistance in the circuit. A small voltage can stir up a large amount of current at that frequency, which is 727 kilohertz. Notice from the last column of figures in the table, figure 19-16, that as frequencies become farther from 727 kilohertz either above or below, greater opposition is set up by the coil-and-capacitor circuit to currents of those frequencies. If we calculate $(X_L - X_C)$ for 730 kilohertz, we find about 4 ohms, but for 800 kilohertz we find 130 ohms, over 30 times as much.

Tuning Devices

If an antenna is connected to a coil and capacitor similar to that shown in figure 19-17, and the antenna is hit by electromagnetic waves from a hundred different broadcasting stations, the coil-and-capacitor tuning circuit conducts only one of those wave frequencies well. Little or no current is circulated between the coil and capacitor at other frequencies. Tuning to another station may be done by changing the capacitance of the capacitor, as in an ordinary radio set. Some receivers select

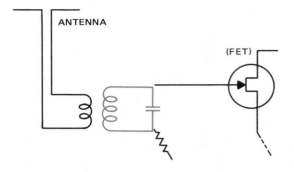

Fig. 19-17 A Simple Tuning Device

stations by moving a ferrite core inside of the coil, thus changing the inductance of the coil. Changing channels in a TV set is done by switching one set of coils out of the circuit and connecting another set in its place. In figure 19-17, the antenna connects to the primary of a transformer, which is generally

Radio Frequencies for Communication

100-550 kHz: Marine communications, air and marine navigation aids

550-1 600 kHz: Commercial AM broadcasting

1 600-50 000 kHz (1.6-50 mHz): Aircraft, amateurs, fire and police, highway and railroad, Loran, government and international, beam relays, etc.

50-54 MHz: The 6-meter amateur band

54-88 MHz: TV, channels 2-6

88-108 MHz: FM broadcasting

108-174 MHz: Government uses, aero navigation and control, amateurs, police, etc.

174-216 MHz: TV, channels 7-13

216-460 MHz: Government, civil aviation, radio altimeters, weather instruments, amateurs, etc.

460-470 MHz: Citizens' band

470-890 MHz: UHF TV, channels 14-83

890-30 000 MHz: TV relays, aircraft navigation and landing instrument systems, amateur and experimental work, radar, etc.

wound on a non-magnetic core. For very high frequencies, such as TV, the number of turns depends on the type of antenna and transmission line and the type of amplifying circuit, as well as the range of frequencies to be received. At lower frequencies, such as ordinary radio, no external antenna is needed. The part of the circuit shown in color in figure 19-17 acts both as receiving antenna and as tuning circuit. On some receivers the coil is a rectangular arrangement of about 25 or 30 turns of wire spread out on the back panel of the case of the radio. On others, a multi-turn coil is wound on a pencil-size magnetic core.

TRANSMISSION

Electromagnetic waves used for communication are not sound waves. They travel through empty space even more successfully than through air. Their frequencies are far too high to be heard if they were sent into a speaker or headphone. In order to make these waves carry information that can be used to produce sound in a speaker or a picture on a TV tube, changes must be made in the electromagnetic waves many times per second. These changes are made while the wave is generated by the transmitting station. Figure 19-18(A) shows a graph of a steady radio wave. This graph should be compared with figure 19-15 which shows electric and magnetic forces in a portion of a wave only three wavelengths long; figure 19-18(A) shows only the sine-wave variations of the electric field, for about 40 wavelengths. The steady wave of figure 19-18(A) represents the wave sent out by a radio transmitter during periods of silence when no information is transmitted. Seventy years ago, the only way to use the wave for communication was to turn the wave on and then shut it off, as in the key-operated wireless telegraph. Key-operated continuous wave (CW) transmitters are still

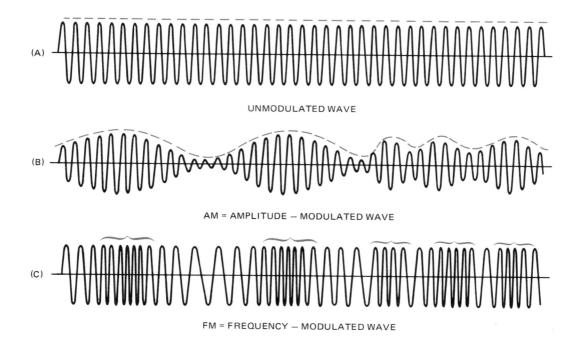

(A)

UNMODULATED WAVE

(B)

AM = AMPLITUDE – MODULATED WAVE

(C)

FM = FREQUENCY – MODULATED WAVE

Fig. 19-18 Graph of Electromagnetic Waves

very much in use in amateur and military communications. Code transmission can be understood under conditions of noise and interference where voice transmission would not be intelligible.

Amplitude Change

The height of the sine wave in figure 19-18(A) is called the *amplitude* of the wave. It represents the strength of the electric field. One way to make the wave carry information is to change the amplitude of the wave produced by the transmitter. Figure 19-18(B) shows a wave whose amplitude is changing frequently. Notice that in the part of the wave at the right, the amplitude changes more often than it does at the left. An ordinary AM radio receiver keeps track of amplitude changes. If the amplitude changes a thousand times per second, the receiver produces a sound of a thousand vibrations per second in the speaker. A slower change in amplitude, such as 200 times per second, produces a lower pitched sound in the speaker, for example, 200 vibrations per second. The amount of amplitude change controls the loudness of sound produced in the speaker. Looking at the color line along the wave tops in figure 19-18(B), notice that the amount of change of amplitude, which is the amount of up-and-down variation, has been increased in the left side of the graph. This greater variation causes the receiver to produce a louder sound than will be produced by the part of the wave at the right. *Amplitude modulated* means changing in amplitude.

These graphs can show only what happens in less than a hundredth of a second; drawing the graph of as much as one second of operation would be too complex. A station broadcasting at 1 000 kilohertz sends out one million waves every second. Perhaps during one of those seconds, they wish to produce the sound of a 500-hertz trumpet note. The 1 000-kilohertz wave increases and decreases in amplitude 500 times per second. To show all this completely, the colored line of figure 19-18(B) would have to rise and fall 500 times. Each hump in the colored line would need to enclose 2 000 little electric sine waves. Amplitude modulation of the TV broadcast controls the formation of the picture in the TV set. One million changes in amplitude per second may be applied to the several million hertz wave used for TV.

Frequency Change

Figure 19-18(C) shows another way to change a wave. The amplitude remains steady, but the frequency often changes slightly. This process is used to carry the sound that accompanies the TV picture as well as carrying sound information for FM radio. Compare graphs (B) and (C) in figure 19-18. Both are intended to convey the same information to the appropriate receiver. In frequency-modulated receivers (FM), the rate of increase and decrease of frequency determines the rate of vibration of the speaker. The amount of frequency change determines the loudness of the sound to be produced.

Ordinary AM radio receivers readily pick up many sorts of interference: crackles, buzzes and howls from lightning, commutator motors, and improperly built TV sets. These sources of interference produce amplitude-modulated waves. A well-built and well-adjusted FM receiver is unaffected by amplitude changes, hence, FM sound can be still clear when an electric mixer is operating nearby. For this reason, FM has developed high-quality musical programs for the enjoyment of its audience.

OSCILLATORS

So far, much has been said about radio frequency (RF) voltages alternating at frequencies of millions of hertz without explaining how they are generated. These frequencies are generated in tube or transistor circuits called *oscillators.* Oscillators are basically amplifiers with part of their output signal connected back into the input. When a public address system is too loud, the system generates its own noise, which is usually painful, because the microphone is listening to the noise from its own speaker. This is amplified and sent to the speaker again, with increased loudness. In radio-frequency oscillators, this feedback is deliberately accomplished in many ways.

Consider the FET radio-frequency amplifier of figure 19-19. An RF signal of the correct frequency, applied to the primary of the input transformer will induce RF currents in L_1 and C_1, causing the voltage of the gate to increase and decrease. The changing gate voltage causes the drain current to rise

and fall. However, the parallel combination of C_2 and L_2 in the drain circuit has a high impedance to the changing current if they are also resonant at the input signal frequency. Ohm's law for ac indicates that a voltage drop must be associated with a changing current through an impedance. L_2 acts like the primary of a transformer. Voltage at the RF frequency is available from the secondary output coil.

Now consider the same circuit modified by connecting the primary of the input coil in series with the drain of the transistor. Figure 19-20 shows that any output current must go through the input transformer primary. L_1 and C_1 experience electron upset when the current is turned on and begin a weak oscillation at their resonant frequencies. This oscillation is amplified by the transistor. Some of the output energy

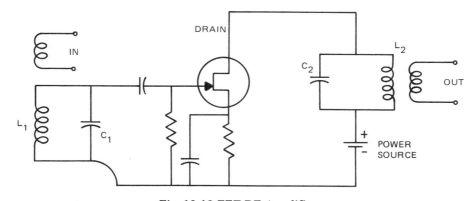

Fig. 19-19 FET RF Amplifier

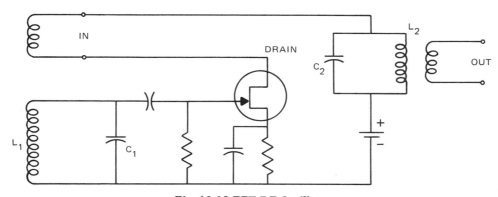

Fig. 19-20 FET RF Oscillator

is sent back into the input primary coil; the rest of the output appears across L_2 and C_2. By listening to itself, the amplifier continuously maintains an RF frequency at its input and output.

There are many different ways of arranging oscillator circuits. One essential in all of them is that energy must be taken from the output circuit of the tube or transistor and fed back into the input circuit (grid and cathode of tube or base and emitter of transistor). This feedback of energy must have the right phase relationship so it will strengthen the pulses of current. The feedback should not have an out-of-step character which cancels out oscillations when they start.

DEMODULATION

After station selection, demodulation is probably the most important function of a radio receiver.

Modulation means changing the radio wave by adding sound-producing information to it. *Demodulation* means changing it back again so that the sound-producing information is obtained from the wave. Demodulation is sometimes called *detection*. For amplitude-modulated signals, the demodulator is a rectifier. Figure 19-21 diagrams a simple receiver consisting of tuned circuit and a diode as a rectifier. Years ago these were called a crystal set, because mineral crystals were used as rectifiers. A germanium diode can be used in building a simple receiver. Unless you live very close to a powerful transmitter, the long antenna is needed to pick up enough energy to operate the headphones. The device has no battery.

Current induced in the tuned circuit applies a voltage to the rectifier and headphone like that diagrammed in figure 19-18(B). Since the diode passes electrons in only one direction, current in the phone will be something like that shown in figure 19-22. One million little pulses each second may be applied to the headphone, but its diaphragm cannot vibrate that rapidly. Figure 19-22, if more accurately drawn, might show a thousand tiny pulses under each rise and fall of the color line that represents the modulation. The thousand little pulses add up to one big push on the headphone diaphragm, one big push that may last only one-thousandth of a second. The headphone can and does vibrate at audio frequencies, responding to the rate at which the waves change in amplitude, figure 19-23.

Occasionally there is an advertisement for a transistor radio that requires no power-line cord and no batteries. The advertiser has rediscovered the circuit of figure 19-21. It can be built with even fewer parts than shown here. The signal from such a radio is very weak and inaudible if it is not operated near a commercial radio station. If there are

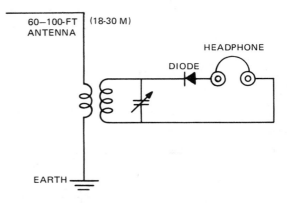

Fig. 19-21 A Crystal Radio Receiver

Fig. 19-22

many radio stations nearby, such a radio will receive a number of them at the same time. Sensitivity and selectivity are severely limited in these radios.

THE AM RECEIVER

To develop a receiver with enough power to operate a loudspeaker from the simple receiver of figure 19-21, amplifier circuits are needed. Amplifiers get their energy from either rectified ac or from a battery. The signal of the antenna will only control energy, not supply it.

Before reading the next few lines, you might have some thoughts about a type of amplifier to use, and where to insert it into figure 19-21. Two possibilities might occur, both of which are used in radio receivers. One is to use an RF amplifier, as in figure 19-19, between the antenna and the diode. Another is to use an audio amplifier between the diode and headphone. To get still greater power output, a whole row of amplifiers, called a *cascade,* can be used. Figure 19-24 shows a plan that will work for connecting amplifiers. It also shows two methods of coupling amplifiers. *Coupling* means passing energy from the output of one amplifier stage to the input of

the next amplifier. The RF amplifiers are transformer coupled. The transformers are usually wound on a nonmetallic core and are part of tuned circuits. Audio amplifiers are usually capacitor coupled because, compared to transformers, it is less expensive to build a network of resistors and capacitors that will meet the requirement of handling the wide range (25 to 16 000 hertz) of audio frequencies without loss of highs or lows. To finally deliver some watts of power to a speaker, an iron-core output transformer is depended upon.

The RF amplifier portion of figure 19-24 is outdated. It is difficult to make an RF amplifier with high sensitivity and selectivity and uniformly high amplification over the entire broadcast band, unless several stages are used. This increases the cost. In the circuit of figure 19-24, two stages of RF amplification are shown. However, three or four might have been used, amplifying the antenna signal at any carrier frequency from 550 to 1 600 kilohertz. A less expensive system called superheterodyne has been devised. This system achieves good reception and high amplifier gain in a small, easily built circuit.

In the superheterodyne receiver, the greatest part of the amplification is done

Fig. 19-23

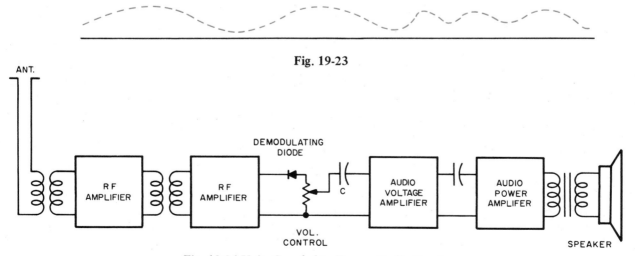

Fig. 19-24 Units Coupled to Form a Radio Receiver

at one new carrier frequency which is produced in the receiver itself, rather than at the frequency received by the antenna. The new carrier frequency is produced by a mixing of two frequencies. *Heterodyning* means frequency mixing.

If the equipment is available, do an experiment to find out what happens when two different frequencies are mixed together. Figure 19-25 suggests two ways of mixing audio frequencies so the results can be heard. In (A), two audio oscillators should be set at the same frequency and adjusted to produce about the same loudness in each pair of phones. While listening to both sounds at once, very slowly turn the frequency control of one back and forth, and adjust to produce about the same loudness in each pair of phones. While listening to both sounds at once, very slowly turn the frequency control of one back and forth. At some points a throbbing sound should be heard. For example, if one is producing a 500 vibrations per second sound and the other is producing 504

vibrations per second, four pulses of loudness each second can be heard. Or in (B), if a matched pair of mounted tuning forks is available, fasten a small weight on a prong of one fork to slow its vibration, then hit both forks. Perhaps each fork originally produced 256 vibrations per second (vps), and one was slowed to 253 vibrations per second. Three pulses per second should be expected. When two frequencies are mixed together, the results of the mixing include a new frequency. The new frequency, which is the difference of the two original vibrations, is sometimes called the *beat frequency*. The throbbing or pulsing sensation of sound is called *beats*.

The part of the receiver circuit in which the frequency mixing is done may be called either a *mixer* or *converter,* figure 19-26. Some amplification of the signal takes place in the converter, but a great voltage gain is made in the intermediate frequency (IF) amplifier which follows the converter. A high-gain amplifier can easily be built for one frequency. Heterodyning makes it possible to get as much gain out of two tubes as one would get from 3 or 4 stages of RF amplifiers built to convert a wide range of frequencies.

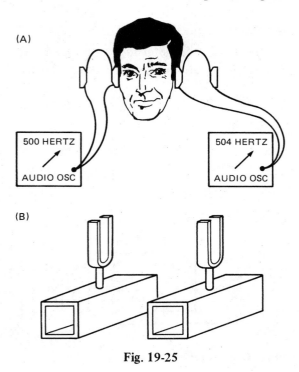

(A)

(B)

Fig. 19-25

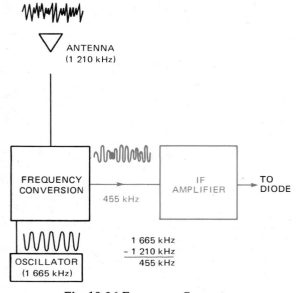

Fig. 19-26 Frequency Converter

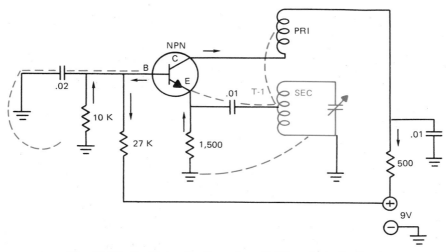

Fig. 19-27 Oscillator Portion of an AM Transistor Radio

Converter Circuit

The variable capacitor in a radio consists of two sections mounted on one shaft. The larger section with more plates tunes the antenna circuit for station selection. At the same time, the smaller section tunes an oscillator circuit to a frequency which is 455 kHz higher than the frequency of the received station. Both of these signals, one from the antenna and one from the oscillator, are fed into the same tube or transistor. Figure 19-26 shows the situation when a receiver is tuned to a broadcast station at 1 210 kHz. The oscillator

at the receiver is incidentally tuned to 1 665 kHz. Mixing of 1 210 and 1 665 produces a new intermediate frequency, 455 kHz which carries the same amplitude modulation as was carried by the wave received at the antenna. If the listener turns his tuning knob from 1 210 to 780 kHz, the oscillator is retuned to 1 235 kHz. The output of the converter circuit is the difference of 1 235 and 780. 1 235 − 780 = 455 kHz.

Figure 19-27 shows the oscillator portion of one type of transistor frequency converter for a small AM radio. Its frequency is set by

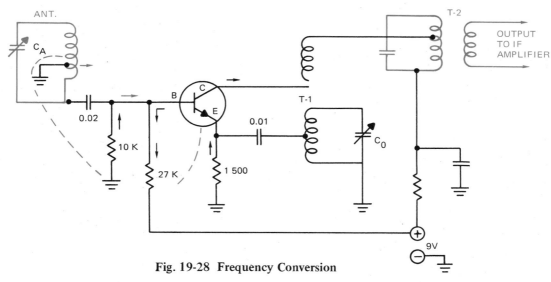

Fig. 19-28 Frequency Conversion

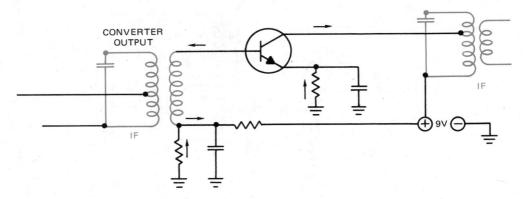

Fig. 19-29 Transistor IF Amplifier

the tuning capacitor across the secondary of T-1. This signal is fed into the emitter through the 0.01-μf capacitor, or in more complete detail, to the emitter-base junction through the 0.01 and 0.02 capacitors and the ground connection. The amplified collector current in the primary of T-1 feeds energy back into the tuned circuit to maintain the oscillation. Resistors in the circuit maintain proper dc voltages at the transistor terminals. Black arrows show the direct currents.

Figure 19-28 shows two additions to the above oscillator circuit. A signal from the antenna and its tuning capacitor is also fed into the base-emitter junction through the base. The collector current now has two controlling factors, for example, an antenna signal at 1 000 and an oscillator signal at 1 455 kHz. The combining of these two frequencies in the collector current develops 455 kHz pulses in the collector current. The resonant circuit of T-2 is tuned to 455 kHz so amplitude-modulated 455 kHz IF is fed into the IF amplifier. C_A and C_O are mechanically mounted together, both retune at the same time when someone changes stations.

IF Amplifier

Figure 19-29 shows a simplified diagram of a transistor IF amplifier. The IF transformers are set for 455 kHz. The signal from the converter is applied to the base-emitter junction in the transistor. The values for resistors depend on the transistor used. Their main function is to maintain proper dc voltages in the various parts of the circuit.

Complete Superheterodyne Receiver

A satisfactory receiver using six transistors and a diode is shown in figure 19-30. The output of the audio voltage amplifier is coupled into a two-transistor arrangement called a push-pull amplifier circuit. This arrangement provides a higher useful power output with less distortion of sound and less wasted energy than any other two-transistor arrangement. In effect, one transistor amplifies the top half of the wave and the other the bottom half of the wave. High power or high-fidelity tube-type audio amplifiers also use a push-pull circuit.

Sound quality obtained from an amplifier depends greatly on the speaker. Often, a tinny-sounding radio or record player can be improved by a new and larger speaker.

POINTS TO REMEMBER

- In the telephone transmitter, sound waves cause variations of dc through carbon grains. Variations of dc in the primary of a transformer induces alternating currents which vibrate the diaphragm in the receiver.

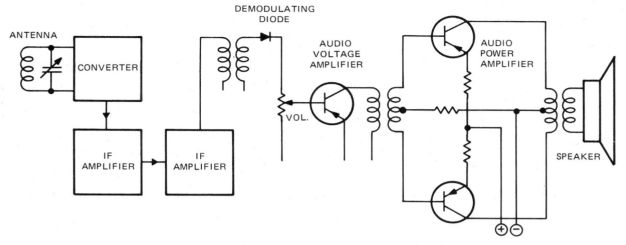

Fig. 19-30 Six-Transistor Receiver

- Electromagnetic waves (radio, TV, light infrared, etc.) consist of alternating electric fields and alternating magnetic fields, moving together at the speed of light.

- For a tuned circuit, resonant frequency may be found from:

$$\frac{1}{2\pi\sqrt{LC}}$$

- Feedback of energy, in proper phase, produces oscillations in a tube or transistor circuit.

- Detection, or demodulation, is the extraction of audio (sound) frequencies from a modulated **RF** wave.

- Heterodyning, the mixing of two different frequencies, produces in the mixture a new frequency which is the arithmetical difference of the two original frequencies.

- Superheterodyne receivers convert the carrier frequency received at the antenna to a new intermediate frequency (IF) to achieve high gain and good selectivity with economy of construction.

REVIEW QUESTIONS

1. Draw a diagram of a simple two-way telephone circuit. Label the parts.

2. Make a sketch of a telephone receiver and a transmitter and explain how each part operates.

3. What is meant by the term *polarized waves?*

4. Why do TV antennas have an arrangement of horizontal conductors?

5. What are electromagnetic waves? Who predicted their discovery? Who discovered and developed them?

6. Explain how the tuning circuit in a radio or TV set is used to select a desired frequency.

7. What are three important elements in a simple crystal radio receiver?

8. Name several important operations in a superheterodyne receiver.

9. Define the term *heterodyning* and explain how it is accomplished.

10. Explain how oscillation is achieved in a circuit. Describe one essential that must be present in all oscillators for them to function properly.

11. What are the advantages of a 2-transistor, push-pull audio output circuit?

12. Define the following terms: amplification, amplitude, modulation, demodulation, and rectification.

13. What is the difference between audio frequency (AF) and radio frequency (RF)? How does an RF transformer differ from an AF transformer?

14. What is the most unique feature of the superheterodyne receiver circuit?

15. What is the purpose of the circuit illustrated in figure 19-28? How does it function?

16. One of the wavelengths assigned for amateur communication is 20 meters. Electromagnetic waves travel through space at a speed of 300 000 000 meters per second. With waves 20 meters apart (figure 19-8 or 19-13), how many waves are sent out each second?

17. A resonant circuit contains a coil with L = 0.1 henry and C = 0.4 microfarads. Calculate its resonant frequency. If ac of this frequency were sent into a headphone, could it be heard?

RESEARCH AND DEVELOPMENT

Experiments and Projects on Electrical Communications

INTRODUCTION

Communication is a vital part of daily life. Modern technology has made it possible to receive news of important events instantly from all over the world. In Chapter 19 you had an opportunity to study about the development of the telegraph, the telephone, and the discovery of electromagnetic waves. You have learned how electromagnetic waves are transmitted, how they travel through the atmosphere, and how they are received.

The suggested experiments and projects are included to provide an opportunity to become familiar with the different kinds of communication devices and to learn how to connect and use them. The radio receivers are included so you can learn more about the operation of vacuum tubes and transistors in electronic circuits.

It is suggested that these projects be assembled from experimental kits or, in the case of take-home projects, from commercial kits, or parts selected to assemble a receiver according to specifications given in manuals or reference books. Many of the parts required to assemble them can be salvaged from old radios or procured through surplus property sources.

EXPERIMENTS

1. Assemble an oscillator and control the frequency of oscillation with capacitors.

PROJECTS

1. Assemble and test a complete telegraph circuit consisting of keys, relays, sounders, and batteries.

2. Wire a simple two-way telephone circuit.

3. Design and build a telephone receiver and a transmitter.

4. Assemble an AM single-tube radio detector set.

5. Assemble a multiple tube AM receiver.

6. Assemble a multiple transistor AM receiver.

7. Assemble an FM receiver.

EXPERIMENTS

Experiment 1 OSCILLATOR

OBJECT

To learn how to assemble an oscillator and to control the frequency.

APPARATUS

1 - Octal tube base
1 - Triode vacuum tube, 6J5 or 6C5
1 - Resistor, 1-megohm, 1/2-watt (R_1)
1 - Capacitor, 0.000 5-μf (C_1)
1 - Capacitor, 0.01-μf (C_2)
1 - Transformer, 6.3-volt (T_1)
1 - Set capacitors, 0.000 1, 0.000 22, 0.000 47, 0.001-μf (C_3)
1 - Set phones (P_1)
1 - Key (K_1)
1 - DC supply, 90-volt
1 - Vector board

PROCEDURE

1. Select the parts and mount them on a base.

2. Wire the circuit with size 0.001-μf capacitor for C_3, figure 19-31, page 476.

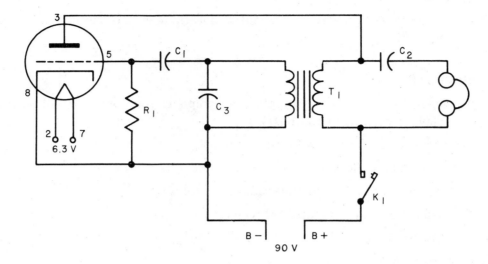

Fig. 19-31

3. Check the circuit and test it. *Note:* If you do not get a tone, reverse the connections to either the primary or secondary of the transformer, but not to both.

4. Check the frequency and record.

5. Remove capacitor C_3 and insert another such as 0.000 22 μf.

6. Check the frequency and record.

7. Use the same procedure for the two remaining capacitors.

OBSERVATIONS

Use a table similar to the one shown here to record your observations.

OBS.	FREQUENCY CHANGE
1	
2	
3	
4	

QUESTIONS

1. From what you have observed, could you plan a circuit that would oscillate at a specific frequency? If you could, explain how you would proceed. If you could not, why not?

2. Explain what happens in the circuit to cause the frequency to change.

3. How could a circuit be arranged so the frequency could be changed without substituting capacitors?

PROJECTS

ASSEMBLE AND TEST A COMPLETE TELEGRAPH CIRCUIT CONSISTING OF KEYS, RELAYS, SOUNDERS, AND BATTERIES

The telegraph developed by J. Henry, was the first dependable device perfected for communication over long distances. It could be used to send messages from one place to another in a matter of minutes. The letters in each word were spelled in a code which consisted of a series of dots and dashes representing the letters of the alphabet. A telegraph circuit is a simple but interesting device to assemble and use.

MATERIALS

2 - Telegraph keys
2 - Telegraph sounders
1 - Telegraph relay
    Annunciator wire
    Dry cells

PROCEDURE

Arrange the component parts on the bench or on a suitable board and wire the circuit, figure 19-32. Adjust the instruments until each operates satisfactorily.

After you have operated the single system, plan and connect a circuit with one or more relays and another key and sounder so messages can be sent back and forth. Record your observations.

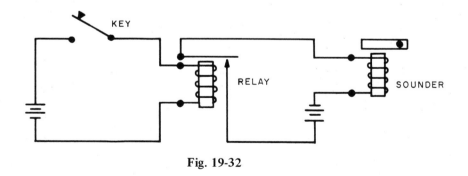

Fig. 19-32

WIRE A TWO-WAY TELEPHONE CIRCUIT

The telephone, one of the most important communication devices, consists in its simplest form of a transmitter, a receiver, and a source of current. The transmitter produces a variable current that has the same characteristics as the sound waves which caused it. The receiver converts this variable current into sound waves which reproduce the original sound. With a few basic parts you can set up an intercommunication telephone system. Use the simple series circuit or set up a call-bell system for signaling the person you wish to call.

MATERIALS

    2 - Telephone transmitters
    2 - Telephone receivers
    2 - Receiver hooks or SPST switches
    4 - Dry cells, 1 1/2-volt or other dc power source
    2 - SPST switches
    2 - Bells or buzzers
        Annunciator wire
        Additional equipment needed for assembling a call-bell system

PROCEDURE

Mount a transmitter and receiver on separate panels. Place the panels thirty or forty feet apart and connect the total circuit in series.

The wire is run from the power source through the transmitter and receiver of one set, to the receiver of the other set, and through the receiver and transmitter back to the power source, figure 19-33.

On-off switches or receiver hook-switches should be placed in the circuit to prevent power loss when the telephones are not in use.

To connect a call-bell system, use conventional wall-type telephones or plan an arrangement consisting of transmitter, receiver, bell, switches, and batteries for each unit. Test and evaluate.

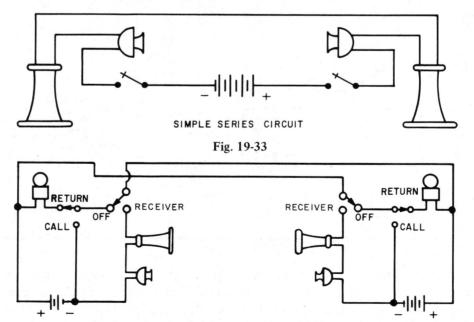

SIMPLE SERIES CIRCUIT

Fig. 19-33

Fig. 19-34 Local Battery-Type Telephone System With Call-Bell

DESIGN AND BUILD A TELEPHONE RECEIVER
AND A TRANSMITTER

This simple telephone receiver and the carbon-button transmitter operate on the same principle as the equipment used in modern telephones. If you are interested in experimenting as the inventors did, you can make a set of instruments that will work.

If you are interested in experimenting as the inventors did, you can make a set of instruments that will work.

The handle, body, and cap of the receiver are made of wood; the diaphragm, of thin tin plate; the permanent magnet, of tool steel; and the electromagnet, of fiber washers and magnet wire.

The body and cap of the transmitter are made of hard wood; the diaphragm, of thin tin plate, and the buttons, of carbon. The carbon electrode from a number six dry cell is ideal material for buttons and granules.

The design for the receiver may be varied according to your equipment and materials. For instance, the body and cap may be made square with rounded corners. Cardboard spacers may be used between the diaphragm and the cap and body. The handle may be made from a curtain rod, the diaphragm from a tin can, the permanent magnet from an old round file and the electromagnet from a discarded transformer. Transmitter designs may be varied in a similar manner.

MATERIALS FOR RECEIVERS

1 - Body, hardwood, 1" x 2 3/4" x 2 3/4" (25 mm x 70 mm x 70 mm)

1 - Cap, hardwood, 3/16" x 3" x 3" (5 mm x 75 mm x 75 mm)

1 - Handle, hardwood, 1 5/8" x 1 5/8" x 4" (41 mm x 41 mm x 101 mm)

1 - Diaphragm, tin plate, 0.006" to 0.010" x 2 1/4" x 2 1/4" (0.15 to 0.25 mm x 57 mm x 57 mm)

1 - Permanent magnet, tool steel, 3/8" diameter x 4 1/6" (9 mm diameter x 103 mm)

2 - Fiber discs, 1/16" x 1 3/16" (1.5 mm x 30 mm) diameter

2 - Machine screws, #6 x 32 x 3/4" (M3.5 x 20 mm L) RH

4 - Machine screw nuts, #6 x 32 (M3.5), hexagonal

2 - Wood screws, #4 x 1/4" FH

3 - Brads, #18 x 1"

Magnet wire, #34 to #36 AWG

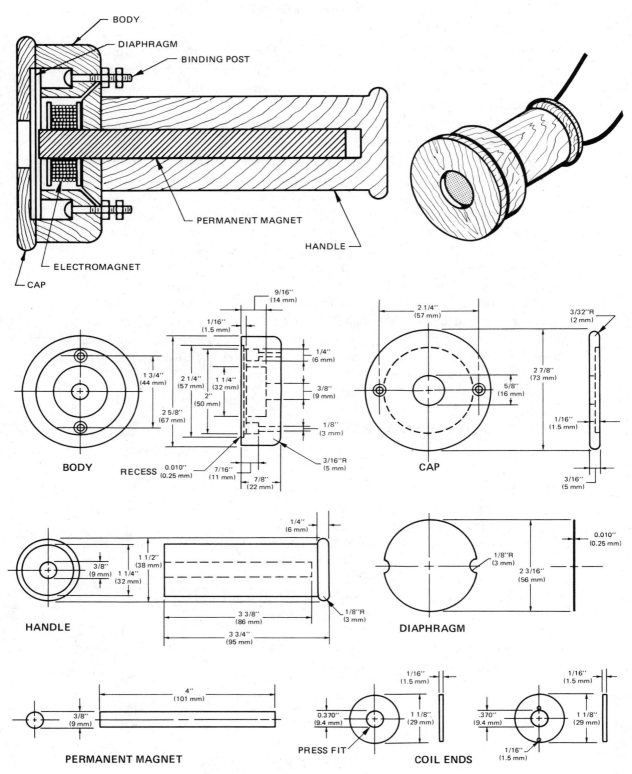

Fig. 19-35 Receiver

MATERIALS FOR TRANSMITTER

1 - Body, hardwood, 1" x 3 1/4" x 3 1/4" (25 mm x 83 mm x 83 mm)
1 - Cap, hardwood 7/8" x 3 1/4" x 3 1/4" (22 mm x 83 mm x 83 mm)
1 - Diaphragm, tin plate, 0.006" to 0.010" x 3 1/8" x 3 1/8" (0.15 to 0.25 mm x 79 mm x 79 mm)
1 - Machine screw, #6 x 32 x 1 1/4" (M3.5 x 30 mm L) RH
1 - Machine screw, #6 x 32 x 3/4" (M3.5 x 20 mm L) RH
1 - Machine screw, #6 x 32 x 5/16" (M3.5 x 8 mm L) RH
5 - Machine screw nuts, #6 x 32 (M3.5), hexagonal
3 - Wood screws, #6 x 1" RH
1 - Pc. carbon, 1/8" x 1" (3 mm x 24 mm) diameter
1 - Pc. carbon, 3/16" x 1" (5 mm x 25 mm) diameter

PROCEDURE

Study the drawing for each unit. Make idea sketches for innovations you wish to develop and have them approved by your instructor.

Obtain the materials and make each part. *Note:* For information regarding coils and magnetizing steel, see Unit 6.

After all the parts are made and finished, assemble the units.

Test and evaluate.

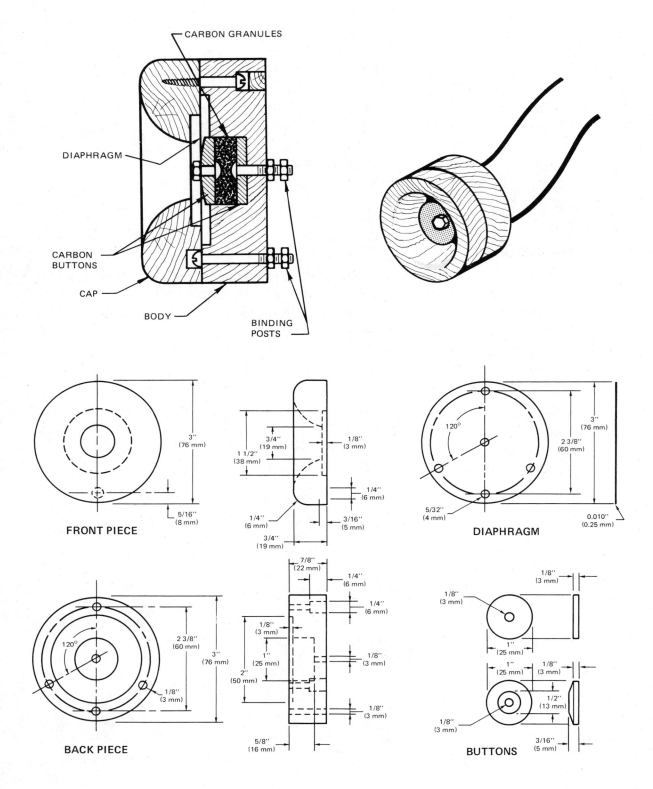

Fig. 19-36 Transmitter

Chapter 20

Solid state AC power control circuits

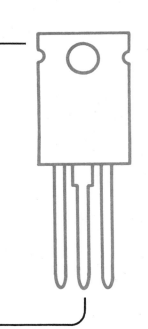

Solid state power control circuits are very efficient in controlling the amount of power used by an electrical device or appliance. Common applications of such circuits include variable-speed electric drills, lamp dimmers, hi-fi color organs, precision heat controllers, timers, battery chargers, alarm circuits, and many others. The term *solid state* is commonly applied to any circuit which depends upon a semiconductor device for its operation. Solid state power control circuits do not represent new applications of electricity. Instead, they represent new ways of accomplishing old tasks. The technician must be familiar with these circuits in order to troubleshoot and repair them. The consumer should have a basic understanding of their principles of operation in order to select and purchase appliances wisely. The present necessity for wise energy utilization and the high reliability of semiconductor devices will encourage wider utilization of such circuits and make knowledge of their principles of operation more valuable.

COMPARISONS WITH RESISTIVE CONTROLS

Analysis of the basic lamp dimming circuits of figures 20-1 and 20-2 provides a better understanding of the solid state power control system. In figure 20-1 a high power rheostat is connected in series with a 300-watt lamp across a 120-volt commercial power source of the type found in homes. Adjusting the rheostat controls the total resistance of the circuit, allowing more or less current through the lamp and therefore controlling its brightness. Figure 20-2 represents a basic full-wave solid state control circuit. In figure 20-2(A), a triac is connected in series with a 300-watt lamp. In addition, this circuit employs a series rheostat and capacitor network between terminals 1 and 2. The gate of the triac is connected to the resistor side of the capacitor through a diac. The diac and triac combination are the semiconductor devices in this circuit. Details describing their behavior will be provided later in this chapter. Adjusting the rheostat controls the amplitude and phase of the voltage waveform appearing across the capacitor. This signal is used to control the conduction of the triac. Hence, the brightness of the lamp in series with it is also controlled. To the circuit user, turning the rheostat knob in either circuit would have an apparently identical effect on lamp brightness. Further analysis beyond simple schematic

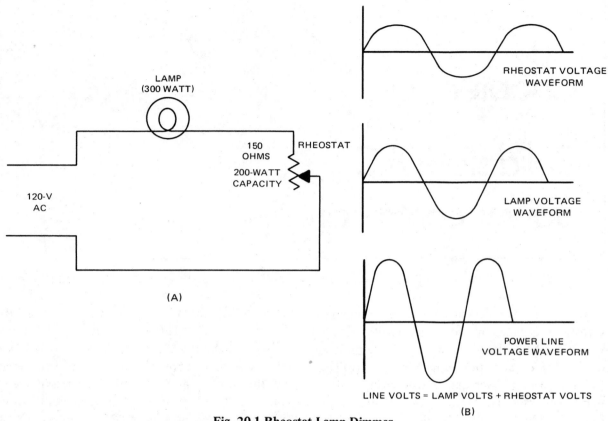

Fig. 20-1 Rheostat Lamp Dimmer

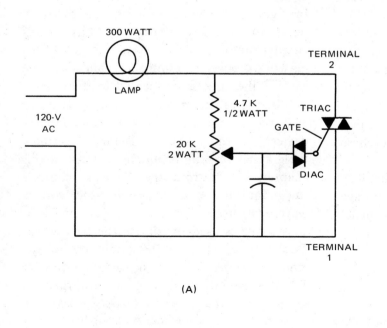

(A)

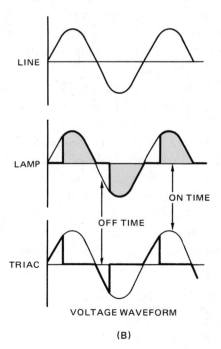

(B)

Fig. 20-2

inspection is necessary to demonstrate the advantage of the solid state method.

If an oscilloscope were connected across the lamp in each circuit and synchronized with the ac power line, one would discover that the voltage waveform would be quite different in the two cases. Isolate the oscilloscope from the ground if this is attempted. In the conventional power control, figure 20-1, the voltage waveform has the shape of a 60-hertz sine wave with an amplitude which could be adjusted into higher or lower magnitude by the rheostat. The maximum value of the voltage amplitude would be 120 volts (RMS) when the rheostat is set at zero ohms. Its minimum value would be approximately 30 volts when the rheostat is set in approximately 150 ohms. These waveforms are illustrated in figure 20-1(B). The reader may connect this circuit and verify these numbers both mathematically and experimentally.

In the case of the solid state dimmer, the waveform is an incomplete sine wave of constant amplitude. The leading part of each half waveform is the place where the wave is expected to cross its center axis going either in a positive or negative direction. This leading part is a short horizontal line representing zero voltage across the lamp for a short time. The voltage across the lamp rises or falls very rapidly to conform with the supply line voltage waveform. Every time the supply line voltage waveform crosses its zero axis, the voltage across the lamp remains at zero for another short time. Adjusting the rheostat controls the time for which the lamp is held in the off state during each half waveform. Adjusting the rheostat to its minimum position will result in very short off times, almost complete line voltage sine waveforms across the lamp, and high lamp brightness. Adjusting the rheostat to its maximum resistance will result in long off

times, very incomplete sine waveforms, and a very dim light that may make the lamp appear to be off.

Oscilloscope analysis of the control element in each circuit, the rheostat in figure 20-1, and the triac in figure 20-2 will reveal why the solid state control circuit is much more efficient. The voltage waveform across the rheostat in figure 20-1 is a sine wave with amplitude adjustable from zero to approximately 90 volts (RMS). If the rheostat were set to deliver one-half power to the lamp (150 watts), it would be set at 48 ohms and produce 150 watts of heat. This energy is used to gain dimming control. (The reader may calculate the 150-watt and 48-ohms figures.) These voltage waveforms are shown with their proper phase relationships under the waveforms for the lamp in figure 20-1(B). The voltage waveform for the triac would look like an incomplete sine wave with low-voltage horizontal lines representing times when it was conducting heavily. Notice that if the points on the voltage waveform across the triac were added to the corresponding points on the lamp waveform, a complete sine-wave image of the line voltage would be created. This result should not be surprising because the sum of the voltage drops around a circuit loop is equal to the voltage source. In practical terms, these two waveforms mean that during the time the lamp is off, instantaneous line voltage appears across the triac which blocks the current. Therefore, little power is used. When the lamp is on, little voltage appears across the triac, line voltage appears across the lamp, and most of the circuit power is developed in the lamp where it is desired. Lamp brightness is controlled by adjusting its on-and-off time durations within each half-cycle. The same result could be achieved with a conventional switch if the switch could be snapped on and off 120 times per second in perfect rhythm with the line voltage. Of course, the switch would soon wear out.

The real advantage of solid state control is its ability to switch circuit states from on-and-off conditions instantaneously and with little expense of control power and no mechanical motion or electric arcing of switch contacts. For the circuit in figure 20-2(A), one may calculate the power lost in the resistor, rheostat, and capacitor when

the rheostat is set at its maximum and minimum values. Ignore the gate connection. Maximum heat generated or lost around the triac is approximately 1 volt (voltage drop when it is conducting) times 2.5 amperes (current when a 300-watt bulb is connected directly to a 120-volt line), which equals 2.5 watts. This power figure represents a small loss in comparison with the resistive control circuit. Triacs utilized in high-power

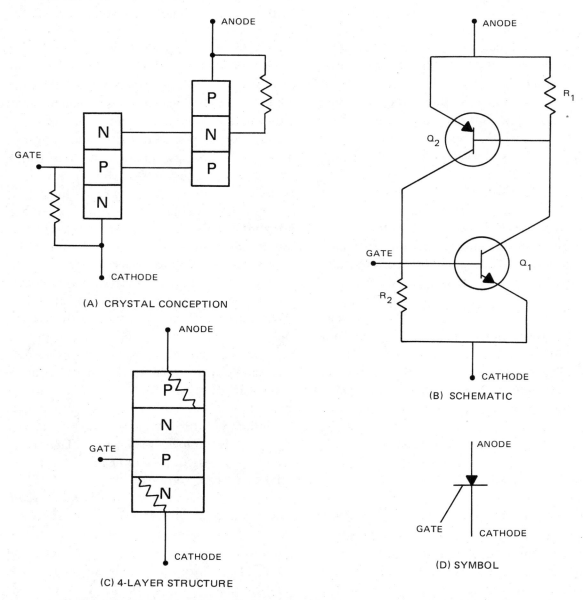

(A) CRYSTAL CONCEPTION

(B) SCHEMATIC

(C) 4-LAYER STRUCTURE

(D) SYMBOL

Fig. 20-3 Symbol, Schematic, and Crystal Diagrams of the Silicon Controlled Rectifier, (SCR)

industrial circuits must be mounted on a heat sink to dissipate the heat they generated. The same general circuits and concepts identified in the lamp dimming application discussed can be used to control the speed of universal motors or to produce varying amounts of heat in incubators, heat lamps, or industrial processes. Control of an ac load current by delaying conduction during each half-cycle is called *phase power control.* This kind of control is similar to that accomplished by thyratrons.

SOLID STATE CONTROL DEVICES

The SCR (silicon controlled rectifier) is an integrated circuit form of a two-transistor switch. It has three terminals: the *anode,* the *cathode,* and the *gate.* The anode and the cathode are the main switch terminals; the gate acts as a controlling element. Figure 20-3(B) shows a schematic of the integrated circuit. Figure 20-4 illustrates the way SCRs actually look.

The integrated circuit of an SCR is made up of two transistors and two resistors directly connected to each other in a way that incorporates all necessary semiconductor areas into 4 layers. If the P-N-P transistor (Q_2) and the N-P-N transistor (Q_1) are shown as semiconductor crystal regions, the schematic converts into the diagrams shown in figure 20-3(A). From this diagram it can be seen that one P- and N-region of each transistor are directly connected together. The two crystals can be merged. The resistances R_1 and R_2 can be incorporated into the crystalline resistances of the two transistor emitters. Figure 20-3(C) shows a cross section of the new four-layer structure of the typical SCR. With this structure the device is very rugged with respect to physical position, shock, and vibration.

In operation, figure 20-3(B), with a positive voltage on the anode and a negative one on the cathode, the device should not conduct because there is no base signal at

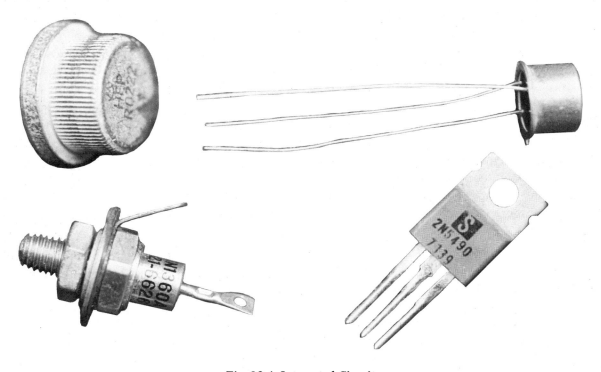

Fig. 20-4 Integrated Circuits

either transistor. It will conduct if voltage is excessive. If a small positive signal is applied to the gate, base of Q_1, current will start in the collector circuit of Q_1. This current causes a voltage drop on R_1 and some of the current is diverted into the base of Q_2. Now Q_2 can conduct and allow electrons up through R_2 and out the anode. A voltage drop appears on R_2, providing more forward bias for the base of Q_1 which now conducts more heavily and the process continues. Current is built up through the device to a level controlled by an outside voltage source and a series load resistor. *NOTE:* If an SCR is connected to a large power source like the ac power line without a series limiting load and is triggered on at the gate, it will build up a current level and heat sufficiently to destroy itself in a split second.

Once turned on, the internal circuit of the SCR will keep the device on as long as the load current remains high enough to develop the forward biases in the crystal resistances. This level of current is called the holding current of the device and will vary from one SCR to another. In ac phase-control applications the device turns itself off each time the line voltage falls to zero.

Specifications of SCRs

There are many specifications related to SCRs which become important in specialized circuits. The most critical and common specifications for applications related to the 60-hertz commercial power line are identified here.

Repetitive reverse voltage is the level of reverse voltage that the SCR can withstand repeatedly (about 200 volts for 120-volt RMS applications).

Forward blocking voltage is the level of forward voltage that the SCR can hold back when there is no signal on the gate. It should be at least 175 volts for 120-RMS applications.

The forward blocking voltage greatly reduces as gate current increases. Hence, gate current controls the turn-on, or firing, levels of the SCR.

Forward current peak is the most current the device could stand for a short time.

Forward current average is the most current the device should carry on the average or continuously with proper heat sink.

Holding current is the small current needed to keep the SCR in an on state.

Gate signal current is the amount of gate current necessary to trigger the device on, with a stated voltage for a power source. Gate current and anode-to-cathode break over voltage (firing voltage) are inversely related.

SCR Power Control in AC

A typical SCR power control circuit is shown in figure 20-5. In such a circuit, the SCR is forward-biased only during positive half-cycles of the line voltage. Therefore, its range of control is limited from the one-half power level downward. In many commercial applications, the SCR is bypassed by a switch which is closed for full power, shunting power current around the SCR.

In using a speed-controlled electric drill, the operator squeezes the trigger switch slightly, turning the drill on. The rheostat is physically connected to the drill trigger and has maximum resistance with the trigger forward. As the trigger is pulled back, the rheostat is adjusted to lower resistances and the drill runs faster. Squeezing the trigger hard closes the bypass switch and allows full line power to operate the drill.

THE FOUR-LAYER DIODE

The four-layer diode is sometimes called a trigger diode or a unilateral switch. It is a miniature SCR which is designed to be

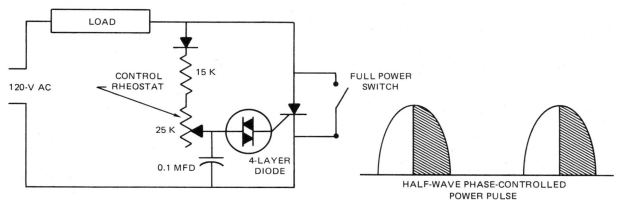

Fig. 20-5 SCR Power Control Circuit

fired by a forward voltage above a given level without any external signal on its gate. In fact, no external gate terminal connection is provided. Once turned on by a sufficient forward voltage, its internal blocking effect is lost, the voltage drop across it will drop to a low value, and it will stay on until the current in it drops below a holding value.

The four-layer diode differs from the zener diode in that there is no switching effect in the zener. It only conducts if the voltage across it remains high enough. While the zener has characteristics necessary for voltage regulation, the four-layer diode is useful for making relaxation oscillators, timing circuits, and blocking gate currents in an SCR circuit until the precise firing time is reached.

In figure 20-5, the four-layer diode isolates the SCR gate from the capacitor voltage until it is built up to about 15 volts. Then it conducts and turns on the SCR. This allows more delay in firing time for low power control as well as extended SCR life spans. The SCR will last longer if it is signaled on by a strong gate pulse rather than turned on by high breakover voltage.

THE TRIAC

The *triac* is an integrated circuit form of two SCRs with opposite polarities connected in parallel. Remember that one SCR can

control power for only one half-cycle. Another SCR could control power on the opposite half-cycle of the ac line if its connections were reversed. Since the SCR is made up of 2 transistors and 2 resistors, the triac must be equivalent to 4 transistors and 4 resistors. Drawing the equivalent circuit of such a device is a challenging exercise. The triac is most commonly showed schematically as in figure 20-2. The integrated circuit of the triac, where directly connected crystalline areas and resistances are merged, is surprisingly simple, figure 20-6.

The three connections to the triac are called terminal 1, terminal 2, and the gate. Gate signal voltages are usually specified with

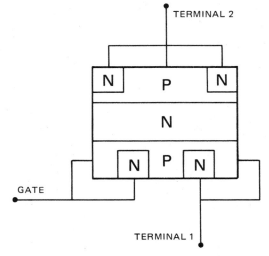

Fig. 20-6 Integrated Circuit Diagram of a TRIAC

respect to terminal 1, which is usually associated with the "common" side of the power circuit. The load is usually connected in series with terminal 2. The gate of a triac is usually fired from a diac or other triggering device for long life and better low power control. The triac can be turned on by any combination of terminal polarity and positive or negative signals on the gate. It is turned off by reducing load current below holding level. This occurs every time the line voltage alternates.

THE DIAC

The *diac* is simply the integrated circuit form of two four-layer diodes with opposite polarities. Therefore, it can fire in either direction. It is also possible to think of it as a miniature triac without a gate lead. One commonly available diac fires at positive or negative 15-volts dc.

TROUBLESHOOTING AND CONSTRUCTION TIPS

The most common problems encountered by the technician involve either a dead circuit or a full-power circuit with no variable control. Isolating the area of difficulty is relatively easy. The techniques suggested apply to either SCRs or triacs.

Dead Circuit

Assuming adequate power is available to the circuit, that is, it is plugged in, turned on, the power lead is good, etc., the line voltage appears either across the load or across the control device (SCR or triac). Since the load is dead, check it for proper operation at full line voltage. Assuming the load is good, measure the voltage across the device cathode-to-anode or terminal 1-to-terminal 2 to verify that the line voltage is available and blocked there. If line voltage

is available, there is no load current for one of two reasons. Either the gate signal is insufficient to turn the device on, or the internal crystal has melted into an open condition. With a pair of jumper leads, connect a 10-K resistor between the anode and the gate. This should provide enough gate signal to turn on any good control device. If the circuit is still dead, replace the SCR or triac. If the circuit comes alive, look for problems in the gate triggering control circuit.

Full Power, No Control

This condition occurs because of higher than normal gate signals or the melting of the internal structure of the device into a conducting mass. If the problem of full power and no control occurs, first disconnect the gate. If the load current stops, the device is still able to control the line voltage. Look for troubles in the gate control circuit. If there is still full power in the load with the gate disconnected, throw away the device.

Solid state devices do not malfunction often. Many beginning technicians accurately determine that a device is burned out, install a new one, and burn it out before realizing that the real root of the trouble may be a shorted load, too heavy a load for the device, or too large a gate signal. Always check all other circuit devices and functions before replacing a semiconductor control device even though it has been determined as "bad."

On many SCRs or triacs the physical distance between the gate terminal and the anode or terminal 2 lead is very small. The inexperienced technician can easily touch both leads at once with a voltmeter probe. If this happens, the gate to the device will be burned out instantaneously.

Always connect some basic fixed resistance, perhaps 4 700 ohms, in series with the rheostat in the control circuit to avoid the possibility of connecting the gate to the high-power source when the control rheostat is set at minimum.

Always use a heat sink when soldering connections to solid state devices. Using a commercial, spring clip heat sink, or holding the lead between the device and the soldering tip with needle nose pliers work equally well.

When connecting test leads to live devices (voltmeters or oscilloscopes) be careful not to accidentally short the gate to the anode (or terminal tube) with the test probe tip. Even a momentary short will destroy the device.

POINTS TO REMEMBER

- Solid state power control produces much less heat in the control circuit than series resistance controls.

- The SCR blocks completely with one polarity and can be switched on when its anode is positive with respect to its cathode. It is a half-wave control.

- The triac can be switched on with either polarity of terminal 1 to terminal 2.

- Diacs, four-layer diodes, neon bulbs, and unijunction transistors are triggering devices.

- In ac power phase control, the control element is turned off every time the line voltage crosses zero.

- In ac power phase control, the control element is turned on (after some delay) during the half-cycle, when polarities forward bias the device.

REVIEW QUESTIONS

1. A solid state control capable of delivering full power to a 500-watt lamp connected to the 120-volt line would have to be able to withstand a current of how many amperes?

2. A solid state device used on a 230-volt RMS system would have to be able to block a reverse voltage of how many volts?

3. Why would a neon bulb firing at 75 volts give less range of control when used as a triggering device than a diac firing at 30 volts?

4. What is the function of a capacitor in a phase control circuit?

5. Why doesn't the control rheostat in a solid state circuit have to be as big physically as the control rheostat in a resistance power control?

6. Draw the two transistor equivalent circuit of an SCR and explain why the device stays switched on as long as there is load current.

RESEARCH AND DEVELOPMENT

Experiments and Projects on Solid State Power Control

INTRODUCTION

Solid state power control is a field of growing importance to industrial technicians as well as common consumers. There are a number of specialized devices on the market across the nation for consumer use. Most of these vendors include a number of suggested circuit applications packaged along with the device purchased. By this time in the study of electricity, the reader should be able to select a device and follow the published project suggestions. Further suggestions for activities are presented below:

EXPERIMENTS

1. Study the basic principles of a half-wave phase control, (SCR).
2. Study full-wave phase control (triac) apparatus.

PROJECTS

1. Obtain an SCR kit from a commercial vendor and experiment with the delayed "on" circuit.
2. Obtain a light-operated SCR (Radio Shack #276-1095) and experiment with its various control possibilities as suggested in its packaging materials. In many ways this device alone can replace a light sensor, amplifier, and a control relay.

EXPERIMENTS

EXPERIMENT 1

OBJECT

To learn the basic principles of a half-wave solid state control.

APPARATUS

1 - Oscilloscope
1 - SCR, RCA #KD2100 or Radio Shack #276-1067
1 - Diac, Radio Shack #276-1050
1 - Capacitor, 0.25-μf
1 - Silicon diode - 200-volts blocking capability
1 - Resistor, 10 000-ohm
1 - Potentiometer, 50 000-ohm
1 - Lamp, 200-watt (load)

PROCEDURE

1. Connect a half-wave phase control circuit as shown in figure 20-5.

2. Isolate the oscilloscope from power line ground by use of an isolation transformer and display patterns of the load voltage and the voltage across the SCR.

3. Adjust the control potentiometer and note the effects on the load voltage waveforms.

4. Substitute a portable electric drill for the lamp and experiment with the apparent range of speed control.

QUESTIONS

1. Increasing the control resistance has what effect on load power?

2. From the oscilloscope analysis, estimate the maximum and minimum power this circuit will allow in the 200-watt lamp.

3. Why does this circuit not allow full power to either the lamp or the drill?

EXPERIMENT 2

OBJECT

To learn about full-wave phase control apparatus.

APPARATUS

1 - Triac, Radio Shack #276-1001
1 - Diac, Radio Shack #276-1050
1 - Capacitor, 0.25-μf
1 - Resistor, 5 000-ohm
1 - Potentiometer, 25 000-ohm

PROCEDURE

1. Connect the full-wave triac circuit as shown in figure 20-6 with the component values listed above.

2. Repeat steps 2 through 4 of Experiment 1 for this full-wave circuit.

QUESTIONS

1. Compare the range of control of the SCR and the triac circuit.

2. Why was the silicon diode left out of the triac control circuit?

Appendix

TABLES

ELECTRICAL PROPERTIES OF METALS AND ALLOYS

PROPERTIES OF MATERIALS	Resistivity in microhm-cm at 20°C	Temp. coeff. of resistance	Melting point (°C)	Specific gravity
Aluminum	2.83	0.004	659	2.67
Alumel	33.3	0.0012		
Aluminum bronze (90% Cu, 10% Al)	12.6-12.7	0.003	1 050	7.5
Brass (various comp.)	6.2-8.3	0.002	880-1 050	8.4-8.8
Bronze (88Cu, 12 tin)	18	0.0005	1 000	8.8
Bronze, "commercial" (Cu, Zn)	4.2	0.002	1 050	8.8
Carbon, amorphous	3 500-4 100	-0.0005	3 500+	1.85
" graphite	720-1 000 (for furnace electrodes at 2 500°C)			
Chromel (Ni, Cr)	70-110	0.0001	1 350	8.3-8.5
Constantan (60%Cu, 40%Ni)	44-49	0.0000	1 190	8.9
Copper, annealed	1.724-1.73	0.004	1 083	8.89
" hard-drawn	1.77	0.0039		
German silver (Cu, Zn, 18%Ni)	33	0.004	1 100	8.4
Gold	2.44	0.0034	1 063	19.3
Invar (65% Fe, 35% Ni)	81		1 495	8
Iron & steel				
Iron, pure	10	0.005	1 530	7.8
cast	60			
soft steel	15.9	0.0016	1 510	7.8
steel, glass-hard	45.7			
" 4% Silicon	51			
" , transformer	11.09			
Lead	22	0.00387	327	11.4
Magnesium	4.6	0.004	651	1.74
Manganin (84%Cu, 12%Mn, 4%Ni)	44-48	0.0000	910	8.4
Mercury	95.7	0.0008	–39	13.6
Monel (Ni, Cu)	42.5	0.00019	1 300	8.9
Nickel	7	0.006	1 452	8.9
Nichrome (Ni, Fe, Cr)	90-112	0.00017	1 350-1 500	8.25
Nichrome V (Ni, Cr)	108	0.00017		8.41
Platinum	11.5	0.003	1 755	21.4
Silver	1.628	0.0038	960	10.5
Sodium	4.4	0.004		
Tin	11.5	0.004	232	7.3
Tungsten	5.51	0.0045	3 410	18.8
at 1730°C	60.			
at 2760°C	100.			
Zinc	6	0.004	419	7.1

Resistivity in microhm-cm is the resistance of a centimeter-cube, in millionths of an ohm.
To obtain ohms per mil-foot, multiply the resistivity figure as given above by 6.015.

ALLOWABLE AMPACITIES OF INSULATED COPPER CONDUCTORS

SINGLE CONDUCTOR IN FREE AIR
Based on Room Temperature of 30°C 86°F.

Size AWG MCM	Temperature Rating of Conductor						
	60°C (140°F.)	75°C (167°F.)	85°-90°C (185°F.)	110°C (230°F.)	125°C (257°F.)	200°C (392°F.)	Bare and Covered Conductor
14	20	20	30 *	40	40	45	30
12	25	25	40 *	50	50	55	40
10	40	40	55 *	65	70	75	55
8	55	65	70	85	90	100	70
6	80	95	100	120	125	135	100
4	105	125	135	160	170	180	130
3	120	145	155	180	195	210	150
2	140	170	180	210	225	240	175
1	165	195	210	245	265	280	205
0	195	230	245	285	305	325	235
00	225	265	285	330	355	370	275
000	260	310	330	385	410	430	320
0000	300	360	385	445	475	510	370
250	340	405	425	495	530	. . .	410
300	375	445	480	555	590	. . .	460
350	420	505	530	610	655	. . .	510
400	455	545	575	665	710	. . .	555
500	515	620	660	765	815	. . .	630
600	575	690	740	855	910	. . .	710
700	630	755	815	940	1 005	. . .	780
750	655	785	845	980	1 045	. . .	810
800	680	815	880	1 020	1 085	. . .	845
900	730	870	940	. . . .	. . . .	. . .	905
1 000	780	935	1 000	1 165	1 240	. . .	965
1 250	890	1 065	1 130	. . . .	. . . .	. . .	. . .
1 500	980	1 175	1 260	1 450	. . . .	. . .	1 215
1 750	1 070	1 280	1 370	. . . .	. . . .	. . .	. . . .
2 000	1 155	1 385	1 470	1 715	. . . .	. . .	1 405

Correction Factors, Room Temps. Over 30°C 86°F.

C F.							
40 104	.82	.88	.90	.94	.95	. . .	. . . .
45 113	.71	.82	.85	.90	.92	. . .	. . . .
50 122	.58	.75	.80	.87	.89	. . .	. . . .
55 131	.41	.67	.74	.83	.86	. . .	. . . .
60 140	. . . .	.58	.67	.79	.83	.91	. . . .
70 158	. . . .	.35	.52	.71	.76	.87	. . . .
75 167	. . . .	. . . .	.43	.66	.72	.86	. . . .
80 176	. . . .	. . . .	.30	.61	.69	.84	. . . .
90 194	. . . .	. . . .	. . . .	.50	.61	.80	. . . .
100 212	. . . .	. . . .	. . . .	. . . .	.51	.77	. . . .
120 248	. . . .	. . . .	. . . .	. . . .	. . . .	.69	. . . .
140 284	. . . .	. . . .	. . . .	. . . .	. . . .	.59	. . . .

These ampacities relate only to conductors described in Table 310-2(a), N.E. Code.

* The ampacities for types FEP, FEPB, RHH and THHN conductors for sizes AWG 14, 12 and 10 shall be the same as designated for 75°C conductors in this table.

Above adapted from the National Electrical Code (NFPA No. 70, Table 310-13)

ALLOWABLE AMPACITIES[†] OF INSULATED COPPER CONDUCTORS

NOT MORE THAN THREE CONDUCTORS IN RACEWAY OR CABLE OR DIRECT BURIAL
Based on Room Temperature of 30°C 86°F

Size AWG MCM	Temperature Rating of Conductor					
	60°C (140°F.)	75°C (167°F.)	85°-90°C (185°F.)	110°C (230°F.)	125°C (257°F.)	200°C (392°F.)
14	15	15	25 *	30	30	30
12	20	20	30 *	35	40	40
10	30	30	40 *	45	50	55
8	40	45	50	60	65	70
6	55	65	70	80	85	95
4	70	85	90	105	115	120
3	80	100	105	120	130	145
2	95	115	120	135	145	165
1	110	130	140	160	170	190
0	125	150	155	190	200	225
00	145	175	185	215	230	250
000	165	200	210	245	265	285
0000	195	230	235	275	310	340
250	215	255	270	315	335	. . .
300	240	285	300	345	380	. . .
350	260	310	325	390	420	. . .
400	280	335	360	420	450	. . .
500	320	380	405	470	500	. . .
600	355	420	455	525	545	. . .
700	385	460	490	560	600	. . .
750	400	475	500	580	620	. . .
800	410	490	515	600	640	. . .
900	435	520	555	. . .	. . .	. . .
1 000	455	545	585	680	730	. . .
1 250	495	590	645	. . .	. . .	. . .
1 500	520	625	700	785	. . .	. . .
1 750	545	650	735	. . .	. . .	. . .
2 000	560	665	775	840	. . .	. . .

Correction Factors, Room Temps. Over 30°C. 86°F.

C F.						
40 104	.82	.88	.90	.94	.95	. . .
45 113	.71	.82	.85	.90	.92	. . .
50 122	.58	.75	.80	.87	.89	. . .
55 131	.41	.67	.74	.83	.86	. . .
60 140	. . .	.58	.67	.79	.83	.91
70 158	. . .	.35	.52	.71	.76	.87
75 167	. . .	. . .	.43	.66	.72	.86
80 176	. . .	. . .	.30	.61	.69	.84
90 194	. . .	. . .	. . .	.50	.61	.80
100 212	. . .	. . .	. . .	. . .	.51	.77
120 248	. . .	. . .	. . .	. . .	. . .	.69
140 284	. . .	. . .	. . .	. . .	. . .	.59

[†]Current-carrying capacity expressed in amperes.

These ampacities relate only to conductors described in Table 310-2(a), N.E. Code

*The ampacities for types FEP, FEPB, RHH and THHN conductors for sizes AWG 14, 12, and 10 shall be the same as designated for 75°C conductors in this table.

Above adapted from the National Electrical Code (NFPA No. 70, Table 310-12)

ALLOWABLE AMPACITIES OF INSULATED ALUMINUM CONDUCTORS

SINGLE CONDUCTOR IN FREE AIR
Based on Room Temperature of 30°C 86°F.

Size AWG MCM	Temperature Rating of Conductor						Bare and Covered Conductor
	60°C (140°F.)	75°C (167°F.)	85° – 90°C (185°F.)	110°C (230°F.)	125°C (257°F.)	200°C (392°F.)	
12	20	20	30 *	40	40	45	30
10	30	30	45 *	50	55	60	45
8	45	55	55 *	65	70	80	55
6	60	75	80	95	100	105	80
4	80	100	105	125	135	140	100
3	95	115	120	140	150	165	115
2	110	135	140	165	175	185	135
1	130	155	165	190	205	220	160
0	150	180	190	220	240	255	185
00	175	210	220	255	275	290	215
000	200	240	255	300	320	335	250
0000	230	280	300	345	370	400	290
250	265	315	330	385	415	. . .	320
300	290	350	375	435	460	. . .	360
350	330	395	415	475	510	. . .	400
400	355	425	450	520	555	. . .	435
500	405	485	515	595	635	. . .	490
600	455	545	585	675	720	. . .	560
700	500	595	645	745	795	. . .	615
750	515	620	670	775	825	. . .	640
800	535	645	695	805	855	. . .	670
900	580	700	750	. . .	. . .	. . .	725
1 000	625	750	800	930	990	. . .	770
1 250	710	855	905	. . .	. . .	. . .	. . .
1 500	795	950	1 020	1 175	. . .	. . .	985
1 750	875	1 050	1 125	. . . .	. . .	. . .	. . .
2 000	960	1 150	1 220	1 425	. . .	. . .	1 165

Correction Factors, Room Temps. Over 30°C 86°F.

C F.							
40 104	.82	.88	.90	.94	.95	. . .	. . .
45 113	.71	.82	.85	.90	.92	. . .	. . .
50 122	.58	.75	.80	.87	.89	. . .	. . .
55 131	.41	.67	.74	.83	.86	. . .	. . .
60 140	. . .	.58	.67	.79	.83	.91	. . .
70 158	. . .	.35	.52	.71	.76	.87	. . . .
75 167	. . .	. . .	.43	.66	.72	.86	. . .
80 176	. . .	. . .	.30	.61	.69	.84	. . .
90 194	. . .	. . .	. . .	.50	.61	.80	. . .
100 212	. . .	. . .	. . .	. . .	.51	.77	. . .
120 248	. . .	. . .	. . .	. . .	. . .	.69	. . .
140 284	. . .	. . .	. . .	. . .	. . .	.59	. . .

These ampacities relate only to conductors described in Table 310-2(a), N.E. Code.

*The ampacities for types RHH and THHN conductors for sizes AWG 12, 10 and 8 shall be the same as designated for 75°C conductors in this table.

Above adapted from the National Electrical Code (NFPA No. 70, Table 310-15)

AMERICAN WIRE GAUGE

B & S Gauge Number	Diameter in Mils	Area in Circular Mils	Ohms per 1 000 Ft. (ohms per 100 m)			Pounds per 1 000 Ft. (kg per 100 m)	
			Copper* 68° F (20°C)	Copper* 167° F (75°C)	Aluminum 68° F (20°C)	Copper	Aluminum
0000	460	211 600	.049 (.016)	.0596 (.0195)	.0804 (.0263)	640 (95.2)	195 (29.0)
000	410	167 800	.0618 (.020)	.0752 (.0246)	.101 (.033)	508 (75.5)	154 (22.9)
00	365	133 100	.078 (.026)	.0948 (.031)	.128 (.042)	403 (59.9)	122 (18.1)
0	325	105 500	.0983 (.032)	.1195 (.0392)	.161 (.053)	320 (47.6)	97 (14.4)
1	289	83 690	.1239 (.0406)	.151 (.049)	.203 (.066)	253 (37.6)	76.9 (11.4)
2	258	66 370	.1563 (.0512)	.190 (.062)	.526 (.084)	201 (29.9)	61.0 (9.07)
3	229	52 640	.1970 (.0646)	.240 (.079)	.323 (.106)	159 (23.6)	48.4 (7.20)
4	204	41 740	.2485 (.0815)	.302 (.099)	.408 (.134)	126 (18.7)	38.4 (5.71)
5	182	33 100	.3133 (.1027)	.381 (.125)	.514 (.168)	100 (14.9)	30.4 (4.52)
6	162	26 250	.395 (1.29)	.481 (.158)	.648 (.212)	79.5 (11.8)	24.1 (3.58)
7	144	20 820	.498 (.163)	.606 (.199)	.817 (.268)	63.0 (9.37)	19.1 (2.84)
8	128	16 510	.628 (.206)	.764 (.250)	1.03 (.338)	50.0 (7.43)	15.2 (2.26)
9	114	13 090	.792 (.260)	.963 (.316)	1.30 (.426)	39.6 (5.89)	12.0 (1.78)
10	102	10 380	.999 (.327)	1.215 (.398)	1.64 (.538)	31.4 (4.67)	9.55 (1.42)
11	91	8 234	1.260 (.413)	1.532 (.502)	2.07 (.678)	24.9 (3.70)	7.57 (1.13)
12	81	6 530	1.588 (.520)	1.931 (.633)	2.61 (.856)	19.8 (2.94)	6.00 (.89)
13	72	5 178	2.003 (.657)	2.44 (.80)	3.29 (1.08)	15.7 (2.33)	4.8 (.71)
14	64	4 107	2.525 (.828)	3.07 (1.01)	4.14 (1.36)	12.4 (1.84)	3.8 (.56)
15	57	3 257	3.184 (1.043)	3.87 (1.27)	5.22 (1.71)	9.86 (1.47)	3.0 (.45)
16	51	2 583	4.016 (1.316)	4.88 (1.60)	6.59 (2.16)	7.82 (1.16)	2.4 (.36)
17	45.3	2 048	5.06 (1.66)	6.16 (2.02)	8.31 (2.72)	6.20 (.922)	1.9 (.28)
18	40.3	1 624	6.39 (2.09)	7.77 (2.55)	10.5 (3.44)	4.92 (.731)	1.5 (.22)
19	35.9	1 288	8.05 (2.64)	9.79 (3.21)	13.2 (4.33)	3.90 (.580)	1.2 (.18)
20	32.0	1 022	10.15 (3.33)	12.35 (4.05)	16.7 (5.47)	3.09 (.459)	0.94 (.14)
21	28.5	810	12.8 (4.2)	15.6 (5.11)	21.0 (6.88)	2.45 (.364)	.745 (.110)
22	25.4	642	16.1 (5.3)	19.6 (6.42)	26.5 (8.69)	1.95 (.290)	.591 (.09)
23	22.6	510	20.4 (6.7)	24.8 (8.13)	33.4 (10.9)	1.54 (.229)	.468 (.07)
24	20.1	404	25.7 (8.4)	31.2 (10.2)	42.1 (13.8)	1.22 (.181)	.371 (.05)
25	17.9	320	32.4 (10.6)	39.4 (12.9)	53.1 (17.4)	0.97 (.14)	.295 (.04)
26	15.9	254	40.8 (13.4)	49.6 (16.3)	67.0 (22.0)	.77 (.11)	.234 (.03)
27	14.2	202	51.5 (16.9)	62.6 (20.5)	84.4 (27.7)	.61 (.09)	.185 (.03)
28	12.6	160	64.9 (21.3)	78.9 (25.9)	106 (34.7)	.48 (.07)	.147 (.02)
29	11.3	126.7	81.8 (26.8)	99.5 (32.6)	134 (43.9)	.384 (.06)	.117 (.02)
30	10.0	100.5	103.2 (33.8)	125.5 (41.1)	169 (55.4)	.304 (.04)	.092 (.01)
31	8.93	79.7	130.1 (42.6)	158.2 (51.9)	213 (69.8)	.241 (.04)	.073 (.01)
32	7.95	63.2	164.1 (53.8)	199.5 (65.4)	269 (88.2)	.191 (.03)	.058 (.01)
33	7.08	50.1	207 (68)	252 (82.6)	339 (111)	.152 (.02)	.046 (.01)
34	6.31	39.8	261 (86)	317 (104)	428 (140)	.120 (.02)	.037 (.01)
35	5.62	31.5	329 (108)	400 (131)	540 (177)	.095 (.01)	.029
36	5.00	25.0	415 (136)	505 (165)	681 (223)	.076 (.01)	.023
37	4.45	19.8	523 (171)	636 (208)	858 (281)	.0600 (.01)	.0182
38	3.96	15.7	660 (216)	802 (263)	1080 (354)	.0476 (.01)	.0145
39	3.53	12.5	832 (273)	1012 (332)	1360 (446)	.0377 (.01)	.0115
40	3.15	9.9	1049 (344)	1276 (418)	1720 (564)	.0299 (.01)	.0091
41							
42	2.50	6.3					
43							
44	1.97	3.9					

*Resistance figures for standard annealed copper. For hard-drawn, add 2%

WIRES PER INCH

Gauge No.	Wires per Inch (wires per cm)		Approx. Wires per Square Inch (Approx. wires per cm²)				Feet per Pound (Meters per kg)		
	S.C.C.	D.C.C.	S.C.C.	D.C.C.	P.E.	Formvar	Bare	P.E.	D.C.C.
8	7.4 (2.9)	7.1 (2.8)	55 (8)	50 (8)	58 (9)		20 (13.4)	19.8 (13.2)	19.5 (13)
9	8.2 (3.2)	7.9 (3.1)	69 (11)	63 (10)					
10	9.3 (3.7)	8.9 (3.5)	86 (13)	98 (12)	92 (14)		31.8 (21.4)	31.5 (21.2)	31 (20.7)
11	10.3 (4)	9.9 (3.9)	108 (17)	98 (15)					
12	11.5 (4.5)	10.9 (4.3)	132 (20)	120 (19)	145 (22)		50.6 (33.9)	50 (33.6)	49 (32.8)
13	12.8 (5)	12.1 (4.8)	166 (26)	148 (23)					
14	14.2 (5.6)	13.5 (5.3)	206 (32)	183 (28)	225 (35)		80.4 (54)	79.4 (53.4)	77 (51.8)
15	15.8 (6.2)	14.8 (5.8)	255 (39)	223 (35)					
16	17.9 (7)	16.5 (6.5)	320 (50)	280 (43)	358 (55)	340 (53)	128 (86)	126 (85)	119 (80)
17	20 (7.9)	18.3 (7.2)	400 (62)	340 (53)					
18	22 (8.7)	21 (8.3)	492 (76)	415 (64)	572 (89)	530 (82)	203 (136)	201 (135)	188 (126)
19	24.5 (9.6)	23.5 (9.2)	625 (97)	510 (79)					
20	27 (10.6)	24.5 (9.6)	770 (119)	625 (95)	875 (136)	800 (125)	323 (217)	319 (214)	298 (200)
21	30 (11.8)	26.7 (10.5)	940 (146)	750 (115)					
22	34 (13)	30.2 (11.9)	1 165 (181)	915 (140)	1 332 (206)	1 200 (185)	514 (346)	507 (341)	461 (311)
23	37.5 (14.8)	32.2 (12.7)	1 400 (217)	1 070 (165)					
24	41.5 (16.3)	35.5 (14)	1 700 (264)	1 260 (195)	2 045 (317)	1 820 (280)	818 (549)	805 (541)	745 (501)
25	45.5 (17.9)	38.5 (15.2)	2 065 (320)	1 495 (230)					
26	50 (19.7)	42 (16.5)	2 510 (390)	1 745 (270)	3 090 (480)	2 700 (420)	1 300 (873)	1 280 (861)	1 118 (752)
27	55 (21.6)	45 (17.7)	3 030 (470)	2 020 (315)					
28	60 (23.6)	48.5 (19.1)	3 654 (565)	2 330 (360)	4 670 (725)	4 000 (620)	2 067 (1 389)	2 030 (1 365)	1 759 (1 182)
29	65 (25.6)	52 (20.5)	4 280 (665)	2 690 (420)					
30	71.5 (28.1)	55.5 (21.8)	5 060 (785)	3 050 (470)	6 860 (1 065)	5 500 (850)	3 287 (2 209)	3 220 (2 165)	2 534 (1 704)
31	77.5 (30.5)	59 (23.2)	6 000 (930)	3 480 (540)					
32	84 (33.1)	62.5 (24.6)	7 050 (1 090)	3 900 (600)	10 050 (1 550)	7 700 (1 200)	5 225 (3 515)	5 120 (3 450)	
33	90 (35.4)	66 (26)	8 100 (1 250)						
34	97 (38.2)	70 (27.6)	9 400 (1 450)		14 250 (2 200)	10 500 (1 600)	8 310 (5 590)	8 160 (5 550)	
35	104 (40.9)	74 (29.1)	10 800 (1 650)						
36	112 (44.1)	78 (30.7)	12 800 (2 000)		20 000 (3 100)	14 900 (2 300)	13 210 (8 880)	12 850 (8 650)	
37									
38	127 (50)	84 (33.1)					21 010 (14 130)		
39									
40	143 (56.3)	90 (35.4)					33 410 (22 470)		

S.C.C. = Single Cotton Covered
D.C.C. = Double Cotton Covered
P.E. = Plain Enamel

FULL-LOAD CURRENTS IN AMPERES, DIRECT-CURRENT MOTORS

The following values of full-load currents are for motors running at base speed.

HP	120V	240V
1/4	2.9	1.5
1/3	3.6	1.8
1/2	5.2	2.6
3/4	7.4	3.7
1	9.4	4.7
1 1/2	13.2	6.6
2	17	8.5
3	25	12.2
5	40	20
7 1/2	58	29
10	76	38
15		55
20		72
25		89
30		106
40		140
50		173
60		206
75		255
100		341
125		425
150		506
200		675

FULL-LOAD CURRENTS IN AMPERES, SINGLE PHASE ALTERNATING CURRENT MOTORS

The following values of full-load currents are for motors running at usual speeds and motors with normal torque characteristics. Motors built for especially low speeds or high torques may have higher full-load currents, and multispeed motors will have full-load current varying with speed, in which case the nameplate current ratings shall be used.

To obtain full-load currents of 208- and 200-volt motors, increase corresponding 230-volt motor full-load currents by 10 and 15 percent, respectively.

The voltages listed are rated motor voltages. Corresponding nominal system voltages are 100 to 120, 220-240, and 440-480.

HP	115V	230V	440V
1/6	4.4	2.2	..
1/4	5.8	2.9	..
1/3	7.2	3.6	..
1/2	9.8	4.9	..
3/4	13.8	6.9	..
1	16	8	..
1 1/2	20	10	..
2	24	12	..
3	34	17	..
5	56	28	..
7 1/2	80	40	21
10	100	50	26

Above adapted from the National Electrical Code (NFPA No. 70, Tables 430-147 and 430-148)

FULL-LOAD CURRENT, THREE-PHASE AC MOTORS

HP	110V	Induction-Type Squirrel-Cage and Wound Rotor Amperes				Synchronous-Type *Unity Power Factor Amperes			
		220V	440V	550V	2300V	220V	440V	550V	2300V
1/2	4	2	1	.8					
3/4	5.6	2.8	1.4	1.1					
1	7	3.5	1.8	1.4					
1 1/2	10	5	2.5	2.0					
2	13	6.5	3.3	2.6					
3		9	4.5	4					
5		15	7.5	6					
7 1/2		22	11	9					
10		27	14	11					
15		40	20	16					
20		52	26	21					
25		64	32	26	7	54	27	22	5.4
30		78	39	31	8.5	65	33	26	6.5
40		104	52	41	10.5	86	43	35	8
50		125	63	50	13	108	54	44	10
60		150	75	60	16	128	64	51	12
75		185	93	74	19	161	81	65	15
100		246	123	98	25	211	106	85	20
125		310	155	124	31	264	132	106	25
150		360	180	144	37		158	127	30
200		480	240	192	48		210	168	40

For full-load currents of 208- and 200-volt motors, increase the corresponding 220-volt motor full-load current by 6 and 10 percent, respectively.

These values of full-load current are for motors running at speeds usual for belted motors and motors with normal torque characteristics. Motors built for especially low speeds or high torques may require more running current, and multispeed motors will have full-load current varying with speed, in which case the nameplate current rating shall be used.

* For 90 and 80 percent P.F. multiply the above figures by 1.1 and 1.25 respectively.

The voltages listed are rated motor voltages. Corresponding nominal system voltages are 110 to 120, 220-240, 440 to 480 and 550 to 600 volts.

Above adapted from the National Electrical Code (NFPA No. 70, Table 430-150)

ELECTRICAL SYMBOLS

ELECTRICAL CONDUCTOR

CONDUCTORS CONNECTED

CROSSING OF CONDUCTORS NOT CONNECTED

JUNCTION OF CONDUCTORS

TERMINAL

GROUND

BATTERY

OR

KNIFE SWITCHES

SINGLE POLE SINGLE THROW

SINGLE POLE DOUBLE THROW

DOUBLE POLE SINGLE THROW

DOUBLE POLE DOUBLE THROW

TRIPLE POLE SINGLE THROW

FIELD DISCHARGE SWITCH

FUSE

THERMAL CUTOUT FLASHER

THERMOSTAT WITH MAKE CONTACT

THREE POLE AIR CIRCUIT BREAKER

THREE POLE OIL CIRCUIT BREAKER

INSTRUMENTS

VOLTMETER

AMMETER

AMMETER WITH SHUNT

WATTMETER

OR

LAMP

RHEOSTAT

OR

FIXED RESISTOR

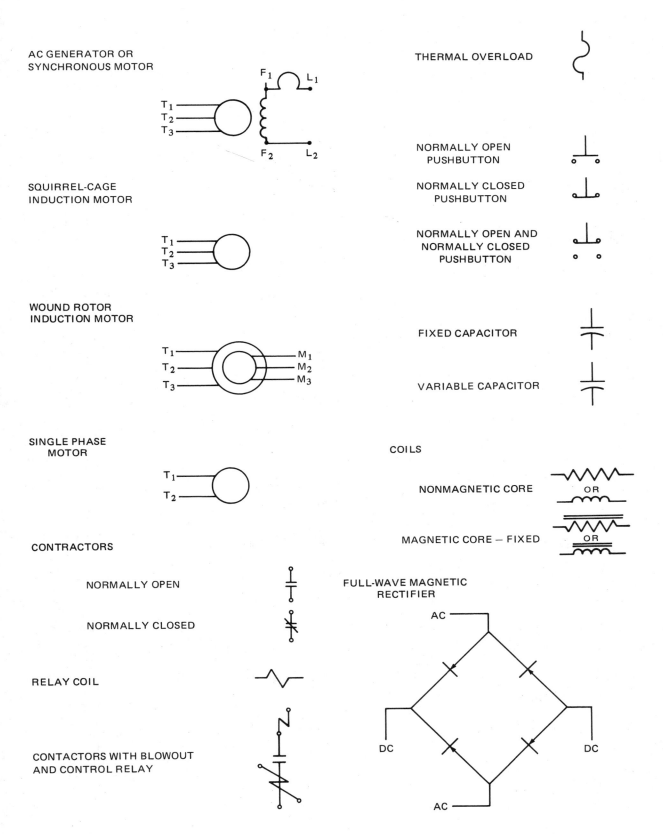

AC GENERATOR OR
SYNCHRONOUS MOTOR

SQUIRREL-CAGE
INDUCTION MOTOR

WOUND ROTOR
INDUCTION MOTOR

SINGLE PHASE
MOTOR

CONTRACTORS

NORMALLY OPEN

NORMALLY CLOSED

RELAY COIL

CONTACTORS WITH BLOWOUT
AND CONTROL RELAY

THERMAL OVERLOAD

NORMALLY OPEN
PUSHBUTTON

NORMALLY CLOSED
PUSHBUTTON

NORMALLY OPEN AND
NORMALLY CLOSED
PUSHBUTTON

FIXED CAPACITOR

VARIABLE CAPACITOR

COILS

NONMAGNETIC CORE

MAGNETIC CORE — FIXED

FULL-WAVE MAGNETIC
RECTIFIER

ELECTRONIC SYMBOLS

ANTENNA	MULTIPLE-DECK CIRCUIT-SELECTOR SWITCH
GROUND	FIXED RESISTOR
BATTERY	ADJUSTABLE RESISTOR
MILLIAMMETER	TAPPED RESISTOR
MILLIVOLTMETER	POTENTIOMETER
FREQUENCY METER	BALLAST RESISTOR
CONDUCTORS NOT JOINED	INDUCTOR, AIR CORE
	INDUCTOR, IRON CORE
CONDUCTORS JOINED	ADJUSTABLE INDUCTOR
BINDING POST	TAPPED INDUCTOR
LINE PLUG	TRANSFORMER, AIR CORE
FUSE	
KEY	TRANSFORMER, IRON CORE
CLOSED CIRCUIT JACK	
OPEN CIRCUIT JACK	
SINGLE-DECK CIRCUIT-SELECTOR SWITCH	PUSH-PULL TRANSFORMER

POWER TRANSFORMER

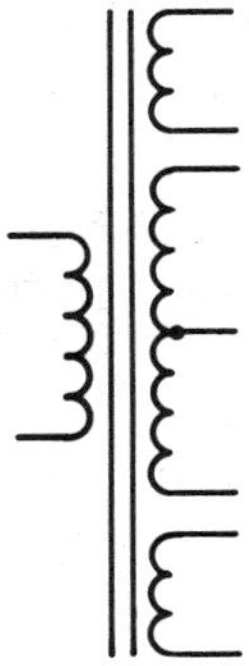

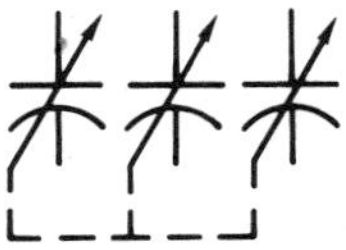

SHIELDED CAPACITOR

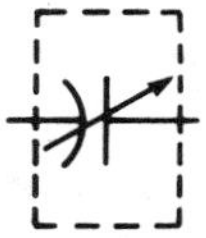

COAXIAL CABLE

EARPHONES

VARIABLE-CORE TRANSFORMER

PERMANENT-MAGNET
DYNAMIC SPEAKER

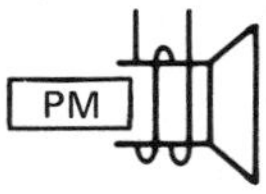

RELAY; CiRCUIT A OPEN
WHEN DE-ENERGIZED

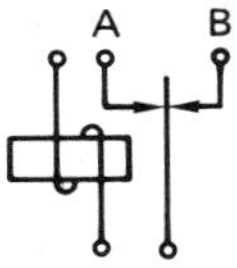

DYNAMIC SPEAKER

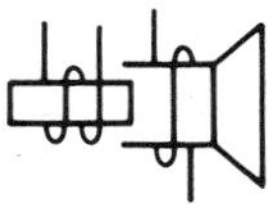

MAGNETIC PHONOGRAPH-
PICKUP

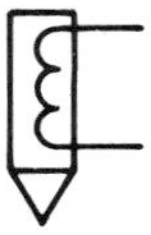

RELAY; CIRCUITS B_1 AND
B_2 CLOSED WHEN
DE-ENERGIZED

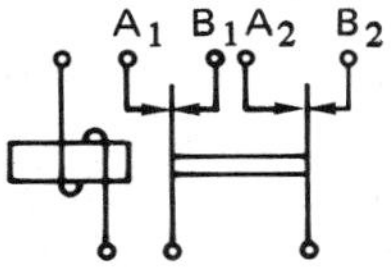

CRYSTAL PHONOGRAPH-
PICKUP

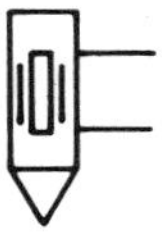

FIXED CAPACITOR

SINGLE-BUTTON CARBON
MICROPHONE

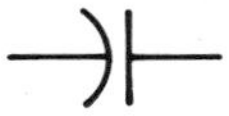

ELECTROLYTIC CAPACITOR

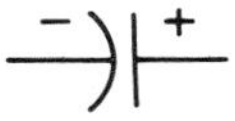

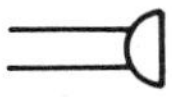

DOUBLE-BUTTON CARBON
MICROPHONE

VARIABLE CAPACITOR

CRYSTAL MICROPHONE

ADJUSTABLE CAPACITOR
(TRIMMER)

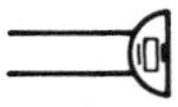

VELOCITY MICROPHONE
(RIBBON)

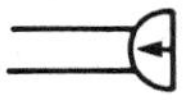

DYNAMIC MICROPHONE
(MOVING COIL)

ADJUSTABLE CAPACITOR
(PADDER)

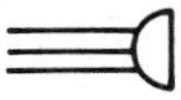

CRYSTAL DETECTOR

CRYSTAL

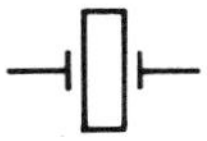

SPLIT-STATOR VARIABLE
CAPACITOR

RECTIFIER, HALF-WAVE

GANGED VARIABLE CAPACITOR
MECHANICAL LINKAGE

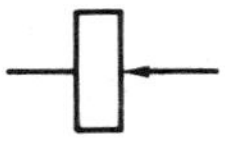

RECTIFIER, FULL-WAVE

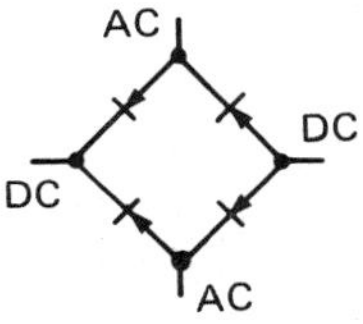

SCHEMATIC SYMBOLS FOR ELECTRON TUBES AND SEMICONDUCTORS

The following are a few of the more common tube symbols used in electronic equipment. More than one tube may be enclosed in one envelope, such as Dual Diode, Dual Triode, Dual Diode-Triode, Triode-Pentode, etc. The addition of a black dot within the symbol indicates a gas-filled tube, such as a Voltage Regulator or Thyratron.

P	Plate	K	Cathode	H	Heater
SG	Screen Grid	F	Filament	D	Diode
SPR	Suppressor Grid	C	Control Grid	OG	Oscillator Grid

VACUUM TUBES

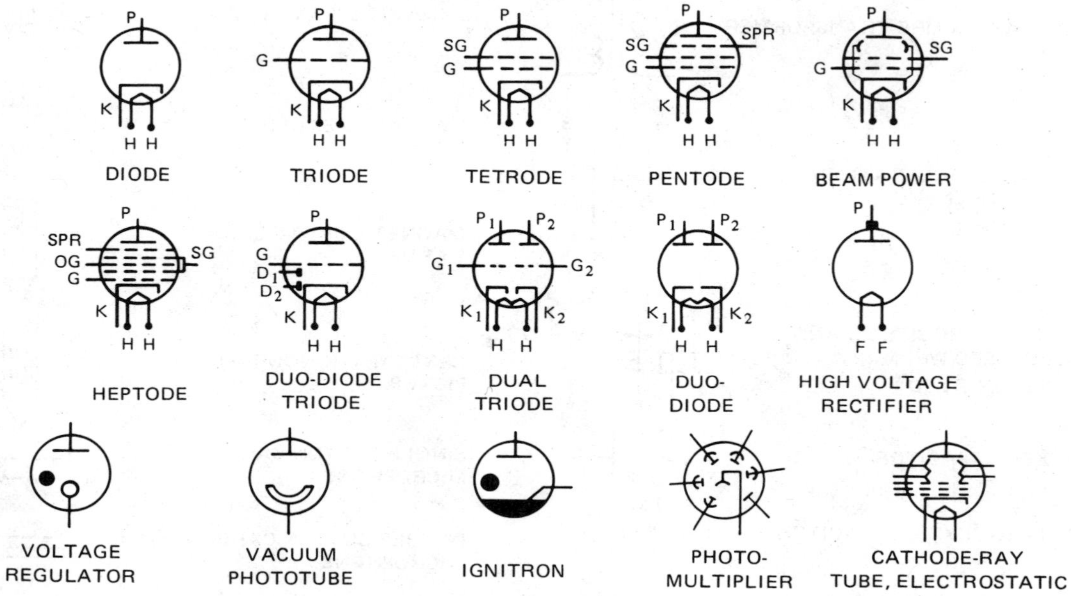

DIODE   TRIODE   TETRODE   PENTODE   BEAM POWER

HEPTODE   DUO-DIODE TRIODE   DUAL TRIODE   DUO-DIODE   HIGH VOLTAGE RECTIFIER

VOLTAGE REGULATOR   VACUUM PHOTOTUBE   IGNITRON   PHOTO-MULTIPLIER   CATHODE-RAY TUBE, ELECTROSTATIC

SEMICONDUCTORS

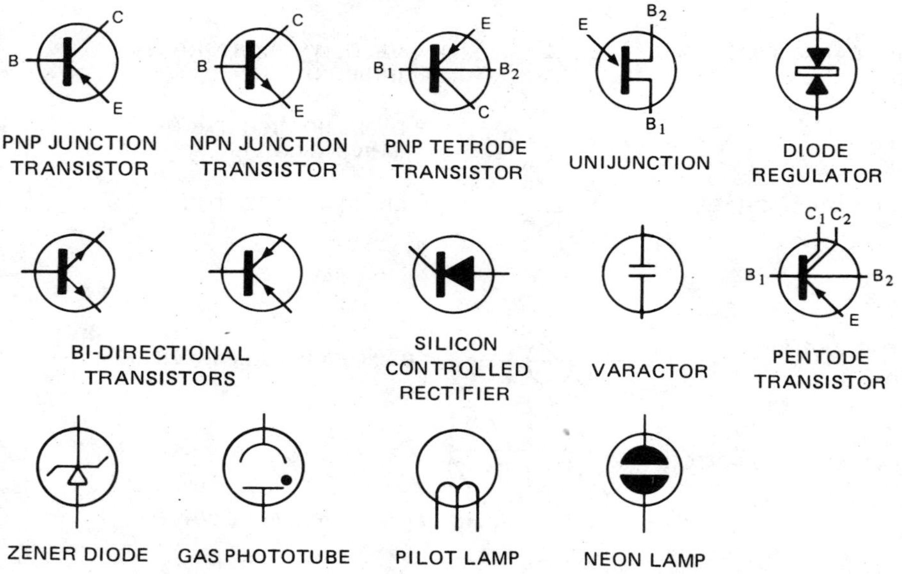

PNP JUNCTION TRANSISTOR   NPN JUNCTION TRANSISTOR   PNP TETRODE TRANSISTOR   UNIJUNCTION   DIODE REGULATOR

BI-DIRECTIONAL TRANSISTORS   SILICON CONTROLLED RECTIFIER   VARACTOR   PENTODE TRANSISTOR

ZENER DIODE   GAS PHOTOTUBE   PILOT LAMP   NEON LAMP

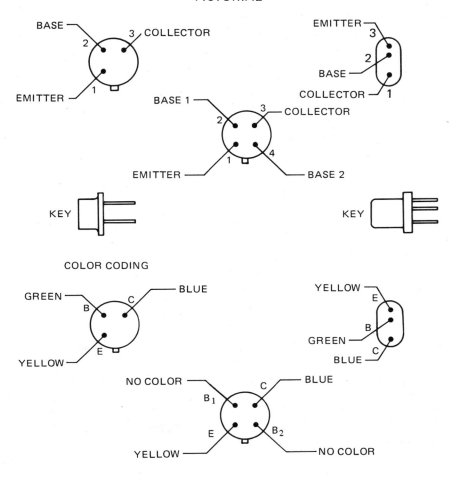

PICTORIAL

COLOR CODING

COLOR CODE FOR CHASSIS WIRING

The standard colors to be used in chassis wiring for the purpose of circuit identification of the equipment shall be as follows:

BLACK	Grounds, grounded elements and returns
BROWN	Heaters or filaments, ac isolated from source (24 VDC or below)
RED	Power supply B plus
ORANGE	Screen grids
YELLOW	Cathodes
GREEN	Control grids
BLUE	Plates
VIOLET (Purple)	Power supply B minus
GRAY	AC power lines (from source)
WHITE	Miscellaneous, above or below ground returns, AVC, etc.

CAPACITOR COLOR CODE MARKING
(MIL-STD CAPACITORS)

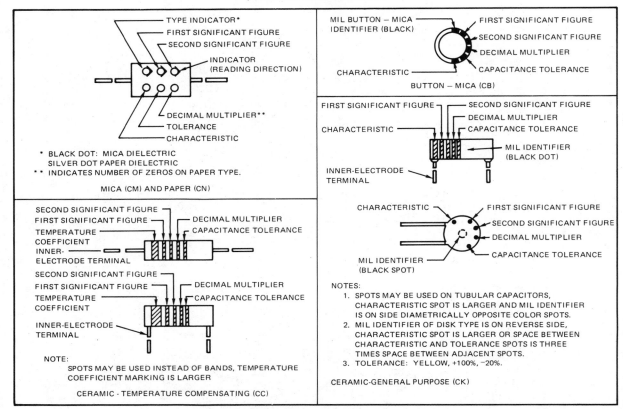

CAPACITOR COLOR CODE

COLOR	SIG FIG.	MULTIPLIER		CHARACTERISTIC[1]				TOLERANCE[2]					TEMPERATURE COEFFICIENT (UUF/UF/°C)
		DECIMAL	NUMBER OF ZEROS	CM	CN	CB	CK	CM	CN	CB	CC		
											OVER 10UUF	10UUF OR LESS	CC
BLACK	0	1	NONE		A			20	20	20	20	2	ZERO
BROWN	1	10	1	B	E	B	W				1		− 30
RED	2	100	2	C	H		X	2		2	2		− 80
ORANGE	3	1 000	3	D	J	D			30				− 150
YELLOW	4	10,000	4	E	P								− 220
GREEN	5		5	F	R						5	0.5	− 330
BLUE	6		6		S								− 470
PURPLE (VIOLET)	7		7		T	W							− 750
GRAY	8		8			X						0.25	+ 30
WHITE	9		9								10	1	− 330 (±500)[3]
GOLD		0.1						5		5			+100
SILVER		0.01						10	10	10			

1. LETTERS ARE IN TYPE DESIGNATIONS GIVEN IN MIL-C SPECIFICATIONS.
2. IN PERCENT, EXCEPT IN UUF FOR CC-TYPE CAPACITORS OF 10 UUF OR LESS.
3. INTENDED FOR USE IN CIRCUITS NOT REQUIRING COMPENSATION.

RESISTOR COLOR CODE MARKING

AXIAL — LEAD RESISTORS
(INSULATED)

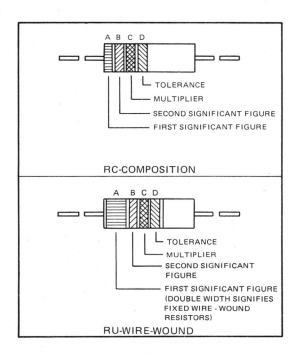

RC-COMPOSITION

RU-WIRE-WOUND

RESISTOR COLOR CODE

BAND A		BAND B		BAND C		BAND D	
COLOR	FIRST SIGNIFICANT FIGURE	COLOR	SECOND SIGNIFICANT FIGURE	COLOR	MULTIPLIER	COLOR	RESISTANCE TOLERANCE (PERCENT)
BLACK	0	BLACK	0	BLACK	1	BODY	±20
BROWN	1	BROWN	1	BROWN	10	SILVER	±10
RED	2	RED	2	RED	100	GOLD	± 5
ORANGE	3	ORANGE	3	ORANGE	1 000		
YELLOW	4	YELLOW	4	YELLOW	10 000		
GREEN	5	GREEN	5	GREEN	100 000		
BLUE	6	BLUE	6	BLUE	1 000 000		
PURPLE (VIOLET)	7	PURPLE (VIOLET)	7				
GRAY	8	GRAY	8	GOLD	0.1		
WHITE	9	WHITE	9	SILVER	0.01		

EXAMPLES (BAND MARKING):
    10 OHMS ±20 PERCENT: BROWN BAND A; BLACK BAND B;
    BLACK BAND C; NO BAND D.
    4.7 OHMS ±5 PERCENT: YELLOW BAND A; PURPLE BAND B;
    GOLD BAND C; GOLD BAND D.

ELECTRICAL WIRING SYMBOLS

CEILING OUTLET

WALL BRACKET

LAMPHOLDER WITH PULL SWITCH

FLOOR OUTLET

CEILING OUTLET FOR RECESSED FIXTURE. (OUTLINE SHOWS SHAPE OF FIXTURE)

TELEVISION OUTLET

FAN OUTLET

RANGE OUTLET

SPECIAL PURPOSE OUTLET SUBSCRIPT LETTERS INDICATE FUNCTIONS. I.E. DW — DISHWASHER, CD — CLOTHES DRYER, ETC.

DUPLEX OUTLET

DUPLEX OUTLET (GROUNDING TYPE)

DUPLEX OUTLET, SPLIT CIRCUIT

WEATHERPROOF OUTLET

CONVENIENCE OUTLET OTHER THAN DUPLEX. 1 = SINGLE, 3 = TRIPLEX, ETC.

FLUORESCENT FIXTURE (EXTEND RECTANGLE TO SHOW LENGTH)

S_1 OR S — SINGLE POLE SWITCH

S_D — DOOR SWITCH

S_2 — DOUBLE POLE SWITCH

S_3 — 3-WAY SWITCH

DUPLEX OUTLET, SPLIT CIRCUIT, (GROUNDING TYPE)

S_4 — 4-WAY SWITCH

S_P — SWITCH WITH PILOT

S_{WP} — WEATHERPROOF SWITCH

2-WIRE CABLE OR RACEWAY

3-WIRE CABLE OR RACEWAY

4-WIRE CABLE OR RACEWAY

PUSH BUTTON

BUZZER

BELL

CHIME (ALSO CH)

ANNUNCIATOR

INTERCOM. TELEPHONE

OUTSIDE TELEPHONE

CLOCK (ALSO ⊕)

MOTOR

JUNCTION BOX

GROUND CONNECTION

LIGHTING PANEL

POWER PANEL

ELECTRIC DOOR OPENER

BATTERY

SWITCH LEG INDICATION, CONNECTS OUTLETS WITH CONTROL POINTS

HEATING PANEL

MULTI — OUTLET ASSEMBLY ARROWS SHOW LIMITS OF IN- STALLATION. APPROPRIATE SYMBOL INDICATES TYPE OF OUTLET. SPACING OF OUTLET IS INDICATED BY X INCHES

TRIGONOMETRY — SIMPLE FUNCTIONS

The (sin) of an angle $= \dfrac{\text{opposite side}}{\text{hypotenuse}}$ $\quad \sin A = \dfrac{a}{c}$

Cosine (cos) of angle $= \dfrac{\text{adjacent side}}{\text{hypotenuse}}$ $\quad \cos A = \dfrac{b}{c}$

Tangent (tan) of angle $= \dfrac{\text{opposite side}}{\text{adjacent side}}$ $\quad \tan A = \dfrac{a}{b}$

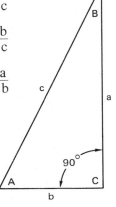
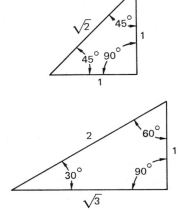

Angle	sin	cos	tan
30°	0.5	$\sqrt{3}/2 = 0.866$	$1/\sqrt{3} = 0.577$
60°	$\sqrt{3}/2 = 0.866$	0.5	$\sqrt{3} = 1.732$
45°	$1/\sqrt{2} = 0.707$	$1/\sqrt{2} = 0.707$	1.0

Law of Sines — In any triangle, the sides are proportional to the sines of the opposite angles. That is,

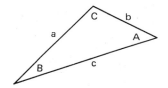

$$\frac{a}{\sin A} = \frac{b}{\sin B} = \frac{c}{\sin C}$$

Law of Cosines — In any triangle the square of any side is equal to the sum of the squares of the other two sides minus twice the product of these two sides and the cosine of their included angle. That is,

$$a^2 = b^2 + c^2 - 2bc \cos A$$

$$\cos A = \frac{b^2 + c^2 - a^2}{2bc}$$

$$\cos B = \frac{c^2 + a^2 - b^2}{2ca}$$

$$\cos C = \frac{a^2 + b^2 - c^2}{2ab}$$

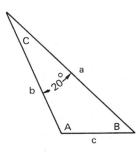

NATURAL SINES AND COSINES

NOTE: For cosines, use right-hand column of degrees and lower line of tenths.

Deg.	°0.0	°0.1	°0.2	°0.3	°0.4	°0.5	°0.6	°0.7	°0.8	°0.9	°1.0	
45	0.7071	0.7083	0.7096	0.7108	0.7120	0.7133	0.7145	0.7157	0.7169	0.7181	0.7193	44
46	0.7193	0.7206	0.7218	0.7230	0.7242	0.7254	0.7266	0.7278	0.7290	0.7302	0.7314	43
47	0.7314	0.7325	0.7337	0.7349	0.7361	0.7373	0.7385	0.7396	0.7408	0.7420	0.7431	42
48	0.7431	0.7443	0.7455	0.7466	0.7478	0.7490	0.7501	0.7513	0.7524	0.7536	0.7547	41
49	0.7547	0.7559	0.7570	0.7581	0.7593	0.7604	0.7615	0.7627	0.7638	0.7649	0.7660	40°
50°	0.7660	0.7672	0.7683	0.7694	0.7705	0.7716	0.7727	0.7738	0.7749	0.7760	0.7771	39
51	0.7771	0.7782	0.7793	0.7804	0.7815	0.7826	0.7837	0.7848	0.7859	0.7869	0.7880	38
52	0.7880	0.7891	0.7902	0.7912	0.7923	0.7934	0.7944	0.7955	0.7965	0.7976	0.7986	37
53	0.7986	0.7997	0.8007	0.8018	0.8028	0.8039	0.8049	0.8059	0.8070	0.8080	0.8090	36
54	0.8090	0.8100	0.8111	0.8121	0.8131	0.8141	0.8151	0.8161	0.8171	0.8181	0.8192	35
55	0.8192	0.8202	0.8211	0.8221	0.8231	0.8241	0.8251	0.8261	0.8271	0.8281	0.8290	34
56	0.8290	0.8300	0.8310	0.8320	0.8329	0.8339	0.8348	0.8358	0.8368	0.8377	0.8387	33
57	0.8387	0.8396	0.8406	0.8415	0.8425	0.8434	0.8443	0.8453	0.8462	0.8471	0.8480	32
58	0.8480	0.8490	0.8499	0.8508	0.8517	0.8526	0.8536	0.8545	0.8554	0.8563	0.8572	31
59	0.8572	0.8581	0.8590	0.8599	0.8607	0.8616	0.8625	0.8634	0.8643	0.8652	0.8660	30°
60°	0.8660	0.8669	0.8678	0.8686	0.8695	0.8704	0.8712	0.8721	0.8729	0.8738	0.8746	29
61	0.8746	0.8755	0.8763	0.8771	0.8780	0.8788	0.8796	0.8805	0.8813	0.8821	0.8829	28
62	0.8829	0.8838	0.8846	0.8854	0.8862	0.8870	0.8878	0.8886	0.8894	0.8902	0.8910	27
63	0.8910	0.8918	0.8926	0.8934	0.8942	0.8949	0.8957	0.8965	0.8973	0.8980	0.8988	26
64	0.8988	0.8996	0.9003	0.9011	0.9018	0.9026	0.9033	0.9041	0.9048	0.9056	0.9063	25
65	0.9063	0.9070	0.9078	0.9085	0.9092	0.9100	0.9107	0.9114	0.9121	0.9128	0.9135	24
66	0.9135	0.9143	0.9150	0.9157	0.9164	0.9171	0.9178	0.9184	0.9191	0.9198	0.9205	23
67	0.9205	0.9212	0.9219	0.9225	0.9232	0.9239	0.9245	0.9252	0.9259	0.9265	0.9272	22
68	0.9272	0.9278	0.9285	0.9291	0.9298	0.9304	0.9311	0.9317	0.9323	0.9330	0.9336	21
69	0.9336	0.9342	0.9348	0.9354	0.9361	0.9367	0.9373	0.9379	0.9385	0.9391	0.9397	20°
70°	0.9397	0.9403	0.9409	0.9415	0.9421	0.9426	0.9432	0.9438	0.9444	0.9449	0.9455	19
71	0.9455	0.9461	0.9466	0.9472	0.9478	0.9483	0.9489	0.9494	0.9500	0.9505	0.9511	18
72	0.9511	0.9516	0.9521	0.9527	0.9532	0.9537	0.9542	0.9548	0.9553	0.9558	0.9563	17
73	0.9563	0.9568	0.9573	0.9578	0.9583	0.9588	0.9593	0.9598	0.9603	0.9608	0.9613	16
74	0.9613	0.9617	0.9622	0.9627	0.9632	0.9636	0.9641	0.9646	0.9650	0.9655	0.9659	15
75	0.9659	0.9664	0.9668	0.9673	0.9677	0.9681	0.9686	0.9690	0.9694	0.9699	0.9703	14
76	0.9703	0.9707	0.9711	0.9715	0.9720	0.9724	0.9728	0.9732	0.9736	0.9740	0.9744	13
77	0.9744	0.9748	0.9751	0.9755	0.9759	0.9763	0.9767	0.9770	0.9774	0.9778	0.9781	12
78	0.9781	0.9785	0.9789	0.9792	0.9796	0.9799	0.9803	0.9806	0.9810	0.9813	0.9816	11
79	0.9816	0.9820	0.9823	0.9826	0.9829	0.9833	0.9836	0.9839	0.9842	0.9845	0.9848	10°
80°	0.9848	0.9851	0.9854	0.9857	0.9860	0.9863	0.9866	0.9869	0.9871	0.9874	0.9877	9
81	0.9877	0.9880	0.9882	0.9885	0.9888	0.9890	0.9893	0.9895	0.9898	0.9900	0.9903	8
82	0.9903	0.9905	0.9907	0.9910	0.9912	0.9914	0.9917	0.9919	0.9921	0.9923	0.9925	7
83	0.9925	0.9928	0.9930	0.9932	0.9934	0.9936	0.9938	0.9940	0.9942	0.9943	0.9945	6
84	0.9945	0.9947	0.9949	0.9951	0.9952	0.9954	0.9956	0.9957	0.9959	0.9960	0.9962	5
85	0.9962	0.9963	0.9965	0.9966	0.9968	0.9969	0.9971	0.9972	0.9973	0.9974	0.9976	4
86	0.9976	0.9977	0.9978	0.9979	0.9980	0.9981	0.9982	0.9983	0.9984	0.9985	0.9986	3
87	0.9986	0.9987	0.9988	0.9989	0.9990	0.9990	0.9991	0.9992	0.9993	0.9993	0.9994	2
88	0.9994	0.9995	0.9995	0.9996	0.9996	0.9997	0.9997	0.9997	0.9998	0.9998	0.9998	1
89	0.9998	0.9999	0.9999	0.9999	0.9999	1.0000	1.0000	1.0000	1.0000	1.0000	1.0000	0°
	°1.0	°0.9	°0.8	°0.7	°0.6	°0.5	°0.4	°0.3	°0.2	°0.1	°0.0	Deg.

NATURAL SINES AND COSINES

NOTE: For cosines, use right-hand column of degrees and lower line of tenths.

Deg.	°0.0	°0.1	°0.2	°0.3	°0.4	°0.5	°0.6	°0.7	°0.8	°0.9	°1.0	
0°	0.0000	0.0017	0.0035	0.0052	0.0070	0.0087	0.0105	0.0122	0.0140	0.0157	0.0175	89
1	0.0175	0.0192	0.0209	0.0227	0.0244	0.0262	0.0279	0.0297	0.0314	0.0332	0.0349	88
2	0.0349	0.0366	0.0384	0.0401	0.0419	0.0436	0.0454	0.0471	0.0488	0.0506	0.0523	87
3	0.0523	0.0541	0.0558	0.0576	0.0593	0.0610	0.0628	0.0645	0.0663	0.0680	0.0698	86
4	0.0698	0.0715	0.0732	0.0750	0.0767	0.0785	0.0802	0.0819	0.0837	0.0854	0.0872	85
5	0.0872	0.0889	0.0906	0.0924	0.0941	0.0958	0.0976	0.0993	0.1011	0.1028	0.1045	84
6	0.1045	0.1063	0.1080	0.1097	0.1115	0.1132	0.1149	0.1167	0.1184	0.1201	0.1219	83
7	0.1219	0.1236	0.1253	0.1271	0.1288	0.1305	0.1323	0.1340	0.1357	0.1374	0.1392	82
8	0.1392	0.1409	0.1426	0.1444	0.1461	0.1478	0.1495	0.1513	0.1530	0.1547	0.1564	81
9	0.1564	0.1582	0.1599	0.1616	0.1633	0.1650	0.1668	0.1685	0.1702	0.1719	0.1736	80°
10°	0.1736	0.1754	0.1771	0.1788	0.1805	0.1822	0.1840	0.1857	0.1874	0.1891	0.1908	79
11	0.1908	0.1925	0.1942	0.1959	0.1977	0.1994	0.2011	0.2028	0.2045	0.2062	0.2079	78
12	0.2079	0.2096	0.2113	0.2130	0.2147	0.2164	0.2181	0.2198	0.2215	0.2232	0.2250	77
13	0.2250	0.2267	0.2284	0.2300	0.2317	0.2334	0.2351	0.2368	0.2385	0.2402	0.2419	76
14	0.2419	0.2436	0.2453	0.2470	0.2487	0.2504	0.2521	0.2538	0.2554	0.2571	0.2588	75
15	0.2588	0.2605	0.2622	0.2639	0.2656	0.2672	0.2689	0.2706	0.2723	0.2740	0.2756	74
16	0.2756	0.2773	0.2790	0.2807	0.2823	0.2840	0.2857	0.2874	0.2890	0.2907	0.2924	73
17	0.2924	0.2940	0.2957	0.2974	0.2990	0.3007	0.3024	0.3040	0.3057	0.3074	0.3090	72
18	0.3090	0.3107	0.3123	0.3140	0.3156	0.3173	0.3190	0.3206	0.3223	0.3239	0.3256	71
19	0.3256	0.3272	0.3289	0.3305	0.3322	0.3338	0.3355	0.3371	0.3387	0.3404	0.3420	70°
20°	0.3420	0.3437	0.3453	0.3469	0.3486	0.3502	0.3518	0.3535	0.3551	0.3567	0.3584	69
21	0.3584	0.3600	0.3616	0.3633	0.3649	0.3665	0.3681	0.3697	0.3714	0.3730	0.3746	68
22	0.3746	0.3762	0.3778	0.3795	0.3811	0.3827	0.3843	0.3859	0.3875	0.3891	0.3907	67
23	0.3907	0.3923	0.3939	0.3955	0.3971	0.3987	0.4003	0.4019	0.4035	0.4051	0.4067	66
24	0.4067	0.4083	0.4099	0.4115	0.4131	0.4147	0.4163	0.4179	0.4195	0.4210	0.4226	65
25	0.4226	0.4242	0.4258	0.4274	0.4289	0.4305	0.4321	0.4337	0.4352	0.4368	0.4384	64
26	0.4384	0.4399	0.4415	0.4431	0.4446	0.4462	0.4478	0.4493	0.4509	0.4524	0.4540	63
27	0.4540	0.4555	0.4571	0.4586	0.4602	0.4617	0.4633	0.4648	0.4664	0.4679	0.4695	62
28	0.4695	0.4710	0.4726	0.4741	0.4756	0.4772	0.4787	0.4802	0.4818	0.4833	0.4848	61
29	0.4848	0.4863	0.4879	0.4894	0.4909	0.4924	0.4939	0.4955	0.4970	0.4985	0.5000	60°
30°	0.5000	0.5015	0.5030	0.5045	0.5060	0.5075	0.5090	0.5105	0.5120	0.5135	0.5150	59
31	0.5150	0.5165	0.5180	0.5195	0.5210	0.5225	0.5240	0.5255	0.5270	0.5284	0.5299	58
32	0.5299	0.5314	0.5329	0.5344	0.5358	0.5373	0.5388	0.5402	0.5417	0.5432	0.5446	57
33	0.5446	0.5461	0.5476	0.5490	0.5505	0.5519	0.5534	0.5548	0.5563	0.5577	0.5592	56
34	0.5592	0.5606	0.5621	0.5635	0.5650	0.5664	0.5678	0.5693	0.5707	0.5721	0.5736	55
35	0.5736	0.5750	0.5764	0.5779	0.5793	0.5807	0.5821	0.5835	0.5850	0.5864	0.5878	54
36	0.5878	0.5892	0.5906	0.5920	0.5934	0.5948	0.5962	0.5976	0.5990	0.6004	0.6018	53
37	0.6018	0.6032	0.6046	0.6060	0.6074	0.6088	0.6101	0.6115	0.6129	0.6143	0.6157	52
38	0.6157	0.6170	0.6184	0.6198	0.6211	0.6225	0.6239	0.6252	0.6266	0.6280	0.6293	51
39	0.6293	0.6307	0.6320	0.6334	0.6347	0.6361	0.6374	0.6388	0.6401	0.6414	0.6428	50°
40°	0.6428	0.6441	0.6455	0.6468	0.6481	0.6494	0.6508	0.6521	0.6534	0.6547	0.6561	49
41	0.6561	0.6574	0.6587	0.6600	0.6613	0.6626	0.6639	0.6652	0.6665	0.6678	0.6691	48
42	0.6691	0.6704	0.6717	0.6730	0.6743	0.6756	0.6769	0.6782	0.6794	0.6807	0.6820	47
43	0.6820	0.6833	0.6845	0.6858	0.6871	0.6884	0.6896	0.6909	0.6921	0.6934	0.6947	46
44	0.6947	0.6959	0.6972	0.6984	0.6997	0.7009	0.7022	0.7034	0.7046	0.7059	0.7071	45
	°1.0	°0.9	°0.8	°0.7	°0.6	°0.5	°0.4	°0.3	°0.2	°0.1	°0.0	Deg.

LETTER SYMBOLS AND ABBREVIATIONS

ac	alternating current	f_r	resonant frequency
dc	direct current	t	time
I	amperes, current	kHz	kilohertz
E	voltage	Ω	ohms
R	resistance	M	meg-, million
W	watts (power)	μ	micro, millionth
X	reactance	μa	microampere
L	inductance	μf	microfarad
X_L	inductive reactance	μh	microhenry
C	capacitor	μv	microvolt
X_c	capacitive reactance	m	milli, thousandth
Z	impedance	ma	milliampere
h	henries (inductance)	mHz	megahertz
M	mutual inductance	mv	millivolt
K	thousand (Kilo-)	mw	milliwatt
P	power	mh	millihenry
PD	potential difference	CW	continuous wave
PF	power factor	AF	audio frequency
emf	electromotive force	IF	intermediate frequency
hp	horsepower	RF	radio frequency
kwh	kilowatt-hour	LF	low frequency
A.H.	ampere hour	HF	high frequency
C.M.	circular mils	VHF	very high frequency
AWG	American Wire Gauge	UHF	ultra high frequency
Hz	hertz (cycles/second)	AM	amplitude modulated
f	frequency	FM	frequency modulated

Acknowledgments

The authors express grateful appreciation to the following educators for their counsel and advice while this material was in the making for the first edition: Nelson S. Mauer, Chairman, Industrial Arts Department, Colonie Central Schools, Albany, NY; Herbert Insley, Instructor of Industrial Arts, Colonie Central High School, Albany, NY; Kenneth Folster, Arlington High School, Poughkeepsie, NY; John Stewart, Coordinator, Industrial Arts, Arlington Central School, Poughkeepsie, NY; Webster G. MacDonald, Instructor, Industrial Arts and Technology, Bethlehem Central High School, Delmar, NY.

Special recognition is due Dr. Clarence A. Cook, Professor of Electronics, State University College, Buffalo, NY, and Thomas W. De Santis, Instructor of Industrial Arts, Narrowsburg Central School, Narrowsburg, NY, who made valuable suggestions regarding experiments and project development.

Index